VOLUME ONE

Developments in Physiology Biochemistry and Molecular Biology of Plants

Developments in Physiology Biochemistry and Molecular Biology of Plants

Bandana Bose and A. Hemantaranjan
Department of Plant Physiology
Institute of Agricultural Sciences
Banaras Hindu University
Varanasi - 221 005
(U.P.) INDIA

NEW INDIA PUBLISHING AGENCY
New Delhi – 110 034

ISBN 978- 81 - 89422- 02- 8

Typeset at:

Laxmi Art Creations
Shiva Mkt., Pitam Pura, New Delhi

Published by:

New India Publishing Agency

101, Vikas Surya Plaza, CU Block, L.S.C. Mkt.,
Pitam Pura, New Delhi- 110 088, (INDIA)
Phone: 011-27341717, Fax: 011-27341616
E-mail: spjain_niph@rediffmail.com
web: www.bookfactoryindia.com

PREFACE

There is no limit in thirst of our knowledge and eagerness and this has led to our concurrence that plant kingdom has always been the integral part of all forms of vitality. To understand the myth of life style of plants, its different types of activities have to be deciphered and this is what we call Plant Physiology. However, physiology deals with all the vital activities of a plant and also explains how it reacts to sustain in natural distress. Inside the plant, what types of physiological actions are going on in chemical level, means what compounds are formed and how these react becomes Biochemistry. It explains how a plant can trap the unlimited solar energy in form of chemical bonding and again at different platforms it utilizes this energy to perform different physiological works. This is the energy cycle, the genesis of whole vital world. But thirst of knowledge does not quench here, and man probed still inside up to molecular and genetic levels of these chemical reactions to understand the nature of biochemical reactions and to control if possible up to the desired level and that is Molecular Biology. We have tried to comprise these three aspects of plant system, *i.e.*, Plant Physiology, Biochemistry and Molecular Biology under one canopy in the form of this book, entitled "***Developments in Physiology, Biochemistry & Molecular Biology of Plants***" contributed with very valuable and relevant chapters by various experts of national and international repute. We trust that the collective wisdom with foresighted vision and dedicated efforts of contributors will fulfil the noble mission of this book in the post genomic era of cyber age. Nevertheless, this book will not only be extremely helpful to satisfy the keenness and inquisitiveness of the students and educators but also enlighten the pavement for researchers of plant science.

Distinguished plant scientists, actively engaged in teaching and research at prominent Universities and ICAR Research Institutes have contributed ten pragmatic review articles, which have been subdivided into six major sections. The *Chapter 1* of *Section I* well deals in part with the recent advances made in plant iron nutrition research meticulously reviewing Fe-uptake and distribution in relation to hormones under Fe deficiency stress followed by physiology and biochemistry of iron nutrition and interactions involving iron on crop yield. The *Section II* of Plant Metabolism consists of three applicable chapters two, three and four. The *Chapter 2* embodies regulation of phospholipid metabolism. The control of phospholipid metabolism in plants is an open area of research, as it is needed in perspective of hormonal effects and environmental stresses. Similarly in *Chapter 3*, the co-author has well emphasized the metabolism of sucrose and starch and their significance. The degradation (breakdown) of starch ensures retaining of important life processes in plants when there is carbon shortage, for instance, during night period in leaves or in resting organs such as potato tubers. Subsequently in *Chapter 4*, experienced co-authors has precisely accentuated the knowledge of nitrate assimilation and nitrate reductase activity in plants, as the understanding of physiology and biochemistry of nitrogen nutrition at molecular level is obligatory and likely to add insight to achieve the hard task goal.

The book is privileged to incorporate *Section III* as Environmental Stresses in Plants, which includes two important reviews by eminent co-authors on plant water relationships and strategies to improve salt tolerance as *Chapter 5* where an effort is made to integrate the physical and chemical properties of water, water potential and factors affecting it, measurement of water potential, water movement from soil to atmosphere through complex plant structure and crop growth, development and yield under abiotic stresses especially salt stress. Afterward, in *Chapter 6*, a team of reputed scientists from three well-known institutions have systematically highlighted the impact of heavy metal pollution in plants illustrating some of the basic facts of this subject. Next to this in *Section IV* on Biological Nitrogen Fixation, acknowledged scientists wonderfully divulged in *Chapter 7* the significance of biological nitrogen fixation in pulses and cereals, which is positively a pre-requisite for sustainability of agriculture and of present importance. Identical to this *Section V* on Sustainable Crop Production consists of two essential reviews initiated with a new group of biofertilizers for sustainable crop productivity in *Chapter 8*. This chapter in depth describes the plant growth-promoting rhizobacteria (PGPR), which presents immense potential and promise as effective substitutes for chemical fertilizers. PGPR enhances plant growth and survival, controls soil-borne fungi, and induces systemic resistance to phytopathogens and is used as inoculants for biofertilization, phytostimulation and biocontrol. This section also focuses on the aim of fruit research strategies as *Chapter 9*. In fact, to achieve the target of required fruit production in future, it is necessary to understand the physiological basis of fruit productivity in order to garner higher productivity of quality fruits per unit area. A well-known ICAR Institute scientist has specified a thorough insight towards an understanding of the factors affecting productivity of fruit crops to enrich the objectives of this book. Eventually, an exclusive review on the role of herbicides in improving crop yields as *Chapter 10* has been integrated in *Section VI* of *Physiology and Biochemistry of Herbicides* and has been left in the original format to honour the distinguished, well experienced and unassuming senior author who along with his associate presented the glorious write up on roles of herbicides in improving crop yields. Certainly, herbicides besides other pesticides are a major and increasing part of the agricultural technology in the decades ahead, which will be needed to provide a greater supply of food, fibre and shelter with greatest cost-effectiveness. Overall, each and every contribution takes us to a sense of accomplishment. We are delighted to state our thankfulness to all distinguished contributors from all over the country. At the same time, with heavy hearts, we pay our homage to our senior colleague, guide, friend and philosopher, Late Professor O. K. Garg, a renowned Plant Physiologist, teacher and excellent human being who co-authored *Chapter 1*, left for heavenly abode in September 2004. While doing this pious work, the first Editor, Dr. (Mrs.) Bandana Bose is deeply overwhelmed to remember her esteemed Ph.D. Guide, Late Professor H. S. Srivastava who is ever indwelling as the source of inspiration in her mind. Furthermore, we are extremely grateful to the admired colleagues, the staff members, and beloved students of the Department of Plant Physiology, Institute of Agricultural Sciences, Banaras Hindu University for their trustworthy moral support. Last but not the least; we acknowledge efficient cooperation of the staff members of the New India Publishing Agency to shape the book near perfection.

April, 2005

Editors:
Dr. (Mrs.) Bandana Bose
Dr. A. Hemantaranjan
B.H.U., Varanasi
(India)

LIST OF CONTRIBUTORS

A. Hemantaranjan
Department of Plant Physiology
Institute of Agricultural Sciences
Banaras Hindu University
Varanasi - *221 005*
E-mail: *hemantaranja@satyam.net.in*

A. K. Bera
Department of Plant Physiology
Bidhan Chandra Krishi Viswavidyalaya
P.O. Krishi Vishwavidyalaya
Dist. Nadia - *741 252*
E-mail: *dr.bera@mantramail.com*

Anita Bera
Institute of Agricultural Sciences,
Calcutta University, 35 Ballygunge
Circular Road
Kolkata - *700 019*
E-mail: *dr.bera@mantramail.com*

B. K. Ghosh
Department of Plant Physiology
College of Agriculture
Orissa University of Agriculture
and Technology
Bhubaneswar - *751 003*
E-mail: *dr_bkghose@yahoo.co.uk*

Bandana Bose, (Mrs.)
Department of Plant Physiology
Institute of Agricultural Sciences
Banaras Hindu University
Varanasi - *221 005*
E-mail : *bandana_bose2000@yahoo.com*

D. N. Tyagi
Department of Plant Physiology
Institute of Agricultural Sciences
Banaras Hindu University
Varanasi - *221 005*
E-mail: *hemantaranja@satyam.net.in*

I. V. Subba Rao
Department of Plant Physiology
College of Agriculture
Acharya N.G.Ranga Agricultural University
Rajendranagar
Hyderabad – *500 030*

J. D. S. Panwar
Division of Crop Physiology,
Biochemistry and Microbiology
Indian Institute of Pulses Research
Kanpur - *208 024*

K. Swarnalakshmi
Division of Crop Physiology,
Biochemistry and Microbiology
Indian Institute of Pulses Research
Kanpur - *208 024*

M. Kar
Department of Plant Physiology
College of Agriculture
Orissa University of Agriculture
and Technology
Bhubaneswar - *751 003*
E-mail: *dr_bkghose@yahoo.co.uk*

O. K. Garg
Department of Plant Physiology
Institute of Agricultural Sciences
Banaras Hindu University
Varanasi - *221 005*
E-mail: *hemantaranja@satyam.net.in*

Padmanabh Dwivedi
Department of Botany
Arunachal University
Rono Hills
Itanagar - *791 111*
E-mail: *pdwivedi25@rediffmail.com*

Pramod Kumar
Senior Scientist (Plant Physiology)
Crop Improvement Division
Vivekananda Parvatiya Krishi Anusandhan Sansthan
Almora - *263 601*

Samadrita Barman Roy
Chinusurah Rice Research Station
Govt. of West Bengal
P.O.Chinsurah
Dist. Hooghly - *712 103*
E-mail: *dr.bera@mantramail.com*

T. Y. Madhulety
Department of Plant Physiology
College of Agriculture
Acharya N.G.Ranga Agricultural University
Rajendranagar
Hyderabad – *500 030*

V. K. Singh
Fruit Physiology Laboratory
Central Institute for
Subtropical Horticulture
(ICAR), Lucknow - *227 107*

Vijay Laxmi
Division of Crop Physiology,
Biochemistry and Microbiology
Indian Institute of Pulses
Research
Kanpur - *208 024*

Vinod Goyal
Department of Botany and
Plant Physiology
CCS Haryana Agricultural
University
Hisar - *125 004*
E-mail: *goyal2973@yahoo.com*

*****

CONTENTS

	Preface	*v*
	List of Contributors	*vii*
	Section I: Recent Advances in Micronutrient Research	
1.	Physiological and Molecular Advances in Plant Iron Nutrition *- A. Hemantaranjan, Bandana Bose, O. K. Garg, and D. N. Tyagi*	1-19
	Section II: Plant Metabolism	
2.	Phospholipid Metabolism in Plants *- Pramod Kumar*	21-36
3.	Metabolism of Sucrose and Starch in Plants: Synthesis and their Degradation *- Padmanabh Dwivedi*	37-51
4.	Nitrate Assimilation and Nitrate Reductase Activity *- M. Kar and B.K. Ghosh*	53-80
	Section III: Environmental Stresses in Plants	
5.	Plant Water Relationships and Strategies to Improve Salt Tolerance *-Vinod Goyal*	81-104
6.	Impact of Heavy Metal Pollution in Plants *-A. K. Bera, Anita Bera and Samadrita Barman Roy*	105- 124
	Section IV: Biological Nitrogen Fixation	
7.	Biological Nitrogen Fixation in Pulses and Cereals *-J. D. S. Panwar and Vijay Laxmi*	125-158
	Section V: Sustainable Crop Productivity	
8.	PGPR: A New Group of Biofertilizers for Sustainable Crop Productivity *-J. D. S. Panwar and K. Swarnalakshmi*	159-180
9.	Towards An Understanding of the Factors Affecting Productivity of Fruit Crops *-V. K. Singh*	181-201
	Section VI: Physiology and Biochemistry of Herbicides	
10.	Role of Herbicides in Improving Crop Yields *- I.V.Subba Rao, and T.Y.Madhulety*	203-287
	Colour Plates	289-290
	Index	291-302

Section I: Recent Advances in Micronutrient Research

Developments in Physiology, Biochemistry and Molecular Biology of Plants, 2005
Eds.: Bandana Bose and A. Hemantaranjan
Vol., 1, pp. 1-19, New India Publishing Agency, New Delhi
E-mail: spjain_niph@rediffmail.com web: www.bookfactoryindia.com

CHAPTER - 1

PHYSIOLOGICAL AND MOLECULAR ADVANCES IN PLANT IRON NUTRITION

A. HEMANTARANJAN, BANDANA BOSE, O. K. GARG, AND D. N. TYAGI

INTRODUCTION

Iron in the terrain's crust is more oxidized now than it was when life evolved. It has been hypothesized that dissolved ferrous iron could have been a convenient oxygen acceptor and that deposition of oxidized iron in sediments must have taken place before oxygen could have entered the atmosphere in significant quantities. Obviously, during early evolution on this planet, the essential elements are present only in reduced form. Today a number of elements are present only in a highly oxidized form and considerable energy must be expanded to reduce them to a form similar to that on early earth which plants can utilize.

Iron represents essential elements which are usually available in oxidized form but must be reduced, often using metabolic energy, before they can be utilized. For the first time, Gris (1844) observed yellowing of grapevine leaves in absence of iron in the nutrient solution in which these plants were growing. Later, iron was reported to be essential for diverse group of plants. Due to a great variation in iron requirement by different plant species, it is difficult to decide whether iron is a macro or micronutrient.

Iron functions in a plant in many ways, but a lack of iron in the growing medium most often evidenced by a yellowing of the leaves commonly referred to as iron chlorosis. Although iron is abundant in nearly every soil, its availability may be so low (particularly in calcareous soils) that plants are unable to absorb enough to sustain normal growth and development. Iron chlorosis involves a reduction in the chlorophyll content of leaves. The resultant decrease in photosynthesis directly affects plant growth and development and reduces productivity for economic uses by human beings. Iron chlorosis affects the production and well being of many kinds of plants including field crops, nursery stock, large and small fruits, as well as forage and turf plants. Iron chlorosis symptoms have often been confused with effects due to other interacting stresses in the terminal leaves of sensitive plants. Classical Fe-deficiency stress is believed to cause chlorosis by limiting chlorophyll formation in the chloroplasts (Platt-Aloia *et al.,* 1983; Miller *et al.,* 1984).

Iron is involved in energy producing and utilizing processes in the plant and is important in many redox reactions. It functions in both heme and non-heme forms in many enzymatic reactions. Photosynthesis and respiratory processes supply the metabolic energy required for plant physiological functions. Iron nutrition interactions arise out of need to bioregulate biochemical oxidation and reduction processes. Electron transport systems largely evolved around Fe as a key redox factor (Clark *et al.,* 1981).

Besides these, for legumes, iron deficiency may inhibit symbiotic nitrogen fixation by affecting the growth and survival of -(brady)*rhizobia*, nodule formation, nodule function or through effects on the growth of host plants. Iron is essential for the constitution of the nitrogen-fixing enzyme nitrogenase as well as leghaemoglobin and ferredoxin, present in the nodule. Presence of active iron would improve their status for taking part in nitrogen fixation to improve the total dry matter content of the plant (Evans and Russel, 1971).

IRON INSUFFICIENCY – A CRITICAL EVALUATION

Non-availability of iron in the rooting medium results in the appearance of characteristic visible effects of the deficiency of the element as chlorosis of young emerging leaves (Hewitt, 1963; 1983). The "die back" of young shoots has been reported in many fruit trees under severe iron deficiency (Wallace, 1961; Chapman, 1973). Zouari *et al.* (2001) critically evaluated two mutants of tomato and their corresponding wild-type genotypes, T*fer*/TFER and *chloronerva*/Bonner Beste, were grown in nutrient solution under conditions leading to iron (Fe) deficiency. Iron deficiency caused decreases in growth, leaf chlorosis, and changes in the morphology of roots. Ferric chelate reductase activities of whole roots were generally lower in Fe-deficient plants than in control, Fe-sufficient plants. Plants grown for 7 days without Fe, however, had transient increases in whole root ferric chelate reductase activity after the addition of small amounts of Fe (2μM) to the nutrient solution. Also, adding sequential 0.5μM Fe pulses to the nutrient solution led to high whole root ferric chelate reductase activities. Similar results were obtained with a protocol using excised root tips instead of whole root systems to measure ferric chelate reductase activities. The protocol using root tips generally gave higher ferric chelate reductase rates than the method using whole roots, due to the localized expression of the enzyme in the distal root zones (Grusak, 2001).

Further studies on response of iron-deficient barley plants to manganese in nutrient solution were made in the recent years by Alam *et al.* (2001). In that investigation barley (*Hordeum vulgare* L. cv. Minorimugi) plants were cultured in iron (Fe)-deficient nutrient solutions at three levels (0.25, 0.025, and 0.0025μM) of manganese (Mn). Iron-deficient plants with high Mn (0.25μM) had brown spots on older leaves and stems characteristic of Mn toxicity. In the nutrient solution at 0.0025μM Mn, the younger leaves had typical symptoms of Mn deficiency. Iron-deficient plants with 0.025μM Mn had shoot Mn concentration statistically similar to Fe-sufficient plants with 0.25μM Mn. Tissue Fe and Mn concentrations were reciprocally related in shoots and roots of plants grown in 0.25μM Mn with and without Fe. The Mn concentration of Fe-deficient plants decreased with decreasing Mn concentrations in the nutrient solution. In contrast, the Mn levels in the nutrient solution did not significantly affect the Fe concentrations of Fe-deficient plants with identical Fe deficiency symptoms. The highest level of phytosiderophore (PS) released from roots of Fe-deficient plants occurred with 0.025μM Mn and the least with 0.0025μM Mn. A deficiency of Fe decreased the translocation of Fe and Cu from roots to shoots, but had little or no effect on the translocation of Mn and Zn.

It is generally believed that when plants are green they do not respond to application of iron. If this were the case, the '*Sufficiency Value*' for iron would be 1.00. When plants show visual symptoms of iron deficiency, it is often concluded that there is no harm if the symptoms are slight, but considerable harm when the symptoms are severe. Leaf analysis for iron of plants grown under controlled conditions is a good indicator of iron status (Wallace and Jones, 1993) from which '*Sufficiency Value*' can be estimated. Analysis of field grown plants, however, is less sure because primarily of dust contamination (Wallace and Jones, 1993).

In most oxygenated soils, Fe exists predominantly in the ferric form (Fe^{3+}), being partitioned between soluble and insoluble fractions depending on local soil conditions. Soluble Fe (*i.e.*, available Fe) can be found as various chelated species (bound to organic ligands), as inorganic species, and as the free ferric ion (Hinsinger, 1998). When soil pH is high and/or when soil organic matter is low (10^{-8} M) and this can lead to deficiency conditions for plants growing in such locations. Therefore, in order to acquire adequate Fe, plants must increase the availability of Fe in the rhizosphere, as well as increase their uptake capacity for Fe and Fe species.

Iron deficiency in plants may be corrected either by fertilizer applications (foliar or soil) or by selection of plants, which are less susceptible. Foliar applications of iron are supposed to be of limited use to plants. However, in the last few years, studies made on foliar applications of iron and other micronutrients were found to be of immense use especially at critical deficiency period (Hemantaranjan, 1999; Sahu, 2000). Crop plants prior to their reproductive growth development have greater need for ferrous (physiologically active) form of iron for various synthetic processes in plants, ultimately leading to flowering and reproduction for optimum yield and yield attributes. In case of unavailability of iron, which easily converts from ferrous to ferric (unavailable form) in soil or plant system, hampers considerable amount of reduction in a number of biosynthetic and other metabolic processes that iron controls? Therefore, quick availability of ferrous iron could be mediated through the foliar application of iron especially, during the pre- or post- flowering stage, which is the true time of need for iron in plants. For a rapid availability of Fe, the subject of direct injection of liquids in to woody plants has been already reviewed (Wallace and Wallace, 1986).

Numerous experiments have shown that not only HCO^-_3 but also NO^-_3 may induce Fe-chlorosis (Chen and Barak, 1982). In alkaline and calcareous soils, no major NH^+_4 accumulation occurs because of the high soil pH, which causes a deportation of NH^+_4 and the resulting NH^+_3 can be easily lost by volatilization. This indicates that on soils with a higher pH level plants are mainly, if not exclusively, nourished by NO^-_3 (Mengal and Geurtzen, 1986). The high pH values in the apoplast may inhibit iron transport across the plasma membrane. NH^+_4 nutrition which is supposed to decrease the apoplastic pH should then increase the mobilization of iron and induce a regreening of the chlorotic tissue. However, regreening of chlorotic leaves occurred during a lag phase of two to three days. The pH in the apoplast of the plant tissues depends also on plasmalemma located ATPase, which pumps H^+ from the cytoplasm into the apoplast. Indole-3-acetic acid (IAA) and fusicoccin (FC) are known to stimulate the plasmalemma located ATPase. If the hypothesis is correct that too high an apoplastic pH is the cause of iron-chlorosis, application of FC or IAA should result in a regreening of chlorotic leaves. It was found that FC or IAA sprayed on the chlorotic leaves caused a regreening of leaves after a lag phase of five days. Why high pH levels in the apoplast induce chlorosis still needs elucidation. Since alkaline nutrition may increase the nutrient solution by about two pH units (from 5-7) during a period of 24 hours (Mengel *et al.*, 1983), it is feasible that in the limited space of the leaf-apoplast extremely high pH values may occur that could bring about a precipitation of Fe-III hydroxyoxides. High phosphate concentrations in the apoplast might favour this kind of iron immobilization. Fe-III is reduced at plasmalemma and then translocated across the plasma membrane.

Ellsworth *et al.* (2000) reported that iron (Fe) chlorosis continues to be a problem in the world where crops are grown in calcareous soils. Although it is possible to remedy this problem with fertilizer application, selection of cultivars resistant to Fe deficiency chlorosis is a more economical and practical solution. In addition to field chlorosis rating studies, several methods have been developed to screen cultivars. One of the most promising is quantifying individual plant response mechanisms induced during Fe deficiency stress, such as Fe reduction or hydrogen ion (H^+) extrusion by the roots. Eight dry bean (*Phaseolus vulgaris* L.) and 11 soybean (*Glycine max* L.) cultivars of varying resistance to Fe deficiency chlorosis were grown in solutions of low (0.05 mg L^{-1}) and no (0 mg L^{-1}) Fe, respectively. Beginning at day 2 after imposition of low Fe treatments, plant roots were incubated for one hour in a solution to collect H^+ ions. This solution was then titrated with NaOH to the pH of the control solutions (no plants grown in solution, but air bubbled through it for one hr). The results of six consecutive days were summed and correlated with field chlorosis scores. In dry bean, there was a statistically significant positive relationship between H^+ release and field chlorosis scores, which is opposite of the theoretical relationship, i.e., dry bean cultivars with high chlorosis scores released the largest quantity of H^+ ion and vice versa. Soybean exhibited a statistically significant negative relationship between quantity of H^+ release and field chlorosis scores, but Fe reduction measurements provided better screening ability than H^+ ion release. Adjusting the measurements for fresh root weight did not alter these relationships; thus, eliminating the need to weigh roots in a breeding nursery. Combining the effects of H^+ release and Fe reduction using standardized scores did not improve the selection of Fe efficient cultivars over Fe reduction alone in either dry bean or soybean. Hydrogen ion quantification was time consuming, tedious, and gave conflicting results in the two species studied. Thus, it is not recommended as a screening technique in either dry bean or soybean.

Franzen and Richardson (2000) made studies on soil factors affecting iron chlorosis of soybean in the Red River Valley of North Dakota and Minnesota. Iron chlorosis tolerant soybeans exhibit chlorosis symptoms in the Red River Valley of North Dakota and Minnesota. Previous research suggests that chlorosis was generally related to high calcium (Ca) carbonate levels in the soil, so a relationship of chlorosis with the presence of Calciaquoll (carcareous at the surface, with very high levels of calcite in the subsurface) and non-Calciaquoll soil types would be expected. Six sites each in 1996, 1997, and 1998 were studied to identify soil factors correlated with the incidence and degree of chlorosis. Gradients between green soybeans and severely chlorotic plants were established, soil type was determined at the green and chlorotic gradient endpoints, and soil samples were taken at the 0–15 cm depth at each gradient location. Iron chlorosis symptoms were filmed using a video recorder. The images recorded were then analysed using an image analyser to give a digital chrome number. The digital values were correlated with soil factors to determine the degree of relationship of symptoms with each factor. Soil type differences were only associated with chlorosis at four of twelve sites. Both Ca carbonate equivalence and soluble salts were most often correlated with chlorosis symptoms. The relationship of iron chlorosis with soluble salts, along with soil carbonate level, appears to be a factor that should be considered in soybean chlorosis resistance breeding for the Red River Valley. A greenhouse experiment was conducted which was unsuccessful in duplicating field chlorosis symptoms when gypsum and Ca carbonate was added to soil obtained from non-chlorotic areas. The study was able to show a decrease in soybean nodule number with increasing Ca carbonate and gypsum levels.

UPTAKE AND DISTRIBUTION OF IRON IN PLANTS

Plants take up iron in ionic form as ferrous or ferric or as iron chelate but the most suitable form of iron supply to plant are usually believed to be the chelated form because of its higher solubility. Suggestions have been made that prior to its absorption by roots, iron is reduced from Fe^{3+} to Fe^{2+} (Olsen and Brown, 1980). Iron is passively transported by mass flow in transpiration stream or iron is translocated by diffusion. Several workers believe that iron uptake is an active process and is controlled metabolically (Epstein, 1960; Mandley, 1981). Later two strategies were proposed for iron acqisition in plants (Marschner *et al.*, 1986; Römheld and Marschner, 1986 a,b; Römheld, 1987; Chaney, 1988). These were referred to as *Strategy I* (in dicots) and *Strategy II* (in monocots).

Since the discovery, classification, and techniques of phytosiderophores collections have been already detailed by Hemantaranjan (1995; 2002), Singh *et al.*(1995) and Vaishampayan (1995) and in recent years, distinctive homeostatic mechanisms have emerged in plants to sustain their micronutrient nutrition, and facilitate them to act in response to micronutrient stress, so this portion will focus attention on mechanisms of iron uptake from siderophores because of the critical role of iron in growth, metabolism and development of plants.

Certain relevant resolutions come out of the overall research in the area of iron nutrition and interactions in plants and/or on phytosiderophores carried out in the last few decades open new vistas to expedite precise work in the new millennium, which are as follows:

1. The identification of Fe-stress-inducible proteins at the molecular level is essential for achieving detailed insight into their regulation.
2. The use of *Arabidopsis* mutants may allow the recognition of key regulatory points in the pathways leading to enhanced uptake of ions.
3. To make further query whether or not the separation between *Strategy I* and *Strategy II* is of importance under conditions of adequate availability of Fe.

4. More physiological and biochemical explanations at molecular level are needed with reference to nicotianamine (NA), mugineic acid and related phytosiderophores, a device for iron supplication under Fe-deficient condition
5. For Fe uptake in higher plants, progress on various fronts is still required to improve our understanding.
6. A better definition of the bioavailable Fe pools is also desirable for economic plants.

The relevant processes include the discharge of substances into the external environment to increase metal solubility and accessibility, the capacity to alter the redox state of the metal, and/or the capability for membrane transport. Numerous facts in relation to root iron acquisition mechanisms constitute the homeostatic processes that help the plant in endeavouring to obtain enough levels of these essential metals, while at the same time preventing excess accumulation and/or toxic damage. In other words, homeostasis is the aptitude of an organism to maintain a relatively constant internal environment, by suitably regulating its physiological processes in response to conditions in the external environment. Many of these processes are highly regulated, exhibiting both spatial and temporal modulation, and as a group are referred to as the micronutrient stress responses. The availability of energy resources, reductants, and/or metabolites may place a supplementary layer of regulation on the functional capacity of expressed homeostatic devices, especially when the plant's wide-ranging vigour is conciliated.

Recently, Schikora and Schmidt (2001) made studies on iron stress-induced changes in root epidermal cell fate, which are regulated independently from physiological responses to low iron availability. Iron-overaccumulating mutants were investigated with respect to changes in epidermal cell patterning and root reductase activity in response to iron starvation. The induction of transfer cells in the rhizodermis appeared to be iron regulated in the pea (*Pisum sativum* L. *cv* Dippes Gelbe Viktoria and *cv* Sparkle) mutants *bronze* and *degenerated leaflets*, but not in roots of the tomato (*Lycopersicon esculentum* Mill. *cv* Bonner Beste) mutant *chloronerva,* suggesting that in *chloronerva* iron cannot be recognized by putative sensor proteins. Experiments with split-root plants supports the hypothesis that Fe (III) chelate reductase is regulated by a shoot-borne signal molecule, communicating the iron status of the shoot to the roots. In contrast, the formation of transfer cells was dependent on the local concentration of iron, implying that this shoot signal does not affect their formation.

In recent years, a number of studies were made regarding the reduction of siderophore and transport of iron, iron transport in phloem-loading and transport vehicles, patterns of phytosiderophore release, immobilized siderophores and synthetic chelates as a source of iron to plants, key enzyme of iron homeostasis as well as the complexities of the phloem transport system with reference to implications for shoot/root interactions and iron homeostasis in plants. All these studies emphasize the significance phytosiderophores production and activity in plants belonging to the groups of dicotyledons and monocotyledons under extreme iron deficient conditions. Iron is an essential trace element for all living cells. In contrast to many other metals, it is difficult to maintain in living systems because of two properties: (a) the ability to easily change its valence under different redox environments, and (b) the dramatic change of solubility of ferrous and ferric compounds at different pH values. Ferric compounds are virtually insoluble at the neutral and weakly alkaline pH range as predominant in living cells, and also in many of the soils (Stephan, 2000: 10th International Iron Symposium, Houstan, Texas, U.S.A.). Roots take up iron from the soil, transport it in radial direction to the central cylinder, and load it into xylem vessels for distribution to all organs with the exception of young, still growing tissues that still lack sufficiently developed vessels. These sites are supplied by the phloem after xylem-phloem-transfer of iron. The root tip, although embedded in the iron-delivering medium, also receives iron via the sieve tubes. Supply by the phloem requires special transport and unloading vehicles.

Interestingly, roots have mechanisms to overcome the poor availability of iron in soils. They increase its solubility by acidification of rhizosphere and by increase of a plasmalemma-bound ferric reductase (*Strategy I*), or by excretion of ferric chelators that are reintroduced after complexation of iron (*Strategy II*). The induction and maintenance of these processes require special metabolic adaptations in the root cells. These activities are regulated by iron sensing of the roots themselves as well as by signal(s) from the shoot transferred by the phloem. Further the transport of iron from the root surface to the xylem vessels may take place apoplastically but it is at least partially a symplastic one in order to bridge the *Casparian strip* that functions as a checkpoint of iron fluxes. During symplast transport suitable chelator molecules to prevent iron precipitations must mask the iron ion. Under reductive environment or in mutants disturbed in their iron homeostasis mechanisms, iron excess can occur. This can cause oxidative stress, which leads to destruction of cell structures. Plants protect themselves by storage compounds, *e.g.*, phytoferritin; also plant-specific iron chelator nicotianamine is discussed to play a role in temporary iron detoxification.

Soils contain siderophores produced by bacteria and fungi, however, the role of siderophores in iron nutrition of plants is uncertain. In recent years, a number of studies were made regarding the reduction of siderophore and transport of iron, iron transport in phloem-loading and transport vehicles, patterns of phytosiderophore release, immobilized siderophores and synthetic chelates as a source of iron to plants, key enzyme of iron homeostasis as well as the complexities of the phloem transport system with reference to implications for shoot/root interactions and iron homeostasis in plants. Responses of Fe deficiency stress have been classified into *Strategy I* and *II*. Iron-efficient cultivars within each strategy exhibit one or more of a combination of specific physiological responses to Fe deficiency stress, while Fe-inefficient cultivars do not. In addition, graminaceous species (grasses) respond quite differently to iron deficiency by increasing sharply the release of nonproteinogenic amino acids ["phytosiderophores"] from the roots. These amino acids mobilize sparingly soluble inorganic Fe (III) in the rhizosphere by the formation of Fe (III) chelates. Plant uptake of iron from chelates is considered to involve a reduction mechanism, in which iron is reduced at the root surface and subsequently taken up as the Fe (II) ion (Chaney *et al.,* 1972; Olsen *et al.,* 1982). At the root surface, iron is released from the chelate in accordance with equilibria relationships governing the chelate and inorganic iron in the soil solution. After deferration, the chelate diffuses away from the plant root and recombines with the iron in solution controlled by solid phase Fe $(OH)_3$. Iron chelates do not increase the solubility of Fe (III) or Fe (II) but only serve to hold iron to the root surface (Lindsay and Schwab, 1982).

Iron chelators are the most effective Fe fertilizers known to date. They are, however, easily leached to depth beyond that of the root zone due to their negative charge. Leaching often creates the need to frequently replenish the rhizosphere with chelated Fe. Binding of the synthetic chelates or microbial siderophores to a solid phase is therefore an attractive alternative.

Mugineic acid (MA), a natural chelator, was isolated from the root washings of barley cultivar "Minorimugi". The discharge of MA by barley roots occurred in the morning, and was much enhanced under Fe-stress (Takagi *et al.,* 1984).

Studies on induction of *in vivo* root ferric chelate reductase activity in fruit tree rootstock by Gogorcena *et al.* (2000) were made. A method has been developed to consistently induce increases in root ferric chelate reductase activity in the fruit tree rootstock GF 677 (*Prunus amygdalopersica*) grown under iron (Fe) deficiency.

Recently, German workers Herbik *et al.*(2002) with their iron uptake studies reported the involvement of a multicopper oxidase by the green algae *Chlamydomonas reinhardtii* has been found. In *Chlamydomonas reinhardtii*, high-affinity uptake of iron (Fe) requires an Fe^{3+}-chelate reductase and an Fe transporter. Neither of these proteins nor their corresponding genes have been isolated. We previously identified, by analysis of differentially expressed plasma membrane proteins, an approximately 150-kD protein whose synthesis was induced under conditions of Fe-deficient growth. Based on homology of internal peptide sequences to the multicopper oxidase hephaestin, this protein was proposed to be a ferroxidase. A nucleotide sequence to the full-length cDNA clone for this ferroxidase-like protein has been obtained. Analysis of the primary amino acid sequence revealed a putative transmembrane domain near the amino terminus of the protein and signature sequences for two multicopper oxidase I motifs and one multicopper oxidase II motif. The ferroxidase-like gene was transcribed under conditions of Fe deficiency. Consistent with the role of a copper (Cu)-containing protein in Fe homeostasis, growth of cells in Cu-depleted media eliminated high-affinity Fe uptake, and Cu-deficient cells that were grown in optimal Fe showed greatly reduced Fe accumulation compared with control, Cu-sufficient cells. Reapplication of Cu resulted in the recovery of Fe transport activity. Together, these results were consistent with the participation of a ferroxidase in high-affinity Fe uptake in *C. reinhardtii* (Herbik *et al.*, 2002).

In plants, the different forms of iron identified are plastid ferritin, lamellar phospholipo-proteins and ribonuclease solubilizable fractions (Terry, 1980). It is generally believed that total iron is not an indicator of iron status of plants because all iron present in the tissue is not in physiologically active form. Various extractants have been used by different workers to extract 'active' iron from plant tissues, which include water, buffers, organic solvents, acids and chelating agents (Pierson *et al.,* 1986). Valenzuela and Romero (1988) characterized Biochemical indicator Index (BI) including catalase, peroxidase, chlorophyll, carotenoid, anthocyanin, active and total iron indices to evaluate the iron status of a plant species.

ROLE OF HORMONES IN THE INDUCTION OF Fe(III) CHELATE REDUCTASE ACTIVITY

To prove whether changes in hormone concentration or sensitivity are involved in the adaptation to suboptimal Fe availability, Schmidt *et al.* (2000) tested 45 mutants of *Arabidopsis* defective in hormone metabolism and/or root hair formation for their ability to increase Fe (III) chelate reductase activity and to initiate the formation and enlargement of root hairs. Activity staining for ferric chelate reductase revealed that all mutants were responsive to Fe deficiency, suggesting that hormones are not necessary for the induction. Treatment of wild-type plants with the ethylene precursor 1-aminocyclopropane-1-carboxylic acid caused the development of root hairs in locations normally occupied by non-hair cells, but did not stimulate ferric reductase activity.

All surveyed mutants appeared to be capable of responding to Fe-deficiency stress by elevated Fe (III) chelate reductase activity. The intensity of staining produced by most Fe-grown mutants was indistinguishable from the wild type, suggesting that hormones are not required for the induction of reduction activity. Mutants exhibiting deviations from the wild type are further characterized by alterations in length and density of root hairs. This was particularly evident for *axr2*, which has almost no root hairs and displayed a small increase in reduction activity, and *eto3*, which represents the other extreme with respect to root hair density and intensity of reductase staining. Thus, reductase activity seems to be associated with root surface area rather than with alterations in hormone levels or sensitivity. Exceptions to this pattern were only noted in cases where the mutation causes a highly reduced root system (*e.g. cri1*, *cyr1*, and *stp1*).

These findings support a model in which, unlike the morphological changes, induction of the physiological responses to Fe deficiency stress is not mediated by hormones. This model is supported by the normal Fe stress response of ethylene insensitive mutants (*ein2-ein7* and *etr*) and other hormone-related mutants. The results presented here are at variance with those reported by Romera *et al.* (1994). However, while the data of the present study do exclude an absolute requirement of hormones for the induction of physiological responses to Fe deficiency, they do not exclude that ethylene can act as enhancer in the transduction of environmental stimuli. Such an enhancer function might be inferred from the relatively low Fe-stress-induced increase in activity of *glabra* and *ttg* phenotypes and the normal response of the *caprice* mutants, which possess fewer root hairs than the wild type under - Fe-conditions.

An essential function for auxin in the induction of ferric reductase activity seems unlikely with regard to the present data. Since inhibitors of polar auxin transport have been suggested to interfere with the induction of Fe-deficiency-induced responses (Landsberg, 1982), the normal response of *tir3-1*, a defect in polar auxin transport, is of particular interest and can be interpreted in this way: The staining of the medium by - Fe-grown roots of the auxin overproducer *sur1* is indistinguishable from the wild type, which may be judged as further evidence against a possible role of auxin in the regulation of root reduction activity.

Function of Root Hairs in Fe Reduction

Determinations of Fe (III) chelate reductase activity *in vivo* are usually based on a root fresh weight basis. This is also true for the half-quantitative visualization of root-mediated Fe(III) reduction in agar medium used in the present study. Since the reduction activity is restricted to young, non-lignified root zones, it may be assumed that an increase in surface area by enhanced root hair formation increases the reduction of Fe (III) in the medium. The present results appear to support this assumption. Further evidence comes from a study by Moog *et al.* (1995) showing that the activity of wild-type Arabidopsis roots was about 2-fold higher than the root-hair-less mutant RM57. With respect to the temporal pattern of induction of the Fe stress syndrome in Arabidopsis, the occurrence of extra root hairs always preceded reduction activity and the development of chlorosis symptoms. Fe-deficiency-induced root hair formation may therefore be regarded as a means to enhance the physiological reactions, and not as a secondary effect of chlorosis, as suggested by Chaney *et al.* (1992). It should be noted, however, that the present data do not support the idea that the formation of root hairs is a prerequisite for the development of enhanced ferric reduction activity. Even mutants with extremely reduced surface area, *e.g. axr2*, showed this response under the conditions in our study.

PHYSIOLOGICAL AND BIOCHEMICAL EFFECTS OF IRON

Photosynthesis

Iron affects various aspects of photosynthetic activities directly or indirectly it is a constituent of a number of heme and non-heme proteins in the chloroplast (Pushnik and Miller, 1989). Suggestions have been made on decreased rate of photosynthesis in iron deficiency, which is due to limited light harvesting and electron transport capacity of Fe-deficient plants (Terry, 1983). Iron deficiency is known to decrease the rate of net photosynthesis. It appears that the decrease in the rate of carbon assimilation due to iron deficiency could be related to the decrease in chlorophyll concentration (Terry, 1983).

Recently Graziano *et al.*(2002) reported that nitric oxide improves internal iron availability in plants. As established, iron deficiency impairs chlorophyll biosynthesis and chloroplast development. In leaves, most of the iron must cross several biological membranes to reach the chloroplast. The components involved in the complex internal iron transport are largely unknown. Nitric oxide (NO), a bioactive free radical, can react with transition metals to form metal-nitrosyl complexes. Sodium nitroprusside, an NO donor, completely prevented leaf interveinal chlorosis in maize (*Zea mays*) plants growing with an iron concentration as low as 10 μM Fe-EDTA in the nutrient solution. *S*-Nitroso-*N*-acetylpenicillamine, another NO donor, as well as gaseous NO supply in a translucent chamber was also able to revert the iron deficiency symptoms. A specific NO scavenger, 2-(4-carboxy-phenyl)-4,4,5,5-tetramethylimidazoline-1-oxyl-3-oxide, blocked the effect of the NO donors.

The effect of NO treatment on the photosynthetic apparatus of iron-deficient plants was also studied. Electron micrographs of mesophyll cells from iron-deficient maize plants revealed plastids with few photosynthetic lamellae and rudimentary grana. In contrast, in NO-treated maize plants, mesophyll chloroplast appeared completely developed. NO treatment did not increase iron content in plant organs, when expressed in a fresh matter basis, suggesting that root iron uptake was not enhanced. NO scavengers 2-(4-carboxy-phenyl)-4, 4, 5, 5-tetramethylimidazoline-1-oxyl-3-oxide and methylene blue promoted interveinal chlorosis in iron-replete maize plants (growing in 250 μM Fe-EDTA). Even though results support a role for endogenous NO in iron nutrition, experiments did not establish an essential role. NO was also able to revert the chlorotic phenotype of the iron-inefficient maize mutants *yellow stripe1* and *yellow stripe3*, both impaired in the iron uptake mechanisms. All together, these results support a biological action of NO on the availability and/or delivery of metabolically active iron within the plant.

Current studies made on expression and biochemical properties of a ferredoxin-dependent heme oxygenase required for phytochrome chromophore synthesis point toward that the *HY1* gene of *Arabidopsis* encodes a plastid heme oxygenase (AtHO1) required for the synthesis of the chromophore of the phytochrome family of plant photoreceptors. To determine the enzymatic properties of plant heme oxygenases, we have expressed the *HY1* gene (without the plastid transit peptide) in *Escherichia coli* to produce an amino terminal fusion protein between AtHO1 and glutathione *S*-transferase. The fusion protein was soluble and expressed at high levels. Purified recombinant AtHO1, after glutathione *S*-transferase cleavage, is a hemoprotein that forms a 1:1 complex with heme. In the presence of reduced ferredoxin, AtHO1 catalyzed the formation of biliverdin IXa from heme with the concomitant production of carbon monoxide. Heme oxygenase activity could also be reconstituted using photoreduced ferredoxin generated through light irradiation of isolated thylakoid membranes, suggesting that ferredoxin may be the electron donor in vivo. In addition, AtHO1 required an iron chelator and second reductant, such as ascorbate, for full activity. These results show that the basic mechanism of heme cleavage has been conserved between plants and other organisms even though the function, subcellular localization, and cofactor requirements of heme oxygenases differ substantially (Muramoto, 2002).

Respiration

In several plant species, the deficiency of iron is reported to decrease the rate of respiration (Hewitt, 1963). Iron is closely involved in the process of respiration evidently because of being a constituent of cytochromes, flavoproteins and enzymes like cytochrome oxidase, succinic dehydrogenase, aconitase and several other FeS centers, which carry out redox reactions in respiratory pathway (Hewitt, 1983).

Carbohydrate Metabolism

The role of iron in carbohydrate metabolism is not yet established. High iron supplies have been shown to increase the synthesis of sugars & starch (Chaturvedi and Singh, 1981).

Nitrogen Metabolism

Nitrogen metabolism of different plant species is affected differently by low iron supply. Iron being the constituent of nitrate reductase its activity is markedly reduced in the deficiency of iron (Hageman and Reed, 1980; Hewitt and Notton, 1980). Iron deficiency is known to decrease the concentration of some amino acids, *viz.*, alanine, arginine, aspartic acid, glutamic acid, aminobutyric acid, proline, lysine, threonine and valine (Mehrotra and Bisht, 1990). The decreased number of ribosomes in iron deficient maize is indicative of the involvement of iron in protein synthesis (Lin and Stocking, 1978).

Iron in Relation to Enzymes

Iron has a specific role as a constituent metal or co-factor and is known to be involved in a number of enzymes of metabolic processes in different plant species. In catalase, iron is bound to porphyrin and protein probably by ionic linkages. Peroxidase is a heme protein with protoheme as prosthetic group. In cytochrome oxidase also iron is present in porphyrin undergoing continuous oxidation-reduction into Fe^{3+} and Fe^{2+}. In aconitase, the bound iron is in ferrous form, whereas succinic dehydrogenase is an Fe-S protein with a tightly bound flavin component. Another iron containing enzyme superoxide dismutase, which catalyses the conversion of superoxide radical (O_2) to H_2O_2, has been described by Salin and Bridges (1980).

Recently Moran *et al.* (2003) reported an iron-superoxide dismutase (FeSOD) with an unusual subcellular localization, VuFeSOD, has been purified from cowpea (*Vigna unguiculata*) nodules and leaves. The enzyme has two identical subunits of 27 kD that are not covalently bound. Comparison of its N-terminal sequence (NVAGINLL) with the cDNA-derived amino acid sequence showed that VuFeSOD is synthesized as a precursor with seven additional amino acids. The mature protein was overexpressed in *Escherichia coli*, and the recombinant enzyme was used to generate a polyclonal monospecific antibody. Phylogenetic and immunological data demonstrate that there are at least two types of FeSODs in plants. An enzyme homologous to VuFeSOD is present in soybean (*Glycine max*) and common bean (*Phaseolus vulgaris*) nodules but not in alfalfa (*Medicago sativa*) and pea (*Pisum sativum*) nodules. The latter two species also contain FeSODs in the leaves and nodules, but the enzymes are presumably localized to the chloroplasts and plastids. In contrast, immunoblots of the soluble nodule fraction and immunoelectron microscopy of cryo-processed nodule sections demonstrate that VuFeSOD is localized to the cytosol. Immunoblot analysis showed that the content of VuFeSOD protein increases in senescent nodules with active leghemoglobin degradation, suggesting a direct or indirect (free radical-mediated) role of the released Fe in enzyme induction. Therefore, contrary to the widely held view, FeSODs in plants are not restricted to the chloroplasts and may become an important defensive mechanism against the oxidative stress associated with senescence (Moran *et al.*, 2003).

Iron in Relation to Reproductive Physiology

Effect of low iron on the development and physiology of pollen in plants has also been observed (Sharma *et al.,* 1984). Iron deficiency causes a marked decrease in pollen producing capacity of anthers, pollen grain size and activities of catalase and peroxidase in maize anthers. If not directly, iron has enough scope to induce reproduction, process of fertilization and seed development even indirectly, as has been observed in the present experiment too.

Nucleic Acid Metabolism

Iron is also known to be involved in the establishment of bond between individual DNA helices and at the same time regulates denaturation of DNA (Goldstein and Gerasimova, 1963). The Fe deficiency changes the concentration of nucleic acids and on spraying the field plants with adenine and guanine, the iron chlorosis was corrected showing indirect of iron on nucleic acid metabolism (Kessler, 1957). In radish leaves Fe deficiency decreased DNA content but had no significant effect on RNA content (Agarwala and Mehrotra, 1984).

Iron and Growth Substances

Coconut milk iron was found to cause stimulation of cell division in cultured plants of tomato, presumably by regulating protein synthesis (Newman and Steward, 1969). Stimulated abscission of citrus fruits by formation of abscission layer by iron application has been suggested to be due to ethylene formation and destruction of IAA (Ben-Yehoshua and Briggs, 1970). Iron deficiency in wheat decreased total auxin, gibberellin and cytokinin content of leaves but increased the ABA content. Similar effects were reported in iron-deficient maize having increased concentration of ABA type growth inhibitors and absence of gibberellin like substance (Stoyanov and Kha, 1981).

INTERACTIONS INVOLVING IRON ON CROP YIELD

Crop yields are controlled by a number of limiting factors which interact in various ways, such as sequential activity, synergism, antagonism and some in-between ways (Wallace, 1990). Iron strongly enters into interactions, and one of the major goals of plant breeding programmes is to eliminate the causes of Fe being a Liebig-type limiting factor (Wallace, 1995).

Productivity of soybean is very low as compared to other crops. The yield potentiality of this crop can be improved to certain degree by manipulating the environmental factors, akin to mineral nutrition. This is for the reason that deficiency of essential mineral nutrients especially the micronutrients in intensive cropping system are of general occurrence. Sulphur deficiency too has been reported in almost every part of the world (Beaton, 1966) and its application improves crop yield and mineral content in legumes (Bentley *et al.,* 1956). Furthermore, S requirement of soybean is more as compared to other leguminous crops. In soils with high pH, Fe availability, *i.e.*, ferrous (Fe^{2+}) or active iron decreases and as a result Fe-chlorosis occurs extensively. In the present days, elemental sulphur has been used to correct Fe-chlorosis with varying success (Kalbasi *et al.,* 1986).

An Fe-efficient and an Fe-inefficient cultivar of soybean grown in calcareous loam soil with FeEDDHA, P and N fertilizers in factorial combination (Wallace, 1990). The results indicate the need for full attention to all variables involved in the production of a crop. With the Fe-efficient cultivar, there was response to Fe alone (40%) and very small responses to P alone (5%), and N alone (13%). When N and P were combined, there was antagonism. The Fe with P and Fe with N both gave sequentially additive interactions and very near. When all three inputs were combined, the responses to Fe, P and N respectively, were 80, 34 and 44%, and the interactions was synergistic. Each nutrient was a Liebig-type limiting factor. With Fe-inefficient soybean, FeEDDHA gave no response when used alone, and N and P singly and together were very antagonistic. When all three nutrients were supplied simultaneously, the interaction was synergistic with responses to Fe, P and N, respectively, 569, 471 and 116%. Even after Fe has been supplied, the N and P were Kiebig-type limiting factors. Yields were higher for the complete treatment with the Fe-inefficient cultivar than with the efficient one. Maximum responses to Fe are possible only if other management practices have been taken care of. In addition, with Fe-inefficient soybean, P was very damaging when used alone. Stresses that decreased yields when applied singly or in pairs resulted in large yield increases when three or four inputs were applied simultaneously.

Field experiments were carried out with soybean (*Glycine max* L. Merr) in sandy loam soil of Varanasi. Foliar application of boron (B) as boric acid and soil applications of individual and combined levels of sulphur (S) (@ 40 and 80 mg kg^{-1} soil) and iron (@ 15 and 20 mg kg^{-1} soil) were made. In general, the foliar application of B showed an increase in morpho-physiological attributes accompanied by total dry matter production and seed yield. B application has been also reported to increase the physiologically active iron (ferrous) in soybean in another experiment (Hemantaranjan, 1999).

Besides these, the soil applied Fe and S revealed significant increase in shoot height, root length, total dry matter production and test weight along with successive increase in oil yield over untreated (control) plants (Hemantaranjan *et al.,* 2000). Soil applied S and Fe @ 80 + 20 mg kg^{-1} soil was optimal amongst individual and combined treatments resulting into enhanced shoot height, root length, number of leaves $plant^{-1}$, leaf area $plant^{-1}$, total dry matter production $plant^{-1}$, test weight (100 seed weight) and seed oil content. Individual S and Fe @ 80 and 20 mg kg^{-1} soil respectively amongst other individual soil applied S and Fe improved all the afore-mentioned characteristics in soybean. Similarly, different combined treatments of S and Fe too stimulated an increase as above in comparison to untreated (control) plants. Increased iron availability helps in maintaining the chlorophyll levels of leaf resulting into greater production of photosynthates and ultimately an increase in the dry matter content. Correspondingly, S fertilization helps to maintain active iron (ferrous iron) level in plants, needed for a number of metabolic processes (Pierson and Clark, 1984).

The overall findings in this part indicate that soil application(s) of S and Fe might be the important factors in improving seed and oil yield of soybean especially in soils having nature and properties similar to the present soil conditions (Hemantaranjan *et al.,* 2000).

REFERENCES

Agarwala, S.C. and Chatterjee, C. 1996. Physiology and biochemistry of micronutrient elements. *Advancements in Micronutrient Research* (Ed. Hemantaranjan, A.). Scientific Publishers, Jodhpur India. 203-365.

Agarwala, S.C. and Mehrotra, S.C. 1984. Iron-magnesium antagonism in growth and mechanism of radish. *Plant Soil,* **80:**355-361.

Alam, S., Kamei, S., Kawai, S. 2001. Response of iron-deficient barley plants to manganese in nutrient solution. *J. Plant Nutr.,* **24:** 12-25.

Beaton, J. D. (1966). Sulphur requirements of cereals, tree fruits, vegetables and other crops. *Soil Sci.,* **101:** 267-282.

Becker, R., Fritz, E. and Manteuffel, R. 1995. Subcellular localization and characterization of excessive iron in the nicotianamine-less tomato mutant *chloronerva*. *Plant Physiol.,* **108:** 269–275.

Bentley, C. F., Garean L. and Renner, R. (1956). Fertilizers and nutritive values of hays. I. Sulphur deficient grey wooded soil. *Canad. J. Agric. Aci.,* **36:** 315-325.

Ben-Yehoshua, S. and Briggs, R.H.1970. Effects of iron and copper ions in promotion of selective abscission and ethylene production by citrus plant fruit and the inactivation of indoleacetic acid. *Plant Physiol.,* **45:** 604-607.

Berkowitz, G.A. and Peters, J.S. (1993) Chloroplast inner-envelope ATPase acts as a primary H^+ pump. *Plant Physiol.,* **102:** 261-267.

Briat, J.F., Fobis-Loisy, I., Grignon, N., Lobréaux, S., Pascal, N., Savino, G., Thioron, S., Van Wirén, N. & Van Wuytswinkel, O. (1995) Cellular and molecular aspects of iron metabolism in plants. *Biol. Cell,* **84**, 69–81.

Chaney, R.L., Brown, J.C. and Tiffin, L.O. 1972. Obligatory reduction of ferric chelates in iron uptake by soybeans. *Plant Physiol.,* **50:** 208-213.

Chaney, R.L., Chen, Y., Green, C.E., Holden, M.J., Bell, P.F., Luster, D.G., Angle, J.S. (1992) Root hairs on chlorotic tomatoes are an effect of chlorosis rather than part of the adaptive Fe-stress-response. *J Plant Nutr.,* **15:** 1857-1875.

Chaney, R.L. 1988. Recent progress aqnd needed research in Fe nutrition. *J. Plant Nutr.,***11:** 1589-1603.

Chatuvedi, G.S. and Singh, A. 1981. Effect of iron and manganese on major biochemical constituents, growth and yield in wheat, *Triticum aestivum* L. *Indian J.Exptl.Biol.,* **19:** 1135-1139.

Chen, Y. and Barak, P. 1982. Iron nutrition of plants in calcareous soils. *Adv.Agron.,***35:** 217-240.

Clark, R.B., Olsen, R.A., Bennett, J.H. and Brown, J.C. 1981. Biological effects on plants. In *EPA Multimedia Criteria for Iron and Compounds.* Eds. V.Finelli, S.S. Que Hee and W.L. Pickens. TDI Inc., Cincinnati, OH (USA).

Evans, H.J. and Russel, S.A. 1971. Physiological chemistry of symbiotic nitrogen fixation by legumes. In: *A Treatise on Nitrogen Fixation.* Ed. J. R. Postgate. 191-244. Plenum, London.

Franzen, D.W. and Richardson, J.L. 2000. Soil factors affecting iron chlorosis of soybean in the red river valley of North Dakota and Minnesota. *J.Plant Nutr.,* **23:** 67-78.

Goldstein, B.I. and Gerasimova, V.V. 1963. The denaturation and fragmentation of deoxyribonucleic acid in animal cells. *Ukrainsky Biokhimicheskij Zhurnal.,* **35:** 3-17.

Graziano, M., Beligni, M.V. and Lamattina, L. 2002. Nitric oxide improves internal iron availability in plants. *Plant Physiol,,* **130:** 1852-1859

Gris, E.1844. Nauvelles experiences sur l'action des composes ferrugineux solubles appliqué a la vegetation at specialement on treatment de la chlorose et da la de' bilite des plantes. *2 C.R. Acad.Sci (Paris),* **19:** 1118-1119.

Grusak, M.A. 2001. Homeostasis and the regulation of micronutrient stress responses in roots. In: *Perspectives on the Micronutrient Nutrition of Crops.* Eds. K.Singh, S.Mori and R.M.Welch. Scientific, Jodhpur, India. 71-90.

Grusak , M.A., Pearson, J.N. and Marentes, E. 1999. The physiology of micronutrient homeostasis in field crops. *Field Crops Res.,* **60:** 41-56.

Hageman, R.H. and Reed, A.J.1980. Nitrate reductase from higher plants. In: *Methods in Enzymology,* **69:** 270-281.

Hemantaranjan, A. 2002. Physiological and molecular advances in phytosiderophores – a contrivance for iron nutrition in plants. In: *Advances in Plant Physiology.* A Treatise, Ed. A. Hemantaranjan, *Section* 4, *Iron Nutrition and Interactions in Plants.* **4**: 287-308.

Hemantaranjan, A. 1995. Phytosiderophore production and activity. *Adv.Iron Nutr.Research,* **1:** 191-213. Ed. A. Hemantaranjan, Scientific Publishers, Jodhpur, India.

Hemantaranjan, A., A. K. Trivedi and Mani Ram. 2000. Effect of foliar applied boron and soil applied iron and sulphur on growth and yield of soybean (*Glycine max* L. Merr.). *Indian J. Plant Physiol.,* **5**(2)New Series: 142-144.

Hemantaranjan, A. 1999. Physiological significance of boron and cobalt on iron activity, nitrogen fixation and yield potentiality of *Glycine max* L. *Plant Physiol.Sustainable Agril.,* **27**: 163-169. Ed. G.C.Srivastava *et al.* Pointer Publishers, Jodhpur.

Herbik, A., Koch, G., Mock, H.P., Dushkov, D., Czihal, A., Thielmann, J., Stephan, U.W., Bäumlein, H. (1999) Isolation, characterization and cDNA cloning of nicotianamine synthase from barley: a key enzyme for iron homeostasis in plants. *Eur J Biochem*., **265:** 231-239

Herbik, A., Bölling, C. and Buckhout, T.J. 2002. The Involvement of a multicopper oxidase in iron Uptake by the Green Algae *Chlamydomonas reinhardtii. Plant Physiol.*, **130**: 2039-2048.

Herbik, A., Giritch, A., Horstmann, C., Becker, R., Balzer, H.J., Bäumlein, H. & Stephan, U.W. (1996) Iron and copper nutrition-dependent changes in protein expression in a tomato wild type and the nicotianamine-free mutant *chloronerva*. *Plant Physiol.,* **111**, 533–540.

Hewitt, E.J. and Notton, B.A.1980. Nitrate reductase systems in eukaryotic and prokaryotic organisms. In: *Molybdenum Containing Enzymes,* Ed. M. Coughlah. 275-325.

Hewitt, E.J. 1983. The effect of mineral deficiencies and excess on growth and composition. In: *Diagnosis of Mineral Disorders in Plants.* Vol.I. Principles (Eds. Late C. Bould, E.J.Hewitt and J.B.D. Robinson) pp. 54-110. H.M.S.O., London.

Hinsinger, P.1998. How do plant roots acquire mineral nutrients? Chemical processes involved in the rhizosphere. *Adv.Agron.,* **64:** 225-265.

Kalbasi, M., Manuchehri, N., and Filsoof, F. 1986. Local acidification of soil as a means to alleviate iron chlorosis in quince orchards. *J. Plant Nutr.*, **9:** 1001-1007.

Kessler, B. 1957. Effect of certain nucleic acid components upon the status of iron, deoxyribonucleic acid and lime induced chlorosis in fruit trees. *Nature,* **179:** 1015-1016.

Landsberg, E.C. (1982) Transfer cell formation in the root epidermis: a prerequisite for Fe-efficiency? *J Plant Nutr.,* **5:** 415-432

Lin, C.H. and Stocking, C.R. 1978. Influence of leaf age, light, dark and iron deficiency on polyribosome levels in maize leaves. *Plant Cell Physiol.,* **19:** 461-470.

Lindsay, W.L. and Schwab, A.P. 1982. The chemistry of iron in soils and its availability to plants. *J.Plant Nutr.,* **5:** 812.

Liu, D.H., Adler, K. & Stephan, U.W. (1998) Iron-containing particles accumulate in organelles and vacuoles of leaf and root cells in the nicotianamine-free tomato mutant *chloronerva*. *Protoplasma,* **201**: 213–220.

Mandley, S.L. 1981. Iron uptake and translocation by *Macrocystis pyrifera. Plant Physiol.,* **68:** 914-918.

Marschner, H., Römheld, V. and Kissel, M. 1986. Different strategies in higher plants in mobilization and uptake of iron. *J.Plant Nutr.,* **9**: 695-713.

Mehrotra, S.C. and Bisht, S.S. 1990. Genotype differences in concentration of free amino acids in susceptible and tolerant cultivars of chick-pea under iron deficiency stress. *Ind.J.Exptl.Biol.,* **28:** 893-894.

Mengel, K. and Geurtzen, G. 1986. Iron chlorosis on calcareous soils. Alkaline nutritional conditions as the cause for the chlorosis. *J.Plant Nutr.,* **9:** 161-173.

Mengel, K., Robin, P. and Salsac, L. 1983. Nitrate reductase activity in shoots and roots of maize seedlings as affected by the form of nitrogen nutrition and the pH of the nutrient solution. *Plant Physiol.,* **71:** 618-622.

Miller, G.W., Pushnik, J.C. and Welkie, G.W. 1984. Iron chlorosis, a world wide problem, the relation of chlorophyll bisynthesis to iron. *J.Plant Nutr.,***7:** 1-22.

Moog, P.R., van der Kooij, T.A.W., Brüggemann, W., Schiefelbein, J.W., Kuiper, P.J.C. (1995) Responses to iron deficiency in *Arabidopsis thaliana*: the turbo iron reductase does not depend on the formation of root hairs and transfer cells. *Planta,* **195:** 505-513.

Moran, Jose F., James, E. K., Rubio, M. C., Sarath, G., Klucas, R.V. and Becana, M. 2003. Functional characterization and expression of a cytosolic iron-superoxide dismutase from cowpea root nodules. *Plant Physiol.,* **133**: 773-782.

Muramoto, T., Tsurui, N., Terry, M.J., Yokota, A. and Newman, Kohchi, T. 2002. Expression and biochemical properties of a ferredoxin-dependent heme oxygenase required for phytochrome chromophore synthesis. *Plant Physiol.*, **130**: 1958-1966.

Olsen, R.A. and Brown, J.C. 1980. Factors related to iron uptake by dicotyledonous plants. *J.Plant Nutr.,* **2:** 629-660.

Olsen, R.A., Brown, J.C., Bennett, J.H. and Blume, D. 1982. Reduction of Fe^{+3} as it releases to Fe chlorosis. *J.Plant Nutr.,* **5:** 433-446.

Peters, J.S., Berkowitz, G.A. 1998. Characterization of a chloroplast inner envelope P-ATPase proton pump. *Photosynthesis Res.,* **57:** 323-333.

Pich, A. and Scholz, G. -/1993 The relationship between the activity of various iron-containing and iron-free enzymes and the presence of nicotianamine in tomato seedlings. *Physiol. Plant.,* **88:** 172–178.

Pierson, E. E. and Clark, R. B. 1984. Ferrous iron determination in plant tissue. *J. Plant Nutr.,* **7:** 91-106.

Pierson, E.E., Clark, R.B., Coyne, D.P. and Maranville, J.W. 1986. Iron deficiency stress effects on total iron in various leaves and nutrient solution pH in sorghum and beans. *J.Plant Nutr.,* **9:** 893-908.

Platt-Aloia, K.A., Thomson, W.W. and Terry, N. 1983. Changes in plastid ultrastructure during iron nutrition-mediated chloroplast development. *Protoplasma,* **114:** 85-92.

Pushnik, J.C. and Miller, G.W. 1989. Iron regulation of chloroplast photosynthetic function. Mediation of PSI development. *J. Plant Nutr.,* **12:** 407-421.

Roh, M.H., Shingles, R., Cleveland, M.J., McCarty, R.E. (1998) Direct measurement of calcium transport across chloroplast inner-envelope vesicles. *Plant Physiol.,* **118:** 1447-1454.

Romera, F.J., Alcántara, E., De la Guardia, M.D. (1999) Ethylene production by Fe-deficient roots and its involvement in the regulation of Fe-deficiency stress responses by *strategy I* plants. *Ann Bot.,* **83:** 51-55.

Römheld, V. and Marchner, H. 1986a. Evidence for a specific uptake system for iron phytosiderophores in roots of grasses. *Plant Physiol.,***80:** 175-180.

Römheld, V. and Marschner, H. 1986 b. Mobilization of iron in the rhizosphere of different plant species. In: *Advances in Plant Nutrition,* Vol.II (B.Thinker and A. Lauchli, eds.), 151-204.

Römheld, V. 1987. Existence of two different strategies for the acquisition of iron in higher plants. In: *Iron Transport in Microbes, Plants and Animals* (G.Winkelmann, D.vander Helm and J.B. Neilands, eds.), pp. 353-374. VCH Verlags Gesellschaft, Weinheim, FRG.

Sahu, M. 2000. M.Sc.(Ag.) Thesis, Plant Physiology, Banaras Hindu University, India.

Schikora, A. and Schmidt, W. 2001. Iron stress-induced changes in root epidermal cell fate are regulated independently from physiological responses to low iron availability. *Plant Physiol.,***125:** 1679-1687.

Schmidt, W., Tittel, J. and Schikora, A. 2000. Role of hormones in the induction of iron deficiency responses in *Arabidopsis* roots. *Plant Physiol,,* **122**: 1109-1118.

Sharma, C.P., Sharma, P.N. and Bisht, S.S. 1984. Induction of water stress by boron deficiency. *Actes Proc.,* **4:** 1303-1309.

Shingles, R., McCarty, R.E. (1994) Direct measurement of ATP-dependent proton concentration changes and characterization of a K^+-stimulated ATPase in pea chloroplast inner envelope vesicles. *Plant Physiol.,* **106:** 731-737.

Shingles, R., North, M. and McCarty, R.E. 2002. Ferrous ion transport across chloroplast inner envelope membranes. *Plant Physiol.,* 128: 1022-1030.

Singh, K., Chino, M., Nishizawa, N.K., Mori, S. and Singh, A.K. 1995. Phytosiderophore based iron acquisition in gramineae plants. In: *Advancements in Iron Nutrition Research.* Ed. A. Hemantaranjan. Scientific Publishers, Jodhpur, India. 215-232.

Stephan, U.W. 2000. Intra- and intercellular iron trafficking and subcellular compartmentation within roots. *Proceedings10th International Iron Symposium,* Houstan, Texas, U.S.A.

Stephan, U.W. 1995. The plant-endogenous Fe (II)-chelator nicotianamine restricts the ferrochelatase activity to tomato chloroplasts. *J. Exp. Bot.* **46**: 531–537.

Stoyanov, I. and Kha, Z.T. 1981. Relationships between growth and contents of gibberellin like substances and abscisic acid in maize under the influence of different iron concentrations in the nutrient medium. *Fiziol.Rast.,* **7:** 70-77.

Takagi, S., Nomoto, K. and Takemoto, T. 1984. Physiological aspect of mugineic acid, a possible phytosiderophore of graminaceous plants. *J.Plant Nutr.,* **7:** 469-477.

Terry, N. 1980. Limiting factors in photosynthesis. 1. Usa of iron stress to control photochemical capacity *in vivo. Plant Physiol.,* **65:** 114-120.

Terry, N. 1983. Limiting factors in photosynthesis IV. Iron stress mediated changes in light harvesting and electron transport capacity and its effects on photosynthesis *in vivo. Plant Physiol.,* **71:** 855-860.

Vaishampayan, A. 1995. Siderophore-mediated iron uptake in nitrogen-fixing cyanobacteria. In:*Advancements in Iron Nutrition Research.* Ed. A. Hemantaranjan. Scientific Publishers, Jodhpur, India. 233-254.

Von Wirén, N., Klair, S., Bansal, S., Briat, J.F., Khodr, H., Shioiri, T., Leigh, R.A. and Hider, R.C. (1999). Nicotianamine chelates both Fe III and Fe II. Implications for metal transport in plants. *Plant Physiol.,* **119**: 1107–1114.

Wallace, A. and Jones, Jr. J.B. 1993. Sample preparation and determination of iron in biological materials. In: *Iron Chelation in Plants and Soil Microorganisms.* Eds. L.L.Barton and B.C.Hemming, 447-463. Academic Press, Inc., San Diego, CA.

Wallace, G.A. and Wallace, A. 1986. Correction of iron deficiency in trees by injection with ferric ammonium citrate solution. *J. Plant Nutr.,***9:** 1009-1014.

Wallace, A. 1990. Interaction encountered when supplying iron, phosphorus, and nitrogen fertilizer to two cultivars of soybeans. *J.Plant Nutr.,* **13:** 349-356.

Wallace, A. 1995. Soil and plant procedures to control iron deficiency. *Adv..Iron Nutr.Res.* (Ed. A.Hemantaranjan), 1-20, Scientific Publishers, Jodhpur, India.

Welch, R.M. 1995. Micronutrient nutrition of plants. *Crit. Rev. Plant Sci.,* **14**: 49–82.

Zouari, M., Abadía, A. , Abadía, J. 2001. Iron is required for the induction of root ferric chelate reductase activity in iron-deficient tomato. *J.Plant Nutr.,* **24:** 383-396.

Section II: Plant Metabolism

Developments in Physiology, Biochemistry and Molecular Biology of Plants, 2005
Eds.: Bandana Bose and A. Hemantaranjan
Vol., 1, pp. 21-36, New India Publishing Agency, New Delhi
E-mail: spjain_niph@rediffmail.com web: www.bookfactoryindia.com

CHAPTER - 2

PHOSPHOLIPID METABOLISM IN PLANTS

PRAMOD KUMAR

INTRODUCTION

Phospholipids are the most abundant lipids of biological membranes. They serve primarily as structural components of membranes and are never stored in large amounts. As the name of phospholipid implies, this group of lipids contains phosphorus in the form of phosphoric acid groups. The major phospholipids found in membranes are the phosphoglycerides (phosphatides) (Fig.1). Which have two fatty acid molecules esterified

O
||
O $CH_2—O—C—CH_2—CH_2—CH_2—CH_2—CH_2—CH_2—CH_2—CH_2—CH_2—CH_2—CH_2—CH_2—CH_2—CH_3$
|| |
$^-O—P—O—CH_2—CH—O—C—CH_2—CH_2—CH_2—CH_2—CH_2—CH_2—CH_2—CH_2—CH_2—CH_2—CH_2—CH_2—CH_2—CH_3$
| ||O
^-O

Polar head Nonpolar tail
Phosphatidate (PA), the parent compound of phosphoglycerides

O
||
O $CH_2—O—C—CH_2—CH_2—CH_2—CH_2—CH_2—CH_2—CH_2—CH_2—CH_2—CH_2—CH_2—CH_2—CH_2—CH_3$
|| |
$^+N_3N—CH_2—CH_2—O—P—O—CH_2—CH—O—C—CH_2—CH_2—CH_2—CH_2—CH_2—CH_2—CH_2—CH_2—CH_2—CH_2—CH_2—CH_2—CH_2—CH_3$
| ||O
^-O

Phosphatidylethanolamine (PE)

O
||
O $CH_2—O—C—CH_2—CH_2—CH_2—CH_2—CH_2—CH_2—CH_2—CH_2—CH_2—CH_2—CH_2—CH_2—CH_2—CH_3$
|| |
$(NH_3)^+N—CH_2—CH_2—O—P—O—CH_2—CH—O—C—CH_2—CH_2—CH_2—CH_2—CH_2—CH_2—CH_2—CH_2—CH_2—CH_2—CH_2—CH_2—CH_2—CH_3$
| ||O
^-O

Phosphatidylethanolamine (PC)

O
||
O $CH_2—O—C—CH_2—CH_2—CH_2—CH_2—CH_2—CH_2—CH_2—CH_2—CH_2—CH_2—CH_2—CH_2—CH_2—CH_3$
|| |
$^-OOC—CH—CH_2—O—P—O—CH_2—CH—O—C—CH_2—CH_2—CH_2—CH_2—CH_2—CH_2—CH_2—CH_2—CH_2—CH_2—CH_2—CH_2—CH_2—CH_3$
| ||O
^-O

Phosphatidylethanolamine (PS)

O
||
O $CH_2—O—C—CH_2—CH_2—CH_2—CH_2—CH_2—CH_2—CH_2—CH_2—CH_2—CH_2—CH_2—CH_2—CH_2—CH_3$
|| |
$—O—P—O—CH_2—CH—O—C—CH_2—CH_2—CH_2—CH_2—CH_2—CH_2—CH_2—CH_2—CH_2—CH_2—CH_2—CH_2—CH_2—CH_3$
| ||O
^-O

Phosphatidylethanolamine (PI)

Fig. 1. The common phosphoglycerides

to the first and second hydroxyl groups of glycerol. The third hydroxyl group of glycerol forms an ester linkage with phosphoric acid. Besides this, phosphatides contain a second alcohol, which is also esterified to the phosphoric acid. The second alcohol group is thus located on the polar head of the phosphoglyceride molecule.

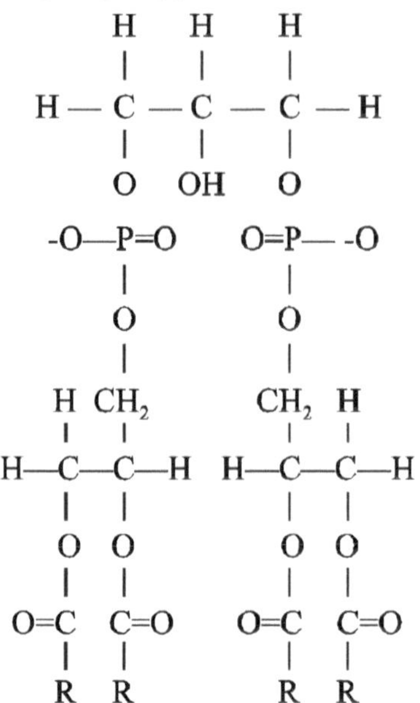

Fig. 2. Cardiolipin (Bisphosphatidylglycerol)

These are several different classes of phosphoglycerides, which differ in their head alcohol group. However, all phospholipids have two nonpolar tails, contributed by their long chain fatty acids. Different types of phosphatides are named according to the alcohol on their polar heads (Fig.1). The parent compound of phosphoglycerides is phosphatidic acid, which has no head alcohol. It is found in free form only in small amounts, but it is an intermediate in biosynthesis of phosphoglycerides. The most abundant phosphoglycerides are closely related phosphatidylethanolamine and phosphatidylcholine, which contain the alcohol ethanolamine and choline, respectively on their polar heads. Each of these may contain different combinations of fatty acids. Other phosphoglycerides include phosphatidylserine, containing serine as the polar compound and phosphatidylinositol, containing the alcohol inositol. Di or Bisphosphatidylglycerol (BPG or DPG) commonly called as cardiolipin, characteristically occurs in the inner membrane of mitochondria, differs from the rest of the phosphoglycerides; it is a "double" phosphoglyceride (Fig.2).

COMPOSITION

The phospholipid lipid composition of numerous plant species, isolated plant organelles and tissue culture has been described and reviewed by several workers (Galliard, 1973; Harwood, 1980; kates *et al.*,1979, Mudd, 1980). Summary of phospholipid composition in higher plants has been described in Table-1.

Table : 1. Phospholipid Composition in Higher Plants.

	PC	PE	PG	PG	PI	DPG	Phospholipid (% of total lipid)	Phospholipid mg/gdw)
Leaves								
Maize, Young	64	27.5	8.5				2.1	0.94
Maize, old	53	22.5	24.5				0.5	0.49
Tissues								
Potato tuber	54.4	26.5	1.5	1.3	13.5	1.3	47.5	0.60
Apple fruit	46.0	30.8	6.6	0.9	12.9	1.5	46.6	0.40
Chloroplast lamellae								
Maize	11.6	76.4	12.0				5.5	61
Spinach	23.3	66.3	10.2				13.7	40
Chloroplast envelopes								
Maize	55	9	18	11.5	6.5	12.2		
Spinach	75.3	4.2	18.6	1.8	33.3			
Endoplasmic reticulum								
Ricinus	45.3	28.7	3.6	2.3	13.1	2.5	76	
Mustard	50	35	8		6	1		300
Mitochondria, Whole								
Mustard	44	34	3		7	11		260
Glyoxysomes								
Ricinus	51.4	27.2	2.7	1.5	9	2.3	78	
Nuclear membrane								
Onion	52.2	25.8	5.2		7.3	1.8	60	

The most common fatty acids associated with phospholipids are palmitate (16:0), Stearate (18:0), oleate (18:1), linoleate (18:2) and linolenate (18:3). A less common fatty acid, trans D^3 hexadecanoic acid (16:1) is found in the phosphatidyl glycerol (PG) molecules associated with chloroplast (Mudd, 1980). Fatty acid composition of plant phospholipids has been depicted in Table-2.

Table : 2. Fatty Acid Composition of Plant Phospholipids.

	Phospholipid	14:0	16:0	16:1	18:0	18:1	18:2	18:3	Unsaturated/saturated fatty acid
Chloroplast lamellae, spinach	PG	1.0	18.3	39.6	trace	0.5	1.6	38.5	4.0
	PC	trace	13.2			7.7	19.8	57.2	6.6
	PI	2.1	42.1	0.5	1	7	14.3	26.3	1.2
Endoplasmic reticulum, castor bean	PC		21		11	18	43	3	1.9
	PE		31		8	9	41	4	1.4
	PI		41		12	9	22	3	0.6
Glyoxysomes, caster bean	PC		20		9	17	43	5	2.2
	PE		26		8	9	51	6	1.9
	PI		35		11	9	38	1	1.0
Mitochondria, caster bean	PC		16		7	17	52	4	3.2
	PE		17		8	8	58	5	2.8
	PI		38		10	9	36	2	1.0

BIOSYNTHETIC PATHWAYS

Making the Backbone: Phosphatidate

The goal of the first enzymes of glycerophospholipid biosynthesis is to make phosphatidate (PA) that serve as precursor for all glycerophospholipids. Sn-Glycerol 3-phophate (glycerol-3-P) provides the 3-carbon backbone of phospholipids, and radioactive glycerol and glycerol-3-P are readily incorporated into these lipids. Sn-Glycerol 3phophate: NAD oxidoreductase (EC 1.1.1.8) has been detected from castor bean endosperm and spinach leaves.

Acyl Units

Harwood (1988) has reviewed the biosynthesis of fatty acids. The majority of evidence indicates the first major fatty acid is palmitic acid (16:0). This synthesis is catalysed by a group of 3 to 7 enzymes collectively known as fatty acid synthase. These enzymes utilize acetyl-ACP (acyl carrier protein) and malonyl-CoA in series of C-2 additions to form longer chain ACP-linked products.

(1)

Acetyl-CoA + ACP-SH → Acetyl-S-ACP +CoA

(2)

Acetyl-S-ACP + Enz(3) → Acetyl-S-Enz(3) +ACP

(3)

Malonyl-CoA+ ACP-SH → Malonyl-S-ACP +CoA

(4)

Acetyl-S-Enz(3) +Malonyl-S-ACP → Acetoacetyl-S-ACP+Enz (3) +CO_2

(5)

Acetoacetyl-S-ACP+NADPH+H^+→ D(-)-β-Hydroxybutryl-S-ACP + $NADP^+$

(6)

D(-)-β-Hydroxybutryl-S-ACP → Δ^2- trans- Crotonyl-S-ACP+ H_2O

(7)

Δ^2- trans- Crotonyl-S-ACP+NADPH+H^+ ®Butyryl-S-ACP+$NADP^+$

(8)

Butyryl-S-ACP+ Enz(3) → Butyryl-S- Enz(3) +$NADP^+$

(9)

Butyryl-S- Enz(3) + Malonyl-S-ACP→ β-Ketohexanoyl-S-ACP+Enz (3) + CO_2 *etc.*

(1) Acetyl transacylase	(4) β - Ketoacyl ACP reductase
(2) Malonyl transacylase	(5) Enoyl ACP hydrase
(3) β- Ketoacyl ACP synthetase	(6) Enoyl ACP reductase

The newly formed butyryl-S-ACP reacts now with another molecule of malonyl-S-ACP and β - ketoacyl-ACP synthetase and the sequence outlined above is repeated until the desired length of fatty acid is attained. The mechanism of termination is still unclear. The precise reaction for palmitate may be written as follows:

$$\text{Acetyl CoA} + 7\ \text{Malonyl CoA} + 14\ \text{NADPH} + 14\ \text{H}^{+} \rightarrow$$

$$\text{Palmitate} + 8\ \text{CoA} + 7\ \text{CO}_2 + 14\ \text{NADP}^{+} + 6\ \text{H}_2\text{O}$$

Several lines of evidence point toward a specific elongase to form 18:0 from 16:0, and other elongases for the addition of further C_2 units. Desaturation generally occurs at the C_{18} stage, with up to three cis- double bonds being introduced. Longer chain unsaturated fatty acids are formed by the addition of C_2 units to the unsaturated C_{18} product.

Acylation

The coenzyme A derivatives of the fatty acids are utilized for the acylation reactions. This requires conversion from ACP derivatives utilized for synthesis to the fatty acyl CoA, or removal from the "ACP track" to the "CoA track" (Shine *et. al.,* 1976). Evidence exists for the sequential addition of fatty acids to glycerol-3-P to form phosphatidiate (PA). Lysophosphatidiate (LPA) is an intermediate, the second addition normally occurs rapidly after the first. Glycerol 3-phosphate acyltransferase is the enzyme, catalyzes this reaction. The reported pH optima for sn- glycerol 3-phosphate acyltransferase (EC 2.3.1.15.) range from 7 to 9.5. Apparent K_ms for palmitol- CoA ranging from 0.006 to 2mM have been obtained; for glycerol-3-P the range is 0.1 to 3 mM (Moore, 1982).

Headgroup

Summary of the pathways for addition of the polar headgroups are outlined in (Fig. 3) and details in (Fig. 4).The central metabolite, PA, may be utilized to form either diacylglycerol (DG) or CDP- diacylglycerol (CDP-DG). From diacylglycerol, PC or PE may be formed, while from CDP- diacylglycerol, PI, PGP, or PS may be formed. BGP is formed from PG. Some interconversion between phospholipid types, for example exchange of serine for ethanolamine of PE to form PS or methylation of PE to form PC, also can occur. The majority of information available on these reactions arises from work on spinach leaf and castor bean endosperm cell fractions, but cauliflower florets, onion root tips, and pea stem sections also have been utilized.

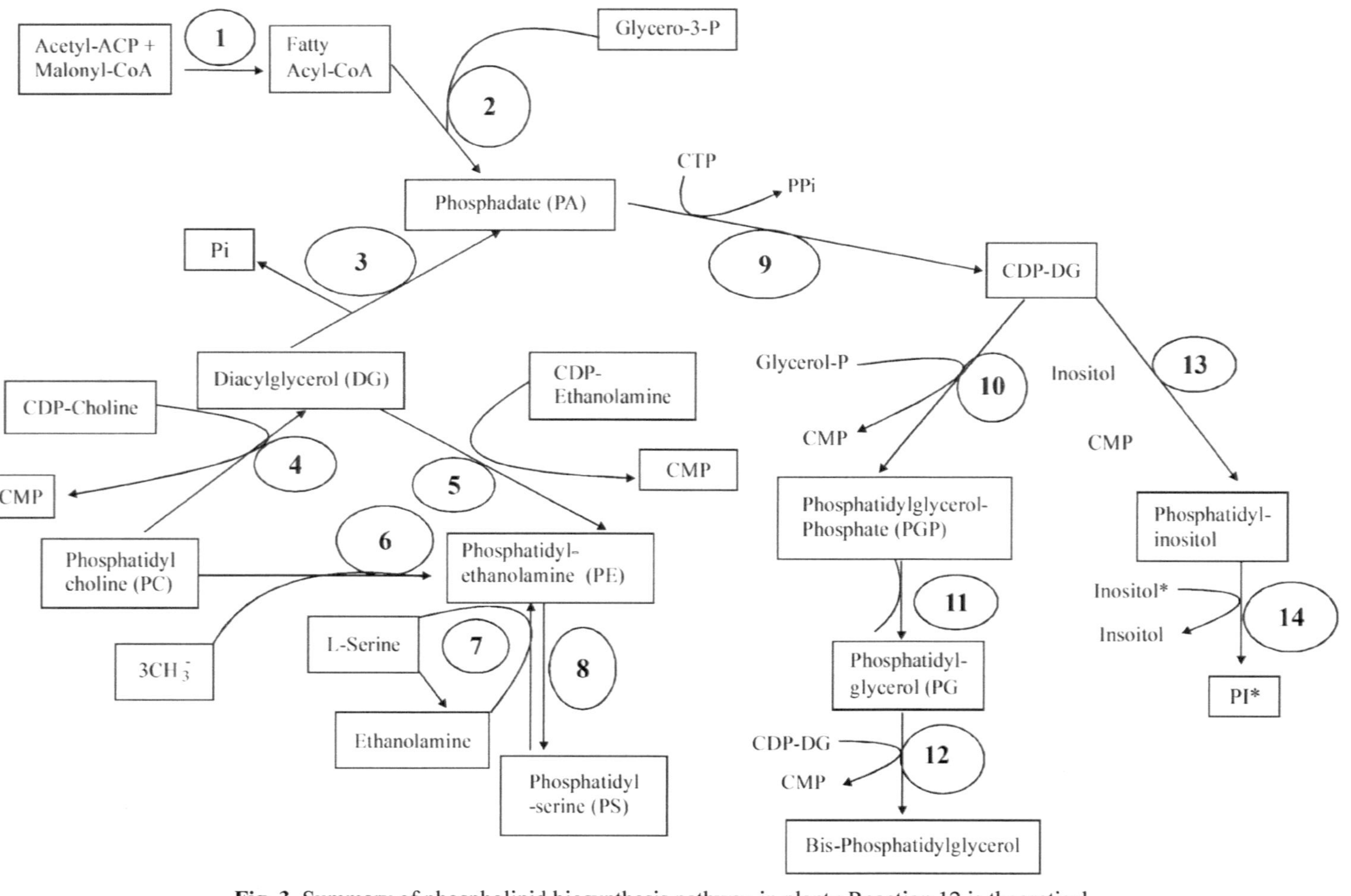

Fig. 3. Summary of phospholipid biosynthesis pathway in plant : Reaction 12 is theoretical

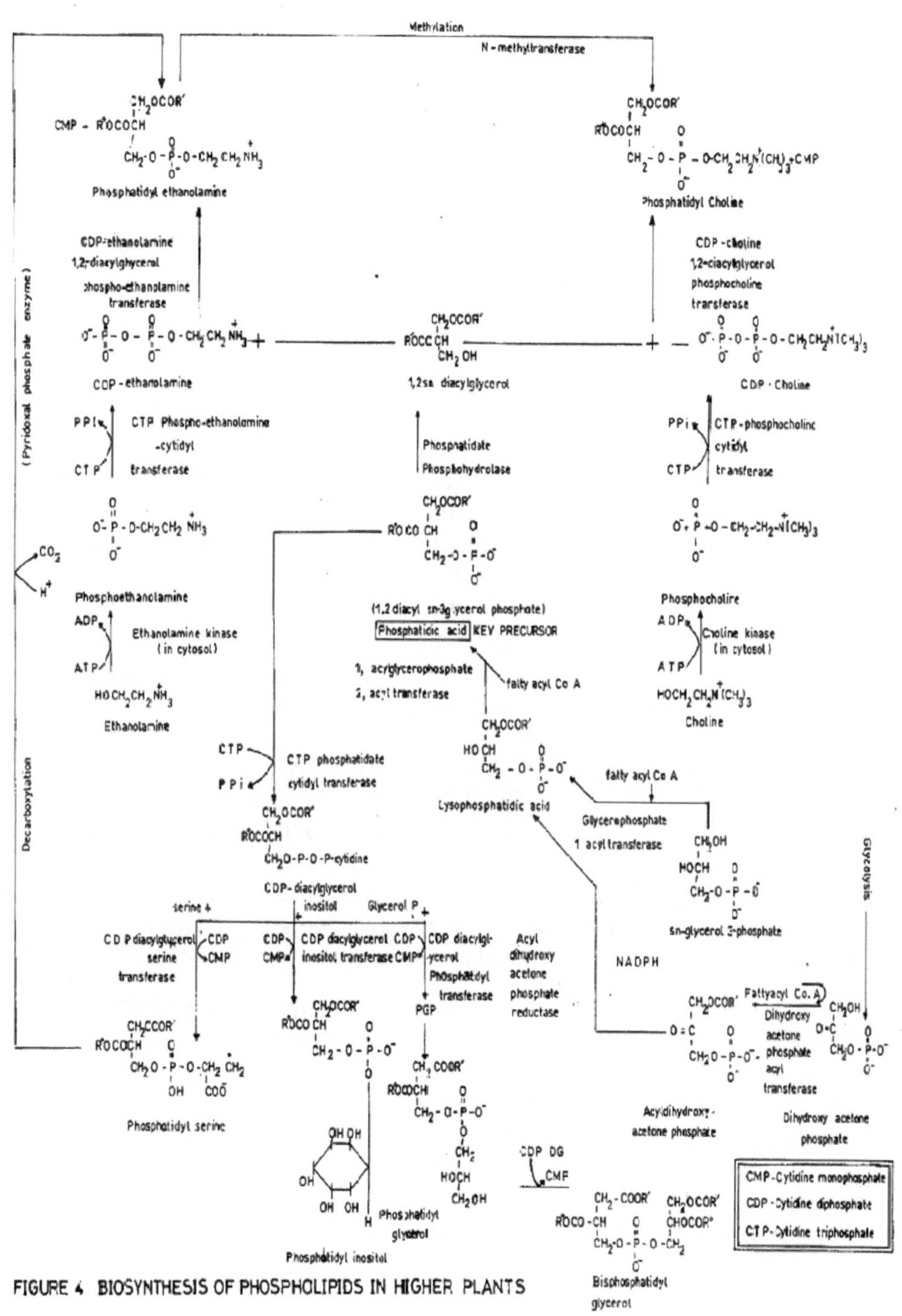

Fig. 4. Biosynthesis of Phospholipids in Higher Plants.

Water-Soluble Precursors

CDP chloine, CDP ethanolamine, myo-inositol, L-serine and glycerol-3-P are the immediate water-soluble substrates for phospholipid headgroup additions. The origin of glycerol-3-P has been discussed above in the "backbone" section.

Serine

This amino acid can be utilized directly for the synthesis of phosphatidylserine (PS) and also serves as a precursor for ethanolamine and choline.

Ethanolamine and choline

The formation of ethanolamine and choline occurs by the pathway:

$$\text{L-Serine} \xrightarrow{\nearrow CO_2} \text{Ethanolamine} \xrightarrow{3CH_3^- \downarrow} \text{Choline}$$

Another interesting source of choline in radish seedlings is sinapine (sinapoylcholine). This compound is stored in the seed and metabolised releasing choline during germination. About 50% of the released choline is utilized for synthesis of PC (Strack, 1981).

Phosphorylation of choline and ethanolamine

Choline kinase (EC2.7.1.32) and ethanolamine kinase (EC 2.7.1) involve in phosphorylation of choline and ethanolamine. Both enzymes have been purified from spinach leaves (Macher and Mudd, 1976) and soybean (Wharfe and Harwood,1979).

$$(CH_3)_3N^+CH_2CH_2OH + ATP \rightarrow (CH_3)_3N^+CH_2CH_2OPO_3^{--} + ADP$$

$$H_2NCH_2CH_2OH + ATP \rightarrow H_2NCH_2CH_2OPO_3^{--} + ADP$$

Choline does not compete with ethanolamine as a substrate for the spinach ethanolamine kinase (Macher and Mudd, 1976).The Km for ethanolamine is 42 μM and for Mg-ATP is 63 μM; free Mg^{2+} also required. The molecular weight of the enzymes is 110,000.

The major choline kinase from soybean has a molecular weight of 36,000, has no ethanolamine kinase activity, and is unstable after purification.

Formation of CDP-Choline and CDP- Ethanolamine

The two enzymes cholinephosphate cytidyltransferase (EC2.7.7.15)

$$(CH_3)_3N^+CH_2CH_2OPO_3^{--} + CTP \rightarrow \text{CDP-Choline} + \text{Ppi}$$

and ethanolaminephospate cytidyltransferase (EC.2.7.7.14)

$$H_2NCH_2CH_2OPO_3^{--} + CTP \rightarrow \text{CDP-ethanolamine} + \text{Ppi}$$

have been reported in plant system for the formation of CDP-Choline and CDP-Ethanolamine.

Head Group Incorporation

Eventhough several reports regarding plant enzymes catalyzing headgroup incorporation into phospholipids exist, no extensive purifications of these enzymes have been reported, and so all the characterizations described below have been performed with membrane preparations. The reaction numbers given with each enzyme are those in (Fig. 3).

Phosphatidate phophatase (E.C2.1.2.4 ; Reaction 3)

Several enzymes are capable of catalyzing this activity in plant cells. The Phosphatidate phophatase of chloroplast envelopes is not active with added phosphatidate (PA), but when PA is generated by the membranes themselves, conversion to diacylglycerol (DG) is obtained. This phosphatase has a pH optimum of 9.0 and appears not to need divalent cations, which may, in fact, inhibit the reaction (Joyard and Douce, 1979).

CDPcholine: 1,2-diacylglycerolcholine phophotransferase (EC2.7.8.2 ; Reaction 4)

This enzymes synthesizes most of the phosphatidyl choline (PC)of plant cells. The microsomal enzymes of caster bean (Moore,1976) and spinach leaf (Devor and Mudd,1971) have apparent K_ms for CDP choline of about 10 μM. Mg^{2+} is the preferred cation, and a pH optimum of 7.5 is obtained with this cation. With Mn^{2+} a pH optimum of 8 is obtained with enzymes from spinach leaf.

CDP-ethanolamine: 1,2-diacylglycerol ethanolamine phosphotransferase (E.C2.7.8.1; Reaction 5)

This enzyme appears responsible for most phosphatidyle -thanolamine (PE) synthesis in plant cells. An exception might occur if certain plants contain the pathway for phosphatidyl serine (PS) synthesis utilizing CDP diacylglycerol and serine, with susequent decarboxylation of PS to PE. The ethanolamine phospho –transferase has apparent K_ms for CDPethanolamine of 6.0 μM in caster bean (Sparace et al, 1981) and 20 μM in spinach (Macher and Mudd, 1974). V_{max} is obtained at 50-60 μM in castor bean and 100 μM in spinach.The pH optimum for castor bean is 6.5 with Mn^{2+} and 7.0-7.5 with Mg^{2+}.

S-Adenosyl-L-methionine: phosphatidylethanolamine N- methyl- transferase (E.C2.1.1.17; Reaction 6)

The formation of PC from PE by sequential methylation of the ethanolamine moiety probably requires at least two enzymes. However, no direct evidence for more than one enzyme has been obtained with plant systems.The enzyme from castor bean endoplasmic reticulum exhibits an apparent Km for S- adenosylmethionine of 31 μM. The optimal pH for the overall reaction is alkaline. The products of the reaction in castor bean are 52% PC and 48% phosphtidyldimethylethanolamine (PDME); in spinach most of the activity is in PC, with small amount in phosphatidylmonomethylethanolamine (PMME) (Moore, 1982).

Phosphatidylethanolamine: L-Serine phosphatidyltransferase (Reaction 7)

This enzyme catalyzes formation of phosphatidylserine (PS) by a Ca^{2+} stimulated exchange of L-serine for ethanolamine of PE. In castor bean, the apparent Km for L-serine is 20 μM. Ethanolaminee compete with L-serine, but D-serine and choline do not. The pH optimum is 7.8. Both PS (48%) and PE (52%) have been observed as products, suggesting the coexistence of PS decarboxylase.

CDPdiacylglycerol serine-O-phosphatidyl transferase (E C2.7.8.8)

Evidence for this enzyme in spinach leaf organelles was reported by Marshall and Kates (1974). The reaction is:

$$\text{L-Serine} + \text{CDPdiacylglycerol} \rightarrow \text{phosphatidylserine} + \text{CMP}$$

Phosphatidylserine decarboxylase (EC 4.1.1.65; Reaction 8)

This reaction occurs in spinach leaves (Marshall and kates, 1974) and indirect evidence also exists for its occurrence in pea seedlings and castor bean. This enzyme has not been charecterized.

CTP:phosphatdate cytidyltransferase (EC2.7.7.41;Reaction 9)

This enzyme catalyzes formation of CDP-diacylglycerol from PA. The pH optima generally are acidic, ranging from 5.6 to 6.5. Mn^2 appears to be the preferred cation when endogenous PA is utilized as the lipid substrate. CTP is required by the enzyme. The apparent Km for CTP varies from 0.5 to 0.7 μM.

Glycerophosphate: CDP diacylglycerol phosphatidyl -transferase (EC 2.7.8.5) plus phosphatidylglycerophosphate phosphohydrolase (EC 3.1.3.27 ; Reactions 10 and 11)

The final product of both reactions (10 & 11) is phosphatidylglycerol (PG), is the major phospholipid of thylakoid membrane (Moore, 1974) and is precursor of bis phosphatidylglycerol (BPG) synthesis in mitochondria. The enzymes of both the endoplasmic reticulum and mitochondria of castor bean endosperm have similar properties (Moore, 1974). The apparent Km S are about 50μM forglycerol-3-P and 2-3 μM for CDPdiacylglycerol. PG is the only product observed, and the pH optima are 7.3. Addition of Mn^{2+} leads to maximum stimulation of both enzymes and it is optimal at 5μM.

Phosphtidylglycerol: CDP diacylglycerol phosphatidyl -transferase (Reaction 12)

There are evidences that cardiolipin synthesis occurs in plant mitochondria (Moore, 1982). The enzyme in potato tuber mitochondria utilizes phosphatidylglycerol and CDP-diacylglycerol as substrates. The reaction requires a divalint cation ($Mg^{2+}$and Co^{2+}). Triton X-100 strongly stimulates the reaction.

CDP diacylglycerol: myo-inositol phosphatidyltransferase (EC 2.7.8.11; Reaction 13)

This enzyme is responsible for the synthesis of phosphatidyl –inositol (PI). It has an apprent Km for myo- inositol of 0.3 μM and for CDP dipalmitoyl glyceride it is 1.35 μM (Sexton and Moore, 1978). A divalent cation is absolutely required, Mn^{2+} is preferred.

Phosphatidylinositol: myo-inositol phosphatidyltransferase (Reaction 14)

This reaction has not been investigated thoroughly in plants. It has been reported only in caster bean endosperm (Sexton and Moore, 1981).The reaction appears not to be a reversal of the CDP diacylglycerol pathway for PI synthesis because of different responses of the enzymes to detergents, different optimal concentrations of Mn^{2-}, the requirement for low levels of substrates necessary to reverse the CDPdiacylglycerol pathway, the low level of CMP in the isolated membrane fractions, and the strong stimulation of the exchange reaction by cytidine nucleotides. This exchange has an apprent Km for myo- inositol of 26 μM. The exchange reaction can be stimulated up to 15-fold by 40 μM CMP, CDP or CTP.

REGULATION

Very little is known about regulation of phospholipid metabolism. Indol- acetic acid, gibberllic acid, kinetin, abscisic acid, cAMP, herbicides, nutrition and "adaptive ageing" or "washing" have been found to influence phospholipid metabolism (Mudd, 1980; Kumar, 1992; Kumar *et al,* 1994; Kumar *et al;* 2001). Metabolite effects, other than substrate concentration, have not been studied.

Phospholipid Transfer

The apparent absence of phospholipid synthesis in certain organelles, especially microbodies and plasma membrane, raises the question of how the phospholipid is transferred from the sites of synthesis to these organelles. Two major mechanisms for phopholipid transfer have been proposed. The first requires membrane- membrane contact, is an integral part of the membrane- flow hypothesis (Morre, 1975), and probably involves transfer of clusters or bulk flow of molecules. The second mechanism relies on carriers for individual molecules, the so-called phospholipid transfer proteins (PTPs) (Kader, 1977). PTPs have been reported from several plant tissues (Moore, 1982). These activities are rapidly inactivated in most cases and so are harder to measure than the mammalian PTPs. The most stable activities appear to be that castor bean endosperm, and PTPs of both PC and PE have been prepared from the tissue. Concern with the origin of PC in the chloroplast has led to tests for PTPs in photosynthetic tissue and their ability to transfer PC to chloroplasts. Such activities have been reported from spinach, oat, and pea leaves, with the capacity to transfer radioactive PC from liposomes to plastids (Moore, 1982).

Compartmentation

A summary of available information on compartmentation of phospholipid metabolism in castor bean endosperm is depicted in Table 3.

CATABOLISM OF PHOSPHOLIPIDS

Phospholipids mobilization is mediated by some important enzymes which is known to be hormonally controlled. Phytohormones like GA, IAA and cytokinin regulate the

Table : 3. Compartmentation of phospholipid biogenesis in castor bean endosperm.

Lipid formed	Percentage of total activity in isolated organelles					
	Reaction no. (Fig.3)	Soluble	Microsomes	Mitochondria	Plastids	Glyoxysomes
Fatty acid	1	0	0		100	0
Phosphadiate	2	0	87	13	0	0
Diacylglycerol	3	present	present	?	?	0
Phosphatidylcholine	4	0	98	2	0	0
Phosphatidylethanolamine	5	0	98	2	0	0
Phosphatidylcholine	6	0	90	10	0	0
Phosphatidylserine	7	0	100	0	0	0
Phosphatidylethanolamine	8	?	present	?	?	?
CDP-Diacylglycerol	9	0	75	25	0	0
Phosphatidylglycerol	10, 11	0	50	50	0	0
BisPhosphatidylglycerol	12	?	?	present	?	?
Phosphatidylinositol	13	0	100	0	0	0
Phosphatidylinositol	14	0	100	0	0	0

metabolic switching.The constituent fatty acids obtained by enzymatic degradation are further broken down by various types of oxidation and provide not only energy through operation of TCA cycle and respiration but also skeleton to the growing embryo. Lipids are converted through glyoxylate cycle, as a by-pass for a portion of TCA cycle. Following phospholipases (enzymes) are involved in phospholipid catabolism.

Phospholipases

Phospholipases constitute a diverse series of enzymes that can be classified into phospholipases D (PLD), C (PLC), A_2 (PL A_2), A_1 (PLA_1) and B (PLB) according to their sites of hydrolysis on phospholipids (Fig. 5).Each class is further divided into subfamilies based on sequence, biochemical properties, or a combination of both. A closely related class of enzyme comprises lysophospholipases (lysoPLAs), which hydrolyzhe products of PLAs (Fig.5).; some phospholipases also comprises lysophopholipases (lysoPLAs), which hydrolyze the products of PLAs (Fig.5), some phospholipases also exhibit this activity. Fatty acids obtained by degradation of phospholipids are further catabolized by various types of oxidation *i.e.* β- oxidation, α- oxidation and ω- oxidation. Phospholipids provide the backbone for biomembranes, serves as rich sources of signaling messenger, and occupy important junctions in lipid metabolism. The activities of phospholipases not only affect the structure and stability of cellular membranes, but also regulate many cellular functions. The past two decades have brought rapid growth in knowledge about the role of phospholipases in cell regulation and signalling processes (Wang, 2001). Activation of phospholipases often is an initial step in generating lipid and lipid-derived second messengers.

Inositol phospholipid turnover acts as a signal transduction mechanism in plant tissues (Morse *et al,* 1987; Cote and Crain, 1993). When an appropriate agonist inter acts with its receptor break down of phosphatidylinositol 4, 5-biphosphate into inositol 1,4,5-triphosphate (IP_3) and diacylglycerol (DAG) is stimulated. Diacylglycerol triggers phosphorylation of

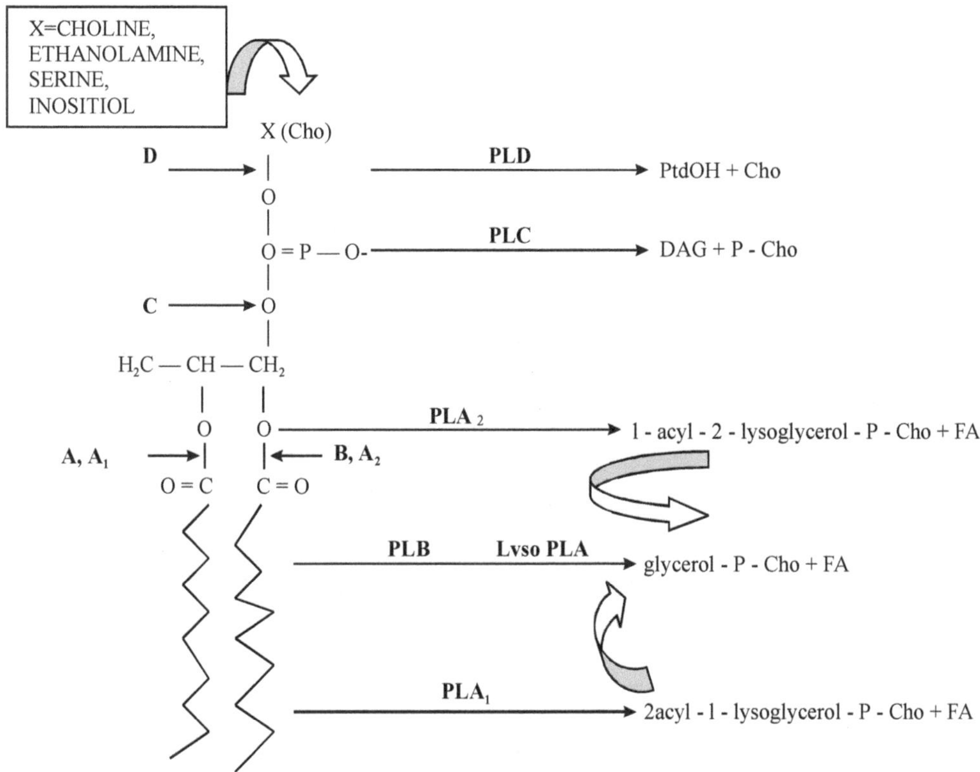

Fig. 5. Hydrolysis of Phosphatidylcholine (PtdCho) by PLD, PLC, PLA2, PLA1, PLB, and lyso PLB and the respective products. Arrow lines indicate their site of hydrolysis. PLA1 hydrolyses the sn-1 aclester bond, whereas lyso PLA removes the last fatty acid from lysophospholpids that can be produced by PLA2 and PLA1, as marked by the curved arrow. PLB sequentially removes two fatty acids from phospholipids, and its final reaction products are the same as those of lyso PLA. Cho, Choline; Pcho, Phosphocholine; FA, fatty acid.

regulatory proteins by the activation of protein kinase C, whereas IP_3 participates in release of Ca^{2+} ions from endoplasmic reticulum during regulated reaction in cell (Fig. 6). Released Ca^{2+} binds with calmodulin (calcium binding proteins) which activates protein kinases and several other enzymes, so that cellular response takes place. Phosphatidylserine also activates the protein kinase C. Altered inositol lipid metabolism has been observed in response to growth hormones (Dureja- Munjal *et al.,* 1992, light Acharya *et al.,* 1991), low temperature (Somolenskasym and Kaeperska, 1996) and "Washing" (Kumar and Pant, 2001).

FUTURE PROSPECTS

Metabolic pathways of major phospholipids in plants are becoming clear, even though, the answers of some questions are still obscure, for instance, mechanism of phosphatidyl serine synthesis. Enzymological studies on phospholipid metabolism in plants have turned

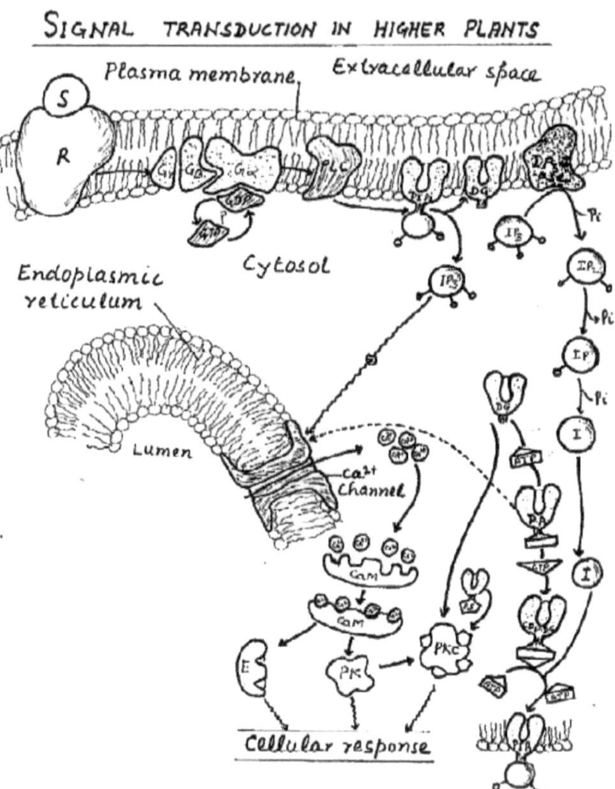

Fig. 6. PIP_2 cycle, when a stimulus (S) binds to its receptor (R) in plasma membrane a cascade of events ensues in a GS protein activates phosphlipase C (PLC). This enzyme hydrolyses PIP_2 to IP_3 and 1,2 diacylglycerol (DG). IP_3 the binds to Ca^{2+} channel embedded in the membrane of endoplasmic reticulum, causing Ca^{2+}ion to be released to the cytosol. IP_3 phosphatase (IP_3ase) converts IP_3 to IP_2, terminating the transmission of the signal. Finally, hydrolysis of IP_2 produces IP which is converted to inositol and incorporated into another P IP_2 molecule Induced cytosolic concentration of Ca^{2+}ion activares calmodolin (CaM) which further stimulates the activity of various enzymes and protein kinases and finally cellular response is occurred.

up no major differences from analogous enzymes in animals. The subcellular location of the enzymes in mitochondria and the endoplasmic reticulum is also similar to animals. Therefore, enzymes associated with the chloroplast are of particular interst, both in synthesis of lysophosphatidic acid and phosphatidic acid. Further studies on chloroplast should reveal important aspects of the phospholipid economy, such as the origin of the phospholipids of the chloroplast envelope and the origin of phosphatidylglycerol in lamellae. Little is known about regulation of phospholipid metabolism. The control of phospholpid metabolism in plants is an open area, and further thorough investigation is needed in context to hormonal effects and environmental stresses. Role of phospholipids in signal transduction in plants has emerged out as an important and interesting area of research, therefore, intensive study is needed in future on the diverse cellular roles of phospholipases in signal transduction.

REFERENCES

Acharya, M.K., Dureja- Munjal, I. and Guha-Mukherjee, S. 1991. Light induced rapid changes in inositol phospholip and phosphatidylchalin in Brassica Seedlings, *Phytochemistry*, **36**, 2895-2897.

Conn, E. E. and Stumpf, P.K., 1985. Outlines of Biochemistry, John Wiley & Sons Inc., New York.

Cote, G.G. and Crain, R.C. 1993. Biochemistry of phosphoinositdes. *Ann. Rev. Plant Physiol. Plant Mol. Biol.*, **44**:333-356.

Devor, K.A. and Mudd, J.B. 1971. Biosynthesis of phosphatidylcholine by enzyme preparations from spinach leaves. *J. Lipid Res.* **12**: 403-411.

Dureja-Munjal, I., Acharya, M.K. and Guha-Mukherjee, S. 1992. Effect of hormones and spermidine on the turnover of inositolphospholipids in Brassica seedlings. *Phytochemistry*, **31**:1161-1163.

Galliard, T. 1973. Phospholipid metabolism in photosynthetic plants. In: *Form and function of phospholipids* (Eds. Ansell, G.B. Dawson, R.M.C. and Hawthorne), Amsterdam, Elsevier: 253-288.

Harwood, J.L. 1980. Plant acyl lipids: Structure, distribution and analysis. In: *The Biochemistry of Plants,* **4**, Lipids: Structure and Function (Eds. Stumpf, P.K.), New York, Academic 1-55.

Harwood, J.L. 1988. Fatty acid metabolism. *Ann. Rev. Plant Physiol. Plant Mol. Biol.*, **39**:101:138.

Kaeder, J.C. 1977. Exchange of phospholipids between membranes. In: *Dynamic Aspect of Cell Surface Organization* (Eds. Poste, G. and Nicolson), Amsterdam, North-Holland, 127-204.

Kates, M. Wilson, A.C. and de la Roche, I.A. 1979. Lipid biosynthesis in plant cell cultures. In: *Advances in the Biochemistry and Physiology of Plant Lipids* (Eds. Appelqvist, L.A. and Liljenberg, C.), Amsterdam, Elsevier/ North- Holland Biomed. Press: 329-342.

Kumar, P. 1992. Impact of "washing" on phospholipid metabolism and ion uptake on corn (Zea mays L.) roots. Ph. D. Thesis, G.B. Pant University of Agriculture and Technology, Pantnagar.

Kumar, P., Pant, R.C. and Singh, K. 1994. Adaptive ageing or "washing" in relation to nutrient uptake and plant productivity. In: *Plant Productivity under Environmental Stress* (Eds. Singh, K. and Purohit, S.S.), Agro Botanical Publishers, Bikaner, India.

Kumar, P. and Pant, R.C. 2001. Inositol phospholipid metabolim in relation to 'washing' induced phosphate uptake in maize (Zea mays L.) roots. *Physiol. Mol. Biol. Plants*, 7:139-144.

Lehninger, A.L. 1982. Principles of Biochemistry, Inc., U.S.A. Worth Publisher.

Macher, B.A. and Mudd, J.B. 1974. Biosynthesis of phosphotidylethanolamine by enzyme preparations from plant tissues. *Plant Physiol.*, **53**:171-175.

Macher, B.A. and Mudd, J.B. 1976. Partial purification and properties of ethanolamine kinase from spinach leaf. *Arch. Biochem. Biophys*. Acta, **260**:558-570.

Joyard, J. and Douce, R. 1979. Characterization of phosphatidate phosphohydrolase associated with chloroplast envelope membranes. *FEBS* Lett., **102**:147-150.

Marshall, M.O.and Kates, M. 1974. Biosynthesis of nitrogenous phospholipids in spinach leaves. *Can. J. Biochem.* **52**: 469-482.

Mathur, J.M.S. and Sharma, N.D. 1989. Plant Lipids. In: *Recent advances in Plant Biochmistry* (Eds. Mehta, S.L., Lodha, M.L. and Sane, P.V.), New Delhi,, ICAR Publication: 341- 396.

Moore, T.S.,. 1974. Phosphatidylglycerol synthesis in castor bean endosperm. Kinetics, requirements and intracellular localization. *Plant Physiol.*, **54:** 164-168.

Moore, T.S., 1975. Phosphatidylserine in castor bean endosperm. *Plant Physiol.*, **56:**177-180.

Moore, T.S.,. 1976. Phosphatidylcholine synthesis in castor bean endosperm. *Plant Physiol.*, **60**:754-758.

Moore, T.S., 1982. Phospholipid biosynthesis. *Ann. Rev. Plant Physiol.*, **33:** 235-259.

Morse, M.J., Crain, R,C. and satter, R.L. 1987. Phosphatidylinositol cycle metabolites in Samanea saman pulvini. *Plant Physiol.*, **83**:640-644.

Mudd, J.B. 1980. Phospholipid biosynthesis. *In: The Biochemistry of Plants*, Vol.4, Lipids: Structure and Function (Eds. Stumpf, P.K.), New York, Academic: 1-55.

Smolenska- Sym, G. and Kaeperska, A. 1996. Inositol 1, 4, 5 - triphosphate formation in leaves of winter oil seed rape plants in response to freezing, tissue water potential and abscisic acid. *Physiol. Plant.*, **96**:692-698.

Sexton, J.C. and Moore, T.S. 1978. Phosphatidylinositol synthesis in castor bean endosperm. Cytidine diphosphate diglyceride: inositol transferase. *Plant Physiol.*, **62**: 978-980.

Sexton, J.C. and Moore, T.S. 1981. Phosphatidylinositol synthesis by a Mn^{++} dependent exchange enzyme in castor bean endosperm. *Plant Physiol.*, **68**:18-22.

Shine, W.E., Mancha, M. and Stumpf, P. 1976. Fate metabolism in higher plants. Differential incorporaion of acyl-coenzyme A and acyl-acyl carrier proteins into plant microsomal lipids. *Arch. Biochem. Biophys.*, **173**:472-479.

Sparace, S.A., Wanger, L.K. and Moore, T.S. 1981. Phosphatidylethanolamine synthesis in castor bean endosperm. *Plant Physiol.*, **67**:922-925.

Strack, D. 1981. Sinapine as a supply of choline for the biosynthes of phosphatidylcholine in Raphanus sativus seedlings. *Z. Naturforsch. Teil C*, **36**:215-221.

Wharfe, J. and Harwood, J.L. 1979. Lipid metabolism in germinating seeds. Purification of ethanolamie kinase from soya bean. *Biochem. Biophys. Acta,* **575**:102-111.

Wang, X. 2001. Plant Phospholipases. *Annu. Rev. Plant Mol. Biol.*, **52**:211-231.

Developments in Physiology, Biochemistry and Molecular Biology of Plants, 2005
Eds.: Bandana Bose and A. Hemantaranjan
Vol., 1, pp. 37-51, New India Publishing Agency, New Delhi
E-mail: spjain_niph@rediffmail.com web: www.bookfactoryindia.com

CHAPTER - 3

METABOLISM OF SUCROSE AND STARCH IN PLANTS: SYNTHESIS AND THEIR DEGRADATION

PADMANABH DWIVEDI

INTRODUCTION

The metabolism of carbohydrates is vital for plant growth and development, and to the yield of major plant products such as starch and sucrose. Starch, sucrose and fructose are the three important carbohydrates which are utilized in most of the higher plants. The majority of sucrose that is synthesized in higher plants during photosynthesis is exported to non-photosynthetic cells (ap Rees, 1992). Starch serves as an important storage for carbohydrate residues. The degradation (breakdown) of starch ensures retaining of important life processes in plants when there is carbon shortage, for instance, during night period in leaves or in resting organs such as potato tubers.

METABOLISM OF SUCROSE

Sucrose exhibits three important inter-related functions in plants. *First*, it is the chief product of photosynthesis; this is evident from the kinetics of labeled sucrose in illuminated leaves exposed to $^{14}CO_2$. Such instances are also available from a range of species covering both C_3 and C_4 indicating that sucrose is the dominant sugar formed in photosynthesis. *Secondly*, sucrose is the major form in which carbon is translocated in plants. This has been demonstrated by analysis of contents of sieve tubes. The studies on distribution of label in petiole and stems when leaf photosynthesis in presence of $^{14}CO_2$ show that most of the exported label is present as labeled sucrose in the phloem (Canny, 1973). *Thirdly*, sucrose is the major soluble storage carbohydrate in higher plants, *e.g.*, sucrose constitutes up to 20% of the dry weight of leaves of snowdrop, and up to 19, 5 and 7.5% of fresh weight of storage tissue of sugarcane, carrot root and red beet, respectively.

Sucrose Synthesis

The ability to synthesize sucrose is a widespread and universal characteristic of higher plant cells. Following enzymes play important role in the sucrose metabolism.

(a) Invertase: This enzyme (β-D-fructofuranoside fructohydrolase; EC 3.2.1.26) catalyzes

$$\text{Sucrose} + H_2O \rightarrow \text{glucose} + \text{fructose}\ (\Delta G = -29.3\ \text{kJ/mol at pH 7.0})$$

Acid invertase has pH optima in the range pH 3.5 - 5.1, and alkaline invertase (or neutral invertase) has pH optima in the range of pH 7.0 - 7.8. Both can be separated by fractionation with ammonium sulphate and their activities vary independently during development. Acid invertase is dominant invertase in most plant tissues. The enzyme is a β-fructofuranosidase, and acts typically on sucrose for which the Km is in the range 2-50 μM, showing the Michaelis-Menten kinetics. This enzyme functions in vacuole and, or, the free space. However, available evidence suggests that acid invertase has a dual location in plant tissues and that the relative proportions in the vacuole and free space vary with the tissue and its state of differentiation.

With regard to alkaline invertase, as a rule, this activity does not reach the high levels characteristic of acid invertase. The Km for sucrose is in the range of 8-25 μM and the enzyme appears to be a β-fructofuranosidase. The enzyme may be located in the cytoplasm; this observation comes from its pH optimum, and the fact that there is a constant ratio between alkaline invertase activity and total protein in extracts of different regions of pea roots (Lyne and ap Rees, 1975).

(b) Sucrose synthase: This enzyme (UDP-glucose: D-fructose-2-α-D-glucosyl-transferase; EC 2.4.1.13) catalyzes

$$\text{UDP-glucose} + \text{fructose} \leftrightarrow \text{sucrose} + \text{UDP}$$

$$(\Delta G = -4.18 \text{ kJ at pH 7.4 and } 37°\text{C}).$$

The enzyme is purified from tubers of Jerusalem artichoke, rice and mung bean. It catalyzes reversible reaction and this distinguishes the enzyme from all other known sugar nucleotide transglucosylases. It also has very wide specificity towards the nucleoside bases like UDP-, ADP-, GDP-, TDP-, glucose. Km for sugar nucleotide ranges from 0.2 to 5.3μM.

Sucrose synthetase has a broad pH optimum for synthesis, pH 7.5 - 8.5. It is located in the soluble phase of the cytoplasm.

(c) Sucrose phosphate synthetase (SPS): This enzyme is UDP-glucose: D-fructose 6-phosphate 2-α-D-glucosyl transferase (EC 2.4.1.14). It catalyzes

$$\text{UDP-glucose} + \text{fructose 6-phosphate} \rightarrow \text{sucrose 6-phosphate} + \text{UDP}$$

$$(\Delta G = -18.7 \text{ KJ at pH 7.5 and } 38°\text{C})$$

It is widely distributed in plants and has been purified from a number of tissues like spinach and maize. Estimates of Km range from 2 to 8 mM for UDPG and from 0.6 to 4 μM for fructose 6-phosphate; this indicates that enzyme needs high concentrations of its substrates from maximum activity. The native enzyme exists as a tetramer (Walker and Huber, 1988; Lunn and ap Rees, 1990). SPS is located in the cytosol, has a neutral pH optimum. The enzyme content increases during the development of leaves, germination of seeds and fruit ripening. This enzyme is subject to regulation by metabolites and is activated by G6P and inhibited by Pi.

(d) Sucrose phosphate phosphatase (SPPase): This enzyme, phosphatase (sucrose 6-phosphate phosphohydrolase; EC 3.1.3.24) catalyzes hydrolysis of sucrose 6-phosphate to sucrose.

Sucrose 6-phosphate + H_2O → sucrose + orthophosphate

SPPase has been partially purified from a variety of plant sources including stem tissue of sugarcane (Hawker and Hatch, 1966), carrot roots (Hawker, 1967), pea shoots (Whitaker, 1984) and rice leaves (Echeverria and Salerno, 1994). It has a mol. wt. of 120 KD (Whitaker, 1984) having two similar subunits each of 55 KD. The enzyme has a pH optimum of 6.7 and Km for sucrose phosphate of between 65 and 250 μM. SPPase reside in cytosol (Hawker *et al.*, 1991). The fact that SPPase is displaced from equilibrium *in vivo* (Krause and Stitt, 1992), it also controls sucrose metabolism.

Pathway for Sucrose Synthesis

Synthesis of sucrose is a general characteristic of both photosynthetic and non-photosynthetic cells of higher plants (Fig. 1). The information on pathway of sucrose synthesis comes from three different types of studies: thermodynamic data, feeding experiments, and measurements of enzyme activities. The recent evidences suggest that sucrose is synthesized via SPS and SPP catalyzed route, and not through an alternative route involving sucrose synthase. This comes from three sources:

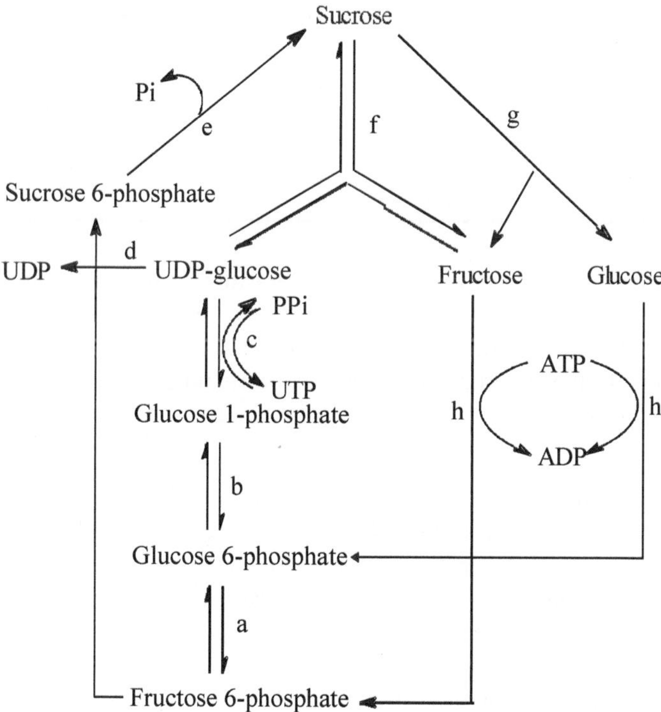

Fig. 1. Pathways of sucrose metabolism. Different letters stand for enzymes catalyzing the reactions: (a) glucose 6-phosphate isomerase; (b) phosphoglucomutase; (c) UDP-glucose pyrophosphorylase; (d) sucrose phosphate synthase; (e) sucrose phosphatase; (f) sucrose synthase; (g) invertase; (h) hexose kinase.

(i) Thermodynamic consideration indicates synthesis via SPS and SPP. The overall change in free energy of reaction catalyzed by sucrose phosphatase strongly favours synthesis even in presence of considerable accumulation of sucrose.

(ii) Kinetics of labelling of hexose phosphates, UDPG, sucrose phosphate and sucrose during photosynthesis in $^{14}CO_2$ indicates the pathway of sucrose synthesis. The fructose 6-phosphate and fructosyl moiety of sucrose are labeled before free fructose labeling in these compounds; this indicates for non-involvement of sucrose synthase in sucrose production (ap Rees, 1984).

(iii) In tissues in which sucrose synthesis is marked, activity of sucrose phosphate synthase exceeds that of sucrose synthase (Table 1). Activity of SPS, and not that of SS, is sufficient to account for rate of sucrose production in some tissues (ap Rees, 1988).

Table : 1. Examples of activities of sucrose phosphate synthase and sucrose synthase in extracts of plant tissues.

Tissues	Activity (nmol min^{-1} g^{-1} fresh wt.)		References
	Sucrose phosphate synthase	Sucrose synthase	
Shoot of Pea	317	4	Bird *et al.* (1974)
Potato tubers	115	8	Pollock and ap Rees (1975)
Stele of pea root	77	460	Lyne and ap Rees (1972)
Leaves of spinach	388	4	Bird *et al.* (1974)

Evidences favour an exclusively cytoplasmic location of sucrose synthesis. The inability of purified isolated chloroplasts to produce labelled sucrose from $^{14}CO_2$ and the impermeability of the chloroplast inner envelope to sucrose, indicate that sucrose is synthesized outside the chloroplasts. Fractionation studies of both photosynthetic and gluconeogenic cells have shown that all the enzymes necessary for this pathway are present in the cytosol.

Studies related to regulation of the pathway are restricted to photosynthetic tissues. Changes in intermediate when the rate of sucrose synthesis is varied indicate that in leaves the synthesis is regulated by sucrose phosphate synthase or sucrose phosphate phosphatase or both (Stitt *et al.,* 1980). On comparing the apparent equilibrium constants of SPS & SPPase with the mass-action ratios of reactants in spinach leaves, it is found that both reactions are far from equilibrium *in vivo* (Krause and Stitt, 1992). Further, studies have shown that SPS exert a major role in control of sucrose synthesis in leaves: (i) a close correlation between rate of sucrose synthesis and extractable activity of SPS (Stitt *et al.,* 1987), which is seen between species and genotypes, (ii) many-fold over expression of maize sucrose phosphate synthase in transgenic tomato plants results in a small increase in leaf sucrose synthesis (Frommer and Sonnewald, 1995).

SPS can exist in two kinetically distinct forms, by having differences in their substrate affinities, sensitivity to inhibition by phosphate and activation by glucose 6-phosphate; the proportion of phosphorylated and dephosphorylated forms of enzyme is decided by relative activity of protein kinase and phosphoprotein phosphatase. Dephosphorylated form of SPS is kinetically active form. This coupled with the variations in the levels of allosteric effectors contribute to regulation of SPS.

Sucrose Degradation

The degradation of sucrose serves two main purposes of providing carbon for polysaccharide synthesis and for respiration. The metabolic fate of sucrose starts with either its hydrolytic cleavage by invertases or its cleavage, via transglycosylation, by sucrose synthase. The contribution of these enzymes to sucrose breakdown can be determined by source of sucrose. Sucrose obtained through translocation can enter a cell via symplast (Patrick, 1990) or the apoplast. In the former case, sucrose is not cleaved into glucose and fructose during transport. The latter route is used by developing seeds in which there are no protoplasmic connections between maternal and embryonic tissues. The other source of sucrose for metabolism is those stored in vacuole; this is mobilized possibly through hydrolysis by acid invertase in vacuole giving rise to hexoses that are released into the cytosol. Studies have shown that level of invertase can determine the amount of sucrose in vacuoles. Sucrose in cytosol can be metabolized by either alkaline invertase or sucrose synthase. The involvement of sucrose synthase is evident by facts like (i) its distribution in different tissues is connected to sucrose breakdown rather than synthesis. Its activity is generally low in photosynthetic and gluconeogenic cells, and is often high in actively growing tissues that rely on sucrose as their respiratory substrate (ap Rees, 1984). (ii) activity of invertases in some tissues are much lower than that of sucrose synthase, and are insufficient to catalyze the observed rates of sucrose metabolism (iii) work with '*shrunken-1*' mutant of maize reveals that reduction in the level of sucrose synthase in developing endosperm restrict the ability of this tissue to metabolize sucrose (Preiss, 1982). The extent of relative contribution of these two enzymes in sucrose breakdown remains poorly understood. This is because the capacities of sucrose synthase and alkaline invertase are sufficient for both to contribute to sucrose metabolism. However, contribution of these enzymes in some instances can be established firstly, by genetic manipulation: the reduction in the extent of starch accumulation, protein content and dry weight of transgenic potato tubers deficient in sucrose synthase confirms that this enzyme is essential for effective sucrose metabolism in the developing tuber (Zrenner *et al.,* 1995). The alternative method to ascertain relative contribution of these two enzymes is to compare rates of metabolism of sucrose and 1-fluorosucrose.

In any case, nevertheless, it is apparent that in most conditions both invertase and sucrose synthase contribute to sucrose breakdown. The free glucose and fructose produced in these reactions are probably phosphorylated by the range of hexose kinases found in plants. These hexose kinases are divided into two classes based on their substrate affinities and specificities: hexokinases and fructokinases. Hexokinase (E.C. 2.7.1.1) may phosphorylate both glucose and fructose, though many hexokinases have a preferential substrate affinity to glucose, like the fraction II hexokinase of pea seeds, with Km values of 0.048 and 10 μM for glucose and fructose, respectively. Fructokinases (E.C. 2.7.1.11) are relatively specific for fructose. For instance, fructokinases showed no activity with glucose compared with fructose, in potato tubers (Renz and Stitt, 1993) and maize kernels (Doehlert, 1989). The extent to which fructokinase uses UTP as a phosphate donor determines the feasibility for the UDP product to be further cycled for cleavage of sucrose by sucrose synthase. Fructokinases from potato tubers (Renz and Stitt, 1993) and sycamore cells (Huber and Akazawa, 1986) show preferred affinity for ATP, compared with UTP, indicating that ATP functions *in vivo* as phosphate donor. Of the two phosphorylated hexoses derived from invertase hydrolysis products, F6P can be further metabolized in glycolysis in the catabolic direction, or in the synthesis of sucrose, whereas G6P can not. G-6-P can enter the pentose phosphate pathway.

The fate of UDP-glucose produced by sucrose, however, is less certain. UDPG may serve directly as substrate for cellulose or callose synthesis via UDP-glucose dehydrogenase (UDPGDH, E.C. 1.1.1.2) (Robertson *et al.,* 1995). Also, UDPG function as substrate for sucrose synthesis via both sucrose synthase and SPS. However, utilization of UDPG for either glycolysis or starch synthesis depends on further metabolism to G1P by the action of UDP-glucose pyrophosphorylase (E.C. 2.7.7.9), catalyzing reversible reaction:

$$\text{UDPG} + \text{PPi} \rightleftharpoons \text{G1P} + \text{UTP}$$

This is the only enzyme known to be present in cytosol of a wide range of tissues at activities sufficient to support the estimated rates of metabolism.

METABOLISM OF STARCH

Starch is a major plant product found to varied extents in most of the organs of almost all higher plants. It constitutes about 45% of the plant matter eaten worldwide. Starch granules contain a mixture of amylase and amylopectin. The former is a linear molecule with 600 - 3000 1,4-α-glucosyl residues, and the latter is generally much larger and more highly branched, containing 6000 to 60,000 glucosyl residues. Most of the higher plants contain two types of starch granules differing in morphology and physiology: (a) relatively small particles with a fast turnover, (b) longer particles which are formed over an extended period of time and are then degraded and disappear. The metabolism of starch is an important and widespread expression of assimilate management, as it is closely interrelated with sucrose metabolism. Starch metabolism is governed by the inter play of metabolic, developmental and environmental influences on genetic expression (Halmer and Bewley, 1982).

Starch synthesis

Starch synthesis from hexose phosphates involves following steps. *Firstly*, ADP-glucose is produced from glucose-1-phosphate by ADP-glucose pyrophosphorylase (E.C. 2.7.7.27), (Fig. 2).

$$\left[\text{Glucose 1-phosphate} + \text{ATP} \rightleftharpoons \text{ADP-glucose} + \text{PPi}\right]$$

The glucosyl unit is then transferred from ADP-glucose to the non-reducing end of an α-glucan primer by starch synthase, forming an additional 1,4-α-glucosidic bond (E.C. 2.4.1.21).

$$[\text{ADP-glucose} + \alpha\text{-glucan}_{(n)} \rightarrow \alpha\text{-glucan}_{(n+1)} + \text{ADP}]$$

Finally, the 1,6-α-glucosyl branch points of amylopectin are introduced by branching enzyme which hydrolyzes a 1,4-α-glucosyl bond and transfers the resulting oligosacchaside to a primary hydroxyl group in a similar glucan chain (E.C. 2.4.1.24).

$$[\text{Linear 1,4-}\alpha\text{-glucan} \rightarrow \text{branched 1,6-}\alpha\text{-1,4-}\alpha\text{-glucan}]$$

The activities of ADP-glucose pyrophosphorylase and starch synthase - the enzymes involved in above pathways are sufficient to account for the rates of starch production in tissues. These enzymes are confined to plastids, the site of starch accumulation. The study of mutant plants also lends support to the pathway of starch synthesis (Okita, 1992). In maize two endosperm mutants namely, '*shruken-2*' and '*brittle-2*' have only about 25% of the normal starch content. This correlates with a reduction in ADP-glucose pyrophosphorylase activity to less than 10% of that found in maize endosperm. In photosynthetic tissues the hexose phosphates required for starch synthesis are provided by CO_2 fixation during photosynthesis, as evinced by the ability of isolated intact chloroplasts to incorporate label from $^{14}CO_2$ into starch. For non-photosynthetic cells the ultimate source of carbon is translocated sucrose. The uptake of hexose phosphate is also provided by the observation that plastids from several non-photosynthetic tissues can incorporate label from [^{14}C]-hexose

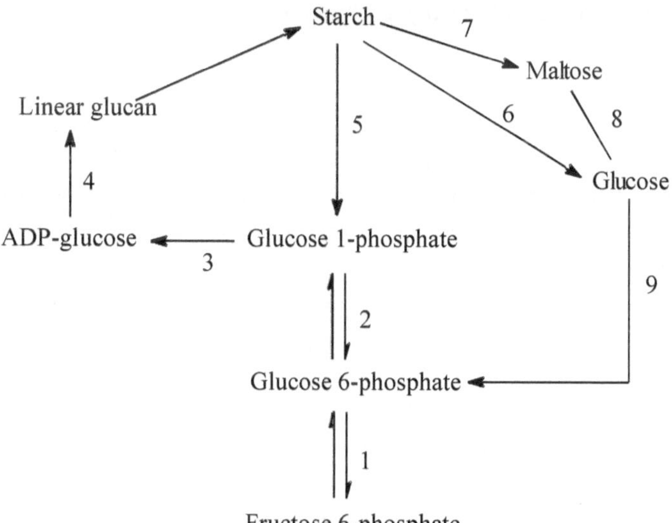

Fig. 2. Pathways of starch metabolism. The enzymes involved include: (1) glucose 6-phosphate isomerase; (2) phosphoglucomutase; (3) ADP-glucose pyrophosphorylase; (4) starch synthase; (5) starch phosphorylase; (6) α-amylase; (7) β-amylase; (8) α-glucosidase; (9) hexose kinase.

phosphates into starch. Further, studies with mutants have shown that in photosynthetic tissues much of the control of pathway of starch synthesis from hexose phosphates is through ADP-glucose pyrophosphorylase. This enzyme is inhibited by phosphate and activated by 3-phosphoglycerate. It is shown that reaction catalyzed by ADP-glucose pyrophosphorylase is far from equilibrium *in vivo* and so flux through this enzyme is sensitive to regulation by allosteric effectors.

The regulating factor in the starch synthesis is likely to be the response of ADP-glucose pyrophosphorylase to 3-phosphoglycerate and phosphate, in chloroplasts. The ratio of these metabolites (3-phosphoglycerate and phosphate) coordinates with the carbon metabolism in the cytoplasm and chloroplasts, thereby regulating the starch metabolism (Preiss, 1982). However, in non-photosynthetic tissues, the mechanism of regulation of starch synthesis is less known. It is suggested that other reactions in the pathway, other than that catalyzed by ADP-glucose pyrophosphorylase, are likely to contribute to the regulation.

Plants contain multiple forms of starch synthase, ADP-glucose pyrophosphorylase and starch branching enzyme, and differing in their molecular and kinetic properties (Preiss, 1991). This feature makes the assessment of relative contribution of each step difficult in regulation of starch synthesis. For example, ADP-glucose pyrophosphorylase is a tetrameric enzyme containing two distinct polypeptides. These isozymes differ in their ability to bind allosteric regulators like 3-phosphoglycerate. Studies with *waxy* mutants of maize, rice and sorghum as well as the corresponding *amylose-free* mutant of potato show the correlation between complete absences of amylose from starch with a massive reduction in the activity of starch synthase. This suggests that starch synthase and starch branching enzyme operate at periphery of starch granule to synthesize amylopectin, which then condense to form matrix of the granule on which the *waxy* protein binds.

Starch degradation

The process of starch breakdown is complex and the information available on the mechanism involved in this process come from studies of germination in starch-storing organs. Prior to starch degradation, the amyloplast membrane must be broken down or permeable to enzymes of starch hydrolysis. There are a number of enzymes which catalyze the breakdown of starch.

(i) α-amylase (E.C. 3.2.1.1): It is responsible for the initiation of dissolution of starch granules. The enzyme degrades both amylose and amylopectin by hydrolyzing non-terminal α- (1, 4) glucosidic bonds, resulting in the production of a mixture of linear and branched oligosaccharides like glucose, maltose, maltotriose and dextrins. The enzyme is a calcium-requiring metallo-enzyme with a pH optimum between 4.5 and 6.0.

(ii) β-amylase (E.C. 3.2.1.2): It hydrolyses α- (1, 4) glucan bonds in both amylose and amylopectin with release of maltose units from the non-reducing ends of the chains. The products include maltose and a β-limit dextrin. This enzyme is a sulphydryl-containing enzyme which can exist in a free, or soluble or insoluble form.

(iii) Starch phosphorylase (E.C. 2.4.1.1): It catalyzes a reversible reaction, producing glucose 1-phosphate from starch and inorganic phosphate.

$$\alpha\text{-glucan(n)} + \text{Pi} \rightleftharpoons \alpha\text{-glucan (n-1)} + \text{glucose 1-phosphate}$$

This enzyme from sweet corn is a dimer containing pyridoxal phosphate and degrades amylopectin *in vitro* (Lee & Braun 1973).

(iv) α-Glucosidase (E.C. 3.2.1.20): Maltose and related oligosaccharides are hydrolyzed to glucose by α-glucosidase.

(v) D-enzyme (E.C. 2.4.1.25): This glycosyl transferase can transfer part of an α- (1, 4) glucan chain to a new 4-position. The enzyme is identified in an α-glycosidase preparation from potato tubers.

$$\alpha\text{-glucan(m)} + \alpha\text{-glucan(m)} \rightleftharpoons \alpha\text{-glucan (m+n-1)} + \text{glucos}$$

(vi) Debranching enzyme (E.C. 3.2.1.10): It removes branch points in starch cleaving 1, 6-α-glucosyl bonds hydrolytically. It releases linear oligosaccharides for further metabolism.

Similar to enzymes of starch synthesis, plants contain multiple isozymes of endoamylase, exoamylase and starch phosphorylase, which are differentially expressed (Sonnewald *et al.,* 1995).

Thus the final products of breakdown can result from either phosphorolysis (glucose1-P) or hydrolysis (glucose, maltose). Endoamylase and debranching enzyme are certainly important to all starch degrading systems; however, the relative importance of other enzymes varies.

The glucose 1-P formed must leave the chloroplast as triose phosphate so that it can be converted to sucrose in the cytosol for export to other plant parts. This pathway of starch breakdown is the predominant, with product export in chloroplast (Jenner, 1982). Glucose and maltose, however, may also be significant end products of chloroplastic starch breakdown. Whereas starch degradation in the cotyledons of germinating legume seeds occur in living cells in which amyloplast membranes appear to lack integrity (Steup, 1988), that in germinating cereal seeds takes place in dead endosperm. In germinating lupin seeds, monosaccharides produced in excess of growth requirements are converted to a branched α-glucan. In the club of *Arum maculatum*, the products of hydrolysis enter the glycolytic pathway via hexokinase (Bulpin and ap Rees, 1978) and finally the reaction of the citric acid cycle.

Among the factors affecting the regulation of starch degradation, α-amylolysis is an important site of regulation. The α- amylase which initiates starch degradation is synthesized *de novo* in germinating cereals following release of gibberellic acid. Calcium availability may regulate the activity of α- amylases. Environmental conditions like pH may influence enzyme activity directly, *e.g.,* the low pH of germinating cereal endosperms correlates with pH optimum of α- amylase (pH 4 – 5), (Fig. 3).

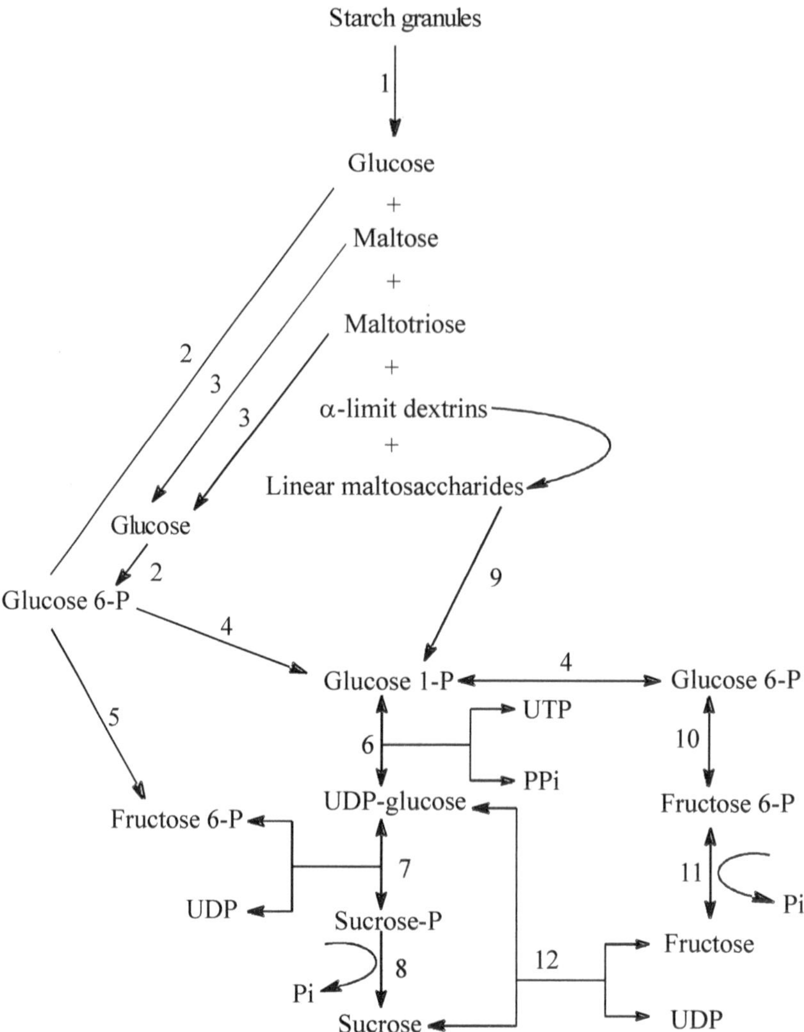

Fig. 3. Some enzymes catalyzing reaction involved in breakdown of starch: (1) α-amylase; (2) hexokinase; (3) α-glucosidase; (4) phosphoglucomutase; (5) glucosephosphate isomerase; (6) UDP-glucose pyrophosphorylase; (7) sucrose phosphate synthase; (8) sucrose phosphatase; (9) phosphorylase; (10) glucosephosphate isomerase; (11) phosphatase; (12) sucrose synthase.

CONTROL OF CARBOHYDRATE METABOLISM BY SUGAR LEVELS

Sugar levels play an important role in controlling source-sink relationships by influencing activity of enzymes of carbohydrate metabolism (Koch *et al*., 1992; Dwivedi, 2000). Sucrose has a key role for the supply of both carbon skeleton and energy making it ideally suited to act as a signal between the source and sink. Sugar regulates the activities of enzymes involved in its own metabolism: exogenous sucrose leads to increase in the activities of acid invertase in root sections of bean (Robinson & Brown 1954) and stem segments of *Avena* (Kaufman *et al.,* 1973), sucrose synthase in detached leaves of eggplant (Claussen *et al.,* 1986), and sucrose-sucrosyl-transferase in detached leaves of *Lolium temulentum* and barley (Housley and Pollock, 1985; Wagner *et al.,* 1986). The role of sugar in modulating gene expression is a well established fact in higher plants (Dwivedi, 2000). A number of studies described either the induction or repression of various plant genes, such as those encoding nitrate reductase (Cheng *et al.,* 1992), ADPG pyrophosphorylase (Muller-Rober *et al.,* 1990), or storage proteins from potato tuber (Hattori *et al.*, 1990) and soybean (Mason *et al.,* 1992). Repression of gene expression by sugar has been observed for several other plant genes such as photosynthetic genes (Krapp *et al.,* 1993), sucrose synthase *Sh1* gene (Koch *et al.*, 1992), α-amylase gene (Yu *et al.,* 1991) and glyoxylate cycle genes (Graham *et al.,* 1994). Studies made so far suggest that sugar affects the expression of genes involved in several processes, such as photosynthesis, respiration, lipid metabolism, nitrogen metabolism, and sucrose and starch metabolism. Sucrose is converted to its hexose components, or more probably a close metabolic derivative, before it becomes the signal and controls gene expression (Jang and Sheen, 1997). Hexokinase may function in the signal transduction pathway of a sugar-sensing mechanism (Table 2).

Table : 2. Examples of enzymes of carbohydrate metabolism whose gene expression is modulated by sugar levels (S = Sucrose, G = Glucose, F = Fructose).

Enzyme	Effectors tested	Effect	Plant	References
Invertase	SGF	Increase	*Chenopodium*	Roitsch *et al.* (1995)
Invertase (Ivr1)	S	Decrease	Corn	Koch and Nolte (1995)
Sucrose synthase	SGF	Increase	*Chenopodium*	Godt *et al.* (1995)
Sucrose synthase (SS2)	SGF	Increase	Rice	Karrer and Rodriguez (1992)
Sucrose synthase (SS1)	SGF	Decrease	Corn	Koch *et al.* (1992)
Sucrose phosphate synthase	G	Increase	Beet	Hesse *et al.* (1995)
ADPG Pyrophosphorylase	S	Increase	Potato	Muller-Rober *et al.* (1990)
Starch synthase	SGF	Increase	Potato	Visser *et al.* (1991)
β-amylase	SGF	Increase	Sweet potato	Nakamura *et al.* (1991)
Starch phosphorylase	S	Increase	Potato	St-Pierre and Brisson (1995)

CONCLUSIONS

Sucrose is central to plant metabolism. It represents the main end product of photosynthesis and the true starting point for sink metabolism. It has a key role in control of metabolism at the level of gene expression. There is a need for a detailed understanding of mechanisms and physiology of starch turnover in order to be better equipped to manipulate plants in service of food production. Starch, on the other hand, can be a quantitatively important part of plant carbon balance, both in source leaves and in storage sinks. It is not qualitatively important. As such there is no evidence which indicates for involvement of starch in control of any plant processes. Its production and breakdown reflect an excess or inadequacy respectively in the supply of carbon for sucrose production in leaves. Its role in metabolism and growth of sinks is less clear. There is a need to understand the mechanisms by which sink tissues metabolize assimilate for growth. This will help in evaluation of control of metabolic pathways along with the significance of tissue and cellular localization of pathways.

The two features of carbohydrate metabolism in plants – its compartmentation and redundancy – pose problem for analysis of plant carbohydrate metabolism. For instance, sucrose and starch are synthesised in cytosol and plastids, respectively. Both these compartments have the enzymic ability to metabolize hexose phosphates. Further, there is often more than one path by which metabolites could be interconvert. These characteristics restrict the finer understanding of regulation of metabolic pathways of carbohydrate.

The use of molecular techniques, however, contributes to our knowledge of carbohydrate metabolism; manipulation of metabolism by altering the activity of specific enzymes (in mutants) presents a powerful way to study regulation of metabolic pathways. A better understanding of factors controlling metabolism and distribution of storage carbohydrates is the main focus of research in recent years as it presents the potential for improvement of crop plants.

REFERENCES

ap Rees, T. 1984. Sucrose metabolism. In: *Storage carbohydrates in vascular plants*. Lewis, D.H. (ed.), Cambridge University Press, Cambridge, 53-74.

ap Rees, T. 1988. Hexose phosphate metabolism by non photosynthetic tissues of higher plants. In: *The Biochemistry of Plants*. Preiss, J. (ed.), Academic Press, San Diego, **14**: 1-33.

ap Rees, T. 1992. Synthesis of storage starch. In: *Carbon Partitioning: Within and Between Organisms*. Pollock, C.J., Farrar, J.F. and Gordon, A.J. (eds.), BIOS Scientific Publishers, Oxford, 115-131.

Bird, I.F., Cornelius, M.J., Keys, A.J. and Whittingham, C.P. 1974. Intracellular site of sucrose synthesis. *Phytochem.*, **13**: 59-64.

Bulpin, P.V. and ap Rees, T. 1978. Starch breakdown in spadix of *Arum maculatum*. *Phytochem*, **17**: 391-396.

Canny, M.J. 1973. *Phloem Translocation.* Cambridge University Press, Cambridge.

Cheng, C.L., Acedo, G.N., Cristinsin, M. and Conkling, M.A. 1992. Sucrose mimics the light induction of *Arabidopsis* nitrate reductase gene transcription. *Proc. Natl. Acad. Sci. USA*, **89**: 1861-1864.

Claussen, W., Loveys, B.R. and Hawker, J.S. 1986. Characterisation of sucrose and hormones on the activity of sucrose synthase and invertase in detached leaves and leaf sections of egg plants (*Solanum melongena*). *Plant Physiol.*, **81**: 345-357.

Doehlert, D.C. 1989. Separation and characterization of four hexose kinases from developing maize kernels. *Plant Physiol.*, **89**: 1042-1048.

Dwivedi, P. 2000. Regulation of root respiration and sugar-mediated gene expression in plants. *Curr. Sci.*, **78**: 1196-1202.

Echeverria, E. and Salerno, G. 1994. Properties of sucrose phosphate phosphatase from rice (*Oryza sativa*) leaves. *Plant Sci.*, **96**: 15-19.

Frommer, W.B. and Sonnewald, U. 1995. *J. Exp. Bot.*, **46**: 587-607.

Godt, D.E., Riegel, A. & Roitsch, T. 1995. Regulation of sucrose synthase expression in *Chenopodium rubrum*: characterization of sugar induced expression in photoautotrophic suspension cultures and sink tissue specific expression in plants. *J. Plant Physiol.*, **146**: 231-238.

Graham, I.A., Baker, C-J. and Leaver, C.J. 1994. Analysis of the cucumber malate synthase gene promoter by transient expression and gel retardation assays. *Plant J.*, **6**: 893-902.

Halmer, P. and Bewley, J.D. 1982. Control by external factors over the mobilization of reserve carbohydrates in higher plants. In: *Encyclopedia of Plant Physiology (NS)*, Loewus, F.A. & Tanner, W. (eds.), Springer, Berlin, **13A**: 748-793.

Hattori, T., Nakagawa, S. and Nakamura, K. 1990. High level expression of tuberous root storage protein genes of sweet potato in stems of plantlets grown *in vitro* on sucrose medium. *Plant Mol. Biol.*, **14**: 595-604.

Hawker, J.S. 1967. Inhibition of sucrose phosphatase by sucrose. *Biochem. J.*, **102**: 401-406.

Hawker, J.S. and Hatch, M.D. 1966. A specific sucrose phosphatase from plant tissues. *Biochem. J.*, **99**: 102-107.

Hawker, J.S., Jenner, C.F. and Niemietz, C.M. 1991. Sugar metabolism and compartmentation. *Aust. J. Plant Physiol.*, **18**: 227-237.

Hesse, H., Sonnewald, U. and Willmitzer, L. 1995. Cloning and expression analysis of sucrose phosphatase synthase from sugarbeet (*Beta vulgaris* L.). *Mol. Gen. Genet.*, **247**: 515-520.

Housley, T.L. and Pollock, C.J. 1985. Photosynthesis and carbohydrate metabolism in detached leaves of *Lolium temulentum*. *New Phytol.*, **99**: 499-507.

Huber, S.C. and Akazawa, T. 1986. A novel sucrose synthase pathway for sucrose degradation in cultured sycamore cells. *Plant Physiol.*, **81**: 1008-1013.

Jang, J-C. and Sheen, J. 1997. Sugar sensing in higher plants. *Trends Plant Sci.*, **2**: 208-214.

Jenner, C.F. 1982. Storage of starch. In: *Encyclopedia of Plant Physiology (NS)*, Loewus, F.A. and Tanner, W. (eds.), Springer, Berlin, **13A**: 700-747.

Karrer, E.E. & Rodriguej, R.L. 1992. Metabolic regulation of rice α-amylase and sucrose genes in plants. *Plant J.*, **2**: 517-523.

Kaufman, P.B., Ghosheh, N.S., Lacroix, J.D., Soni, S.L. and Ikuma, H. 1973. Regulation of inveratse levels in *Avena* stem segments by gibberllic acid, sucrose, glucose and fructose. *Plant Physiol.*, **52**: 221-228.

Koch, K.E. and Notle, K.D. 1995. Sugar-modulated expression of genes for sucrose metabolism and their relationship to transport pathways. In: *Carbon partitioning and source sink interactions in plants*. Modore, M.M., Lucas, W.L. (eds.), *Amer. Soc. Plant Physiol.*, Rockville, MD, 141-155.

Koch, K.E., Nolte, K.D., Duke, E.R., McCarty, D.R. and Avigne, W.T. 1992. Sugar levels modulate differential expression of maize sucrose synthase genes. *Plant Cell*, **4**: 59-69.

Krapp, A. Hofman, B., Schafer, C. and Stitt, M. 1993. Regulation of the expression of *rbcs* and other photosynthetic genes by carbohydrates: a mechanism for the 'sink regulation' of photosynthesis? *Plant J.*, **3**: 817-828.

Krause, K. and Stitt, M. 1992. Investigation of sucrose-6-phosphate levels in spinach leaves and its effects on sucrose-phosphate synthase: No evidence for direct product inhibition during photosynthetic sucrose synthesis. *Phytochem.*, **31**: 1143-1146.

Lunn, J.E. and ap Rees, T. 1990. Purification and properties of sucrose phosphate synthase from seeds of *Pisum sativum*. *Phytochem.*, **29**: 1057-1063.

Lyne, R.L. and ap Rees, T. 1971. Invertase and sugar content during differentiation of roots of *Pisum sativum*. *Phytochem.*, **10**: 2593-2599.

Mason, H., Dewald, D.B., Creelman, R.A. and Mullet, J.E. 1992. Coregulation of soybean vegetative storage protein gene expression by methyl jasmonate and soluble sugars. *Plant Physiol.*, **98**: 859-867.

Muller-Rober, B.T., Kobmann, J., Hannah, L.C. Willmitzer, L. and Sonnewald, U. 1990. One of two different ADP-glucose pyrophosphorylase genes from potato responds strongly to elevated levels of sucrose. *Mol. Gen. Genet.*, **224**: 136-146.

Nakamura, K., Ohto, M., Yoshida, N. and Nakamura, K. 1991. Sucrose-induced accumulation of α-amylase occurs concomitant with the accumulation of starch and sporamin in leaf petiole cuttings of sweet potato. *Plant Physiol.*, **96**: 902-909.

Patrick, J.W. 1990. Sieve element unloading: cellular pathway, mechanism and control. *Physiol. Plant*, **78**: 298-308.

Pollock, C.J. and ap Rees, T. 1975. Effect of cold on glucose metabolism by callus and tubers of *Solanum tuberosum*. *Phytochem.*, **14**: 1903-1906.

Preiss, J. 1982. Biosynthesis of starch and its regulation. In: In: *Encyclopedia of Plant Physiology (NS)*, Loewus, F.A. & Tanner, W. (eds.), Springer, Berlin, **13A**: 397-417.

Preiss, J. 1991. Biology and molecular biology of starch synthesis and its regulation. In: *Oxford Surveys of Plant Molecular and Cell Biology*. Miflin, B.J. (ed.), Oxford University Press, Oxford, 7: 59-114.

Renz, A. and Stitt, M. 1993. Substrate specificity and product inhibition of different forms of fructokinases and hexokinases in developing potato tubers. *Planta*, **190**: 166-175.

Robertson, D., Beech, I. and Bolwell, G.B. 1995. Regulation of the enzymes of UDP sugar metabolism during differentiation of sugarbeet. *Phytochem.*, **39**: 21-28.

Roitsch, T., Bittner, M. and Godt, D.E. 1995. Induction of apoplastic invertase of *Chenopodium rubrum* by D-glucose and a glucose analog and tissue specific expression suggest a role in sink-source regulation. *Plant Physiol.*, **108**: 285-294.

Sonnewald, U., Basner, A., Greve, B. and Steup, M. 1995. *Plant Mol. Biol.*, **27**: 567-576.

Steup, M. 1988. Starch degrdadtion. In: *The Biochemistry of Plants*. Preiss, J. (ed.), Academic Press, San Diego, **14**: 255-296.

Stitt, M., Wirtz, W. and Heldt, H.W. 1980. Metabolite levels in the chloroplast and extra chloroplast compartments of spinach protoplasts. *Biochem. Biophys. Acta*, **593**: 85-102.

Stitt, M., Huber, S.C. and Kerr, P. 1987. Regulation of photosynthetic sucrose synthesis. In: *The Biochemistry of Plants.* Hatch, M.D. & Boardman, W.K. (eds.), Academic Press, New York, **13**: 327-409.

St-Pierre, B. and Brisson, N. 1995. Induction of the plastidic starch-phosphorylase gene in potato storage sink tissue. *Planta*, **195**: 339-344.

Visser, R.G.F., Stolte, A. and Jacobsen, E. 1991. Expression of a chimaeric granule –bound starch synthase – GUS gene in transgenic potato plants. *Plant Mol. Biol.*, **17**: 691-699.

Wagner, W., Wiemken, A. and Matile, P. 1986. Regulation of fructan metabolism in leaves of barley (*Hordeum vulgare* L. cv. Gerbel). *Plant Physiol.*, **81**: 444-447.

Walker, J.L. and Huber, S.C. 1988. Purification and preliminary characterization of sucrose phosphate synthase using monoclonal antibodies. *Plant Physiol.*, **89**: 518-524.

Whitaker, D.P. 1984. Purification and properties of sucrose-6-phosphate from *Pisum sativum* shoots. *Phytochem.*, **23**: 2429-2430.

Yu, S.M., Kuo, Y.H., Sheu, G., Sheu, Y.J. and Liu, L.F. 1991. Metabolic derepression of á-amylase gene expression in suspension-cultured cells of rice. *J. Biol. Chem.*, **266**: 21131-21137.

Zrenner, R., Salanoubat, M., Willmitzer, L. and Sonnewald, U. 1995. Evidences of the crucial role of sucrose synthase for sink strength using transgenic potato plants (*Solanum tuberosum* L.). *Plant J.*, **7**: 97-107.

Developments in Physiology, Biochemistry and Molecular Biology of Plants, 2005
Eds.: Bandana Bose and A. Hemantaranjan
Vol., 1, pp. 53-80, New India Publishing Agency, New Delhi
E-mail: spjain_niph@rediffmail.com web: www.bookfactoryindia.com

CHAPTER - 4

NITRATE ASSIMILATION AND NITRATE REDUCTASE ACTIVITY

M. KAR AND B.K. GHOSH

INTRODUCTION

Among plant nutrients, our soils are particularly low in nitrogen and therefore, require is high doses of nitrogenous fertilizers for getting optimum productivity. Added to this is the problem of poor efficiency of nitrogeneous fertilizers which under normal conditions is only 40 – 50 %, still less for rice ranging from 25 – 35 %, and the residual effect for the succeeding crops is never more than 10 %. The rest of the applied nitrogen is lost either through volatilization, leaching or denitrification (Kashyap, 1978). The high cost of nitrogenous fertilizers and their poor utilization efficiency very much add to the high cost of cultivation. Hence, the need of the hour is to decrease our dependence on chemical fertilizers and increase the efficiency of the fertilizer thus used. The increase in nitrogen efficiency in crop plants can be achieved by reducing denitrification with the use of nitrification inhibitors, reducing the soil compaction, improving drainage conditions, use of soil improving crops, and adopting carefully the irrigation procedures (Carlson, 1980). No doubt exploration of this possibility is a big task, nevertheless, the understanding of physiology and biochemistry of nitrogen nutrition at molecular level is obligatory and is likely to add insight to achieve the hard task goal.

UPTAKE AND METABOLISM OF INORGANIC NITROGEN

Nitrate and Ammonium Uptake in Higher Plants

Nitrate and ammonium are the major sources of inorganic nitrogen taken up by the roots of higher plants. Most of the ammonium has to be incorporated into organic compounds in the roots, whereas nitrate is readily mobile in the xylem and can also be stored in the vacuoles of roots, shoots and storage organs. Nitrate accumulation in vacuoles can be of considerable importance for cation -anion balance, for osmoregulation, particularly in so-called "nitrophilic" species (Smirnoff and Stewart, 1985) and for the quality of vegetable and forage plants. The absorption of nitrate by roots of higher plants is generally thought to be thermodynamically active and to require a significant input of energy. Nitrate uptake in plants is highly regulated and co-ordinated with other transport and metabolic pathways and a number of genes for nitrate uptake and assimilation related have been identified and characterized. Plants absorb nitrate via transporters localized in the root epidermal and cortical cell plasma membrane over a wide nitrate concentration range using several different transport mechanisms (Forde 2002; Coruzzi and Bush, 2001). These include constitutive and nitrate inducible high affinity transport systems, as well as nitrate inducible low affinity transporters. Compared with nitrate reduction and general nitrogen metabolism, the uptake of nitrate, nitrite and ammonium has attracted less interest , probably due to lack of an easy to handle radioactive isotope as well as to many metabolic impacts (Ullrich, 1987). According to Mitchell's (1978) general view of membrane transport anions, nitrate (NO^-_3) in particular, can be taken up by three mechanisms:

(a) Uniport directly energized by an ATPase or redox pump;
(b) Antiport of one anion against another with inverse concentration gradient;
(c) Symport of the anion together with one or more cations having a strong driving force (Mc Clure *et al.,* 1990).

Nitrogen deficiency and duration of N deprivation increased the uptake of NO^-_3–N and NH^+_4–N (Bowman and Paul, 1988). Increase in nitrogen uptake by higher plants is attributed to increased accumulation of soluble carbohydrates in roots following N-starvation (Champigny *et al.,* 1984). Results from other studies of nitrate uptake by plants, however, have led to alternative proposals. In studies of nitrate -starved and nitrate-induced excised maize roots; Thibaud and Grignon (1981) reported that nitrate starved roots excreted protons in the presence of Ca $(NO_3)_2$ while nitrate- induced roots displayed a net H^+ - influx in the same solution. The excreted protons help in activation of proton co-transport system (symport) and exemplify the nitrate uptake from the external solution. Net uptake of nitrate by roots is the difference between concomitant influx and efflux (Glass and Siddiqi ,1995), however the latter has been found in significant when ambient NO^-_3 concentrations are low (Ingmarsson *et al.,* 1987).

Nutrient efficiency of crops may result from increased uptake efficiency or /and utilization efficiency and the former is more important in the soil with lower availability of nutrients (Kalita and Nair, 2001). Plants can cope up with nutrient deficiency by increasing absorbing surface, or by increasing up-take potential per unit root weight or length. In well aerated soil, the predominant form of N available to plant is nitrate regardless of the form in which it is applied to the soil. Thus improvement in nitrate uptake may prove beneficial to improving overall nitrogen efficiency, especially at low level of N availability. Improvement in nitrate uptake *vis-a-vis* heterosis in nitrogen assimilation and utilization and its role in realization of higher yields in hybrids have been studied in various crops like pearl millet, maize and sunflower (Hirel *et al.,* 2001; Stuber 1997; Joshi with *et al.,* 1991) which implicit that the heterosis in nitrogen assimilation is chiefly attributed to nitrogen assimilating enzymes, NR in particular (Shanker *et al.,* 2002).

It has been known for a long time that ammonium suppresses nitrate uptake in many higher plants (Beevers and Hageman, 1983; Ullrich, 1983). In general, there are three ways in which ammonium can become inhibitory:

(a) Repression of the formation of new nitrate carrier and nitrate reductase;
(b) Inhibition of the carrier and /or nitrate reductase either by ammonium or by one of its metabolic products (Syrett, 1987);
(c) Physical inhibition of the nitrate uptake mechanism.

Assimilation of Inorganic Nitrogen in Higher Plants

The capacity for nitrate reduction in higher plants is wide spread (Smirnoff *et al.*, 1984) although habitats differ quite markedly in the extent to which nitrate or ammonium are the available form of nitrogen. Not surprisingly plant species exhibit differences in their capacity to utilize nitrate and ammonium ions as sole nitrogen source (Gigon and Rorison, 1972). Moreover, the efficiency of N uptake, assimilation and redistribution processes may vary with the environment, the level of supplemental nitrogen (Kumar and Singh, 2000) and the growth stages of crop plants (Abdin *et al.*, 1993).

Site of inorganic nitrogen assimilation

In contrast to dinitrogen fixation and ammonium assimilation, which occur in the root system, nitrate reduction can take place in both root and shoot. The proportion of reduction carried out in roots and shoots depends on various factors, including the level of nitrate supply, the plant species, the plant age and the various environmental factors (Marschner, 1995). In general, when the external nitrate supply is low, a high proportion of nitrate is reduced in the roots. With an increasing supply of nitrate, the capacity for nitrate reduction in the roots becomes a limiting factor and an increasing proportion of the total nitrogen is translocated to the shoots in the form of nitrate (Fig. 1).

The reduction and assimilation of nitrate are dependent on the relative levels of enzyme nitrate reductase in different plant parts which also varies in different species and their ages

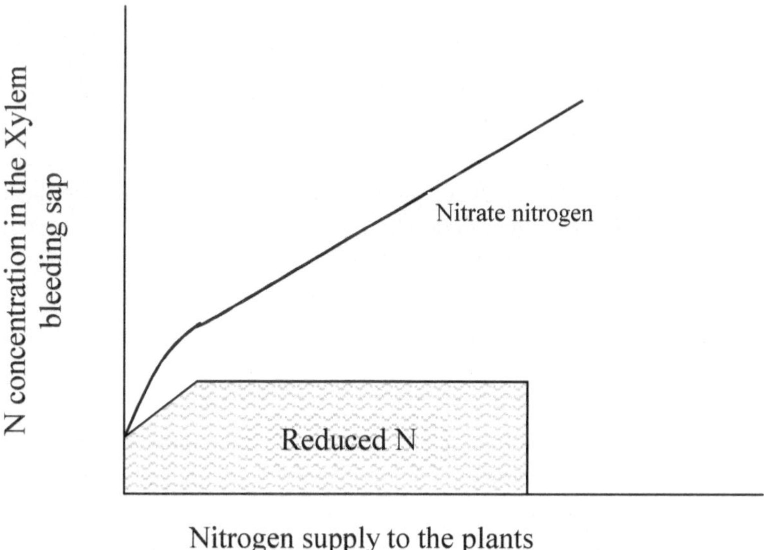

Fig. 1. Schematic representation of the effect of the level of NO_3^- supply in the rooting medium of non-inoculated field pea on the nitrogen compounds in the xylem sap of decapitated plants (Wallace and Pate, 1965).

(Marwaha, 1998; Srivastava, 1975). For example, in wheat and maize, at least two thirds of the nitrate has been reported to be assimilated in the leaf (Beevers and Hageman, 1969), while in pea major amount of nitrate is reduced in the roots (Wallace and Pate, 1965). In potato, the nitrate reduction was about 9-12 times higher in the leaves as compared to roots at different growth stages (Marwaha, 1998). The level of nitrate reduced is directly proportional to the external nitrate concentration *i.e.,* with increasing external supply; there is shift in locale of nitrate reduction from root to shoot (Marschner, 1995; Kar *et al.,* 1994).

The preferential site of nitrate reduction, roots or shoots, may have an impact on carbon economy of plants, and probably also have ecological consequences for the adaptation of plants to low light and high light conditions (Smirnoff and Stewart, 1985). Reduction and assimilation have a high energy requirement and are costly processes when carried out in roots while leaf nitrate assimilation carries a lower energy cost than root assimilation (Sprent, 1980; Pate 1983; Raven, 1985). This suggestion is based upon the use of excess reductant and ATP generated by photosynthesis to assimilate nitrate, and assumes light intensities are high enough to saturate photosynthesis. Unless this condition is met, nitrate and carbon dioxide will compete for reductant and ATP (Canvin and Atkins, 1974) and no energetic advantage will accrue. Other factors favouring leaf assimilation are the cost of synthesizing and translocation sucrose to the root system and the necessity to re-assimilate ammonium in leaves when asparagines and ureide are the products of root nitrate assimilation translocated to the shoot (Stewart *et al.,* 1987).

Nitrate Assimilation and Cation-Anion Balance

Nitrate acts as a strongest osmolytes and thus helps in osmoregulation (Marschner, 1995). So many a times NO^-_3 concentration builds up in cytoplasm of the cell and may impact on ionic balance and intracellular pH. In those plant species where most or all nitrate assimilation occurs in shoot, organic acid anions are synthesized in the cytoplasm and stored in the vacuole (Fig. 2) in order to maintain both cation-anion balance and cytosolic pH (Van Beusichem and Nelemans, 1990) This might lead to osmotic problems if NO^-_3 reduction were to continue after the termination of leaf cell expansion (Raven and Smith, 1976). However, different mechanisms exist for the removal of excess osmotic solutes from the shoot tissue.

- Precipitation of excess solutes in an osmotically inactive form. Synthesis of oxalic acid for charge compensation in NO^-_3 reduction and precipitation as calcium oxalate (Egmond and Breteler, 1972)
- Retranslocation of reduced nitrogen (amino-acids and amides) together with phloem-mobile cations, such as K^+ and Mg^{++} to areas of new growth.
- Retranslocation of organic acid anions, preferentially malate, together with K^+ into the roots and release of an anion (OH^- or HCO^-_3)

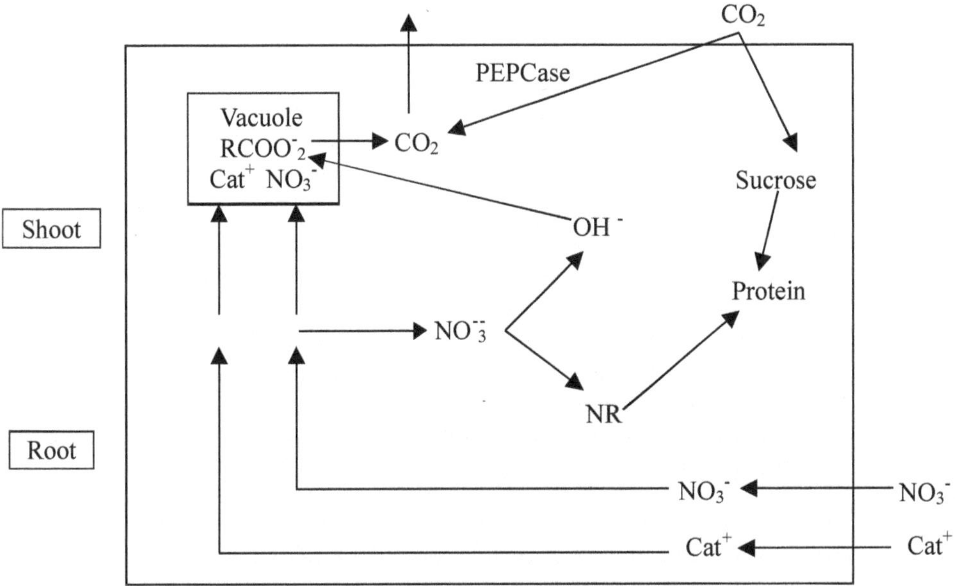

Fig. 2. Model of nitrate assimilation in shoots.

ENZYMES OF INORGANIC NITROGEN METABOLISM

Nitrate Reductase (NR)

Assimilatory NR : structure and function

Function:

Prior to incorporation into amino-acids, nitrate is first reduced to nitrite in the cytosol by nitrate reductase (NR) and then nitrite is reduced to ammonia in plastid by nitrite reductase (Fig 3).The reaction can be written as:

$$AH_2 + NO^-_3 \xrightarrow{NR} A + NO^-_2 + H_2O$$

For the redox couple,

$$NO^-_3 + 2H^+ + 2e^- \xrightarrow{NR} NO^-_2 + H_2O$$

OR,

$$NO^-_3 + NADPH + H^+ \xrightarrow[2e^-]{NR} NO^-_2 + NADP + H_2O$$

The rate limiting and regulating step of nitrate assimilation appears to be the initial reaction catalyzed by NR, as it is the first enzyme of the nitrate assimilatory pathway, is substrate inducible. NR activity is relatively low as compared to other enzymes of the pathway and it is an energy consuming process (Beevers and Hageman, 1969). NR is considered to be a limiting factor for growth, development and protein production in plants and nitrate assimilating organisms (Nair and Abrol, 1982; Nair and Chatterjee, 1990, Selvaraj and Kumari, 1999).

Nitrate reductase is an enzyme that is regulated by several different modes exerted at different levels, namely, enzyme synthesis and degradation, reversible inactivation and concentration of the substrate and effectors (Solomonson and Barber, 1990). The enzyme has a half-life of only a few hours (Beevers and Hageman, 1983), and in plants receiving no nitrate is merely absent (Li and Gresshoff, 1990). Nitrate reductase can be induced within a few hours by addition of NO^-_3 (Oaks *et al.,* 1972) and suppressed by certain amino acids (Oaks, 1991). Nitrate appears to be the major signal leading to induction of higher plant NR. It regulates cellular expression of NR by two mechanisms: *de novo* synthesis of NR and NO^-_3 dependent reversible activation of an inactive NR protein (Pattanayak and Chatterjee, 1997). However, the regulated expression of higher plant NR gene (*nia*) results from a complex interplay of external and internal factors including nitrate, light and phytohormones (Warner and Kleinhofs, 1992; Chatterjee and Naik ,1993; Hoff *et al.,* 1994; Lillo 1994).

The high turnover rate of NR and the distinct modulation of its activity by those effectors (nitrate, light or phytohormones) has initiated many studies using this enzyme as a model for gene regulation by environmental factors in general and nitrate in particular (Creelman *et al.,* 1990). The well known increase in NRA by cytokinin is expressed at the level of NR-mRNA which is increased by cytokinin but suppressed by ABA (Lu *et al.,* 1992). Decrease in the use efficiency of the NR transcript for production of NR- protein is obviously the main factor responsible for the much lower activity of NR in old compared with young leaves (Kenis *et al.,* 1992). The decrease in cytokinin import into older leaves might be involved in this age dependent decline in transcription (Smiciklas and Below, 1992).

Nitrate reduction takes place by the transfer of two electrons from NADH (Fig. 3) to nitrate via three redox centers in NR: Flavin adenine dinucleotide (FAD) heme, and the molybdenum cofactor (MoCo) or molybdo-pterin complex MoCo seems to be the rate limiting step in the reduction of nitrate (Cannons and Solomonson, 1994; Crawford, 1995). The transfer of electrons from heme to the MoCo domain is the likely target for phosphorylation, a post transcriptional mechanism that modulates NR activity rapidly and reversibly (Kaiser and Huber, 1994; LaBrie and Crawford, 1994). Since molybdenum is an important component of MoCo of NR, supplementation of Mo to crop plants could able to increase the NR activities *vis-a-vis* nitrate assimilation (Selvaraj and Kumari *et al.,* 1999; Sagi *et al,* 1997).

It is expected that NRA would be very low in Mo deficient plants. In these the enzyme can be reactivated if the plants are treated with molybdenum containing complexes (Notton and Hewit, 1979). The striking differences in the NR activities in Mo- deficient and Mo-sufficient plants and the rapid response to Mo supply can be used to determine the Mo nutritional status of plants (Witt and Jungk, 1977).

There is ample of evidences that NR can be induced in all parts of the plant (Wallace, 1987), its amount and activity reflecting the flux of NO^-_3 into the cell. While hormonal and

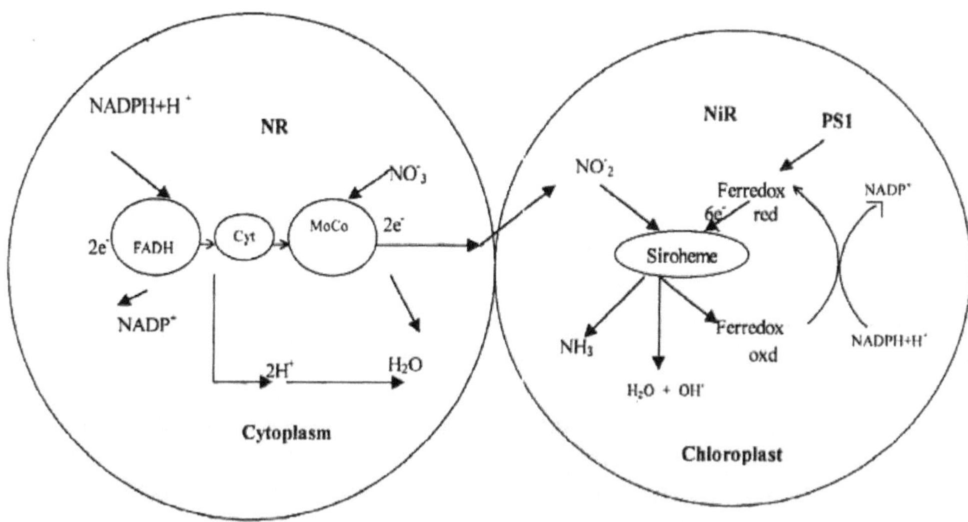

Fig. 3. Schematic representation of nitrate assimilation in leaf cells.

environmental factors have some influence on the level of NO^-_3 assimilation, the activity of NR is most likely to be limited by the availability of NADH and nitrate (Brunetti and Hageman, 1976). Carbohydrate limitation is likely under experimental conditions in the laboratory (sub-optimal light, but optimal NO^-_3). In the field, nitrate is more likely to be the limiting factor, especially during summer and at the late vegetative stages of plant growth (Nelson-Schreiber and Schweitzer, 1986). Moreover, the capacity of different genotypes to extract soil NO^-_3 -N also limits the NR activity in plant tissues (Kalita and Nair, 2001). Hence, it is important to understand the influence of nitrogen source, dose, season and genotypes and their interaction on the NR activity as this is one of the important traits used for crop improvement in India where the crop seasons vary significantly for temperature, rainfall *etc*. (Kumar and Singh, 2001).

Structure:

The native and subunit molecular weight of *chlorella* nitrate reductase are 360-380 kD and 90-100 kD respectively (Solomonson *et al.,* 1986). Howard and Solomonson, 1982). It contains one each of the prosthetic groups FAD, heme and Mo per subunit. The heme classified as a β-type cytochrome (b_{557}) (Solomonson and Vennesland, 1972), exhibits a visible spectrum identical to hepatic sulphite oxidase (Cohenand Fridovich, 1971). EPR analysis of the oxidized enzyme revealed a paramagnetic species with g-values of 2.97 and 2.25 attributable to a low spin ferric heme (Solomonson *et al.,* 1984 a).

Molybdenum, present as a pterin-Mo complex similar to the Mo-cofactor isolated from other molybdoenzymes (Johnson *et al.,* 1980), exhibits an X- ray absorption edge and EXAFS spectroscopy that indicate its environment in *Chlorella* nitrate reductase closely resembles that in sulphite oxidase (Cramer *et al,* 1984). Molybdenum in oxidized NR has two terminal oxygens at 1.7 A⁰ as well as two or three sulphurs at 2.4 A⁰ . A single terminal oxygen at 1.7 A⁰ and a set of sulphurs at 2.4 A⁰ were found upon NADH reduction. Partial reduction of NR generated a triad of Mo^V EPR signals, gav approximately 1.97 with precise g-values and line shapes dependent on pH and the presence of anions (Solomonson and Barber, 1987). The Mo^v signal was abolished by reduction with NADH, reoxidation with nitrate and reaction of the reduced enzyme with CN^- (Barber and Solomonson, 1986). This latter observation suggests that CN^-, a possible physiological regulator of nitrate reductase acts by binding to reduced Mo reversible, thus inactivating the enzyme. *Chlorella* NR also contains a tightly bound FAD prosthetic group (Solomonson *et al.,* 1975). Visible spectra of NR reduced with NADH or with dithionite in the presence of NAD^+ showed an increase in absorbance at long wavelengths (> 600 nm) which was not observed when the enzyme was reduced in the absence of NADH or NAD^+ (Solomonson and Barber, 1984) The EPR spectrum of NADH-reduced NR (pH 7.0) showed a free radical species (g = 2.004, line width 1.4 mT). Idential spectra were obtained at pH 9.0 and for 2H- substituted nitrate reductase indicating the flavin species formed in the presence of NAD^+ is the anionic semiquinone, flavine adenine dinucleotide free radical (FAD^{*-}).

Nitrate reductase exhibits a number of partial activities utilizing one or more of the prosthetic groups as is evident from the study of its kinetic parameters (Solomonson and Barber, 1987). NADH – dehydrogenase activities utilize NADH as the electron donor, involve the FAD prosthetic group, and require the participation of an essential sulphahydryl group for catalytic activity, whereas nitrate reducing activities utilize alternate electron donors and require a functional molybdenum centre. Initial velocity studies showed that increased ionic strength stimulated NADH: NRA by increasing both V_{max} and K_m for nitrate (Kelly *et al.,* 2000). Comparison of the rates for the partial activities indicated electron transfer from heme to Mo to be the rate limiting step in enzyme turn over. The steady-state kinetic parameters for nitrate reductase are summarized in (Fig. 4).

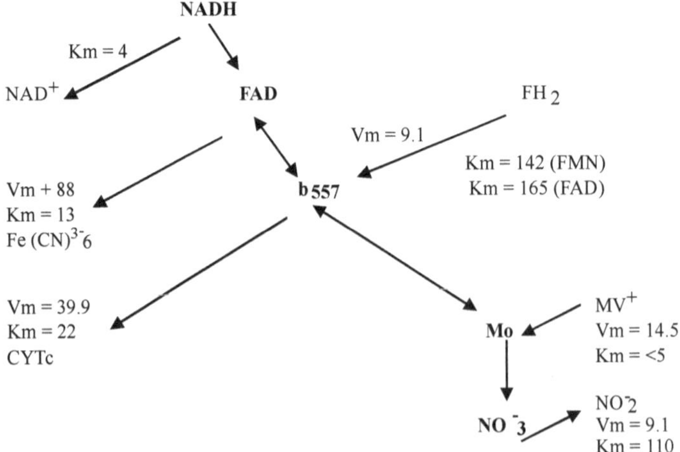

Fig. 4. Summary of the kinetics constants obtained for the partial activities of NR. Apparent V_{max} values are given as μ mol 2 electrons min^{-1} $nmol^{-1}$ heme, while apparent K_m values are given in μ M (Solomonson and Barber 1987).

Photoregulation of nitrate assimilation

Nitrate reduction in green organisms is a basic photosynthetic process since it uses, either directly or indirectly, a high proportion of the total of the reducing power photo generated in the chloroplast. Furthermore, both the intensity and the quality of light may control NO^-_3 assimilation by modulating either NO^-_3 uptake (Calero *et al.,* 1980) or NO^-_3 reduction through regulation of the synthesis and activity of the enzyme NR (Losada *et al.*, 1981). As has been reported the levels of extractable NR (Hewitt *et al* 1976; Beevers and Hageman, 1969) and NiR (Gupta and Beevers, 1984; Sawney and Naik, 1972) are photoregulatory and hold a key to NO^-_3 metabolism. Activities of NR (Bowsher *et al.,* 1991) and NiR increased markedly when NO^-_3 supplied to etiolated or green seedlings kept in dark are exposed to light (Hageman and Flesher, 1969). Although nitrate is the primary inducer of nitrate reductase, light is also required for the maximum effect (Taiz and Zeiger, 1991; Somers *et al.*, 1983). The effect of white light can be seen at the level of mRNA followed by *de novo* synthesis of NR protein. NR activity can be reversibly regulated by red and far red light indicating control the phytochrome system (Hopkins, 1999). With the aid of modern molecular techniques, it has been clearly established that phytochrome does control gene expression and that it can do at the transcription level. However, it is not clear, how P_{fr} regulates gene transcription. Some studies have suggested that P_{fr} may activates another protein (a second messenger?) that binds to certain DNA sequences called light regulated elements (Moses and Chua, 1988). Light regulated elements are short DNA sequences located in the promoter region downstream from, the gene itself. The possible binding of the regulatory protein to the light regulated elements apparently stimulates the transcription of NR gene; transcription will not proceed in its absence. Recent investigation suggests that phytochrome mediated NR expression may be via the activation of protein kinase type of activity through phosphoinositide cycle (Chandok and Sapory, 1992).

Maldonado and Aparico (1987) investigated photoregulation of nitrate assimilation in eukaryotic organisms. They reported that spinach NR which has been in activated by incubation with NADH and CN^- is reactivated in minutes when it is irriadiated with white or blue light. Red light or dark treatments are completely ineffective. The action spectrum for photo reactivation of inactive NR from *Neurospora crassa* has been determined which resembles the absorption spectrum of a flavin (Roldan and Butler, 1980). Hence, the enzyme bound FAD appears to act as photoreceptor. The role of flavins as photosensitizers greatly enhances the rate of photo-reactivation. The activating action of the blue light plus the flavin system on CN^- - inactivated NR has been also shown in green algae (Aparicio and Maldonado, 1979), in the fungus (Fritz and Ninnemann, 1985) and other higher plants (Aryan *et al.,* 1983; Echevarria *et al.,* 1984). Light dependent reduction of nitrate in the pea chloroplast (House and Anderson 1980), in maize (Oaks *et al.,* 1982) and wheat leaves (Sawhney *et al* 1978 b) in the presence of nitrate reductase has also been documented. However, Sehtiya and Goyal (2002) stated that light *per se* is not obligatory for NO^-_3 assimilation rather carbohydrate supply may mimic light.

Earlier reports suggest that light is not absolutely essential for induced synthesis of nitrate reductase. Travis *et al.* (1970) and Travis and Key (1971) have reported a close relationships between polyribosome contents and synthesis of NR. In etiolated corn seedling transferred to light, continuous increase in the enzyme activity as well as polyribosome contents of leaves was noticed. They proposed that light exerts its effect through supply of energy to maintain integrity of polyribosome. They also showed that synthesis of NR occurred even in dark if the energy requirement was fulfilled by exogenous supply of carbohydrates. Aslam *et al.* (1976) reported that etiolated leaves of barley possessed a very low NR activity even though these accumulated large amount of nitrate. On transferring them to light, rapid increase in enzyme activity was accompanied by appearance of nitrate in metabolically active pool (Cheng *et al.*, 1991). It was therefore, contemplated that in dark endogenous nitrate is confined in storage pool and illumination facilitates its transfer to metabolic pool which then acts as inducer for nitrate reductase (Sawhney and Naik, 1988). Lips and Roth-Bejerano (1969) and Roth Bejerano and Lips (1970) have suggested that light is necessary for maintaining proper hormonal balance conducive for synthesis of nitrate reductase. In another study, Nasrulhaq- Boyce and Jones (1977) found that treatment of either etiolated or green seedlings of barley with levulinic acid, an inhibitor aminolevulinic acid dehytrase, suppressed light dependent increase in NR. Since NR is a heme-protein, it was thought that in chlorophyllous tissues light acts by maintaining an adequate supply of porphyrins for its synthesis.

Either directly or indirectly, light plays a cardinal role for the synthesis/ activation of nitrate reductase and thus indispensable for nitrate assimilation. The evidence available till date univocally opine the existence of a strong and positive correlation between light and NR functioning, the effect of light being multifaceted. Nevertheless, in which way the light exactly regulates the NR activity *vis-à-vis* the nitrate assimilation is debatable till date. Undoubtedly, the results obtained so far with regards to photo induction of NR would be highly useful and would add insight to work in depth in future so far as the photoregulation of NR activity is concerned.

Genetic control of nitrate assimilation

The genetic analysis of mutant strains indicates that at least six or seven genes are involved in the synthesis of an active nitrate reductase complex (Fernandez and Matagne, 1984), one or two for the apoproteins of the complex, one for regulation of its synthesis and four for the molybdoprotein cofactor production and elaboration.NR in higher plants is encoded by *nia* gene (Pattanayak and Chatterjee, 1997). Much of the knowledge about NR gene structure and function came from studying the mutant deficient in NR apoenzyme. So far *nia* genes or *nia* cDNA, from at least 10 different plant species, have been cloned. These are: squash (Crawford *et al.*, 1986), tobacco (Vaucheret *et al.*, 1989), Arabidopsis (Cheng *et al.*, 1988), tomato (Daniel-Vedel *et al.*, 1989), rice (Hamat *et al.*, 1989), bean (Jensen *et al.*, 1996), barley (Schnorr *et al.*, 1991) and petunia (Salanoubat and Dang, 1993). Comparison of nia gene sequences shows that their overall structure is well conserved among plants. There is striking homology in NR gene structure and organization with some exceptions, among plants, suggesting that one structural model can describe most of the eukaryotic assimilatory NR (Pattanayak and Chatterjee, 1997).

The genes involved in nitrogen assimilation and its regulation in *Neurospora crassa (N. crassa)* are *nit-3, nit-2, nit-4, nmr-1* and the genes that encode for the enzymes for MoCo biosynthesis and its regulation (Crawford and Arst, 1993). The *nit-3* gene encodes for the NR apoenzyme, and mutation in this gene leads to complete loss of NR activity (Okamoto *et al.,* 1991). The *nit-3* transcript is 3.1 kb in length and encodes a protein of 982 amino-acid residue with molecular mass of 108 kDa.

Nit-4 gene encodes a transacting positive regulatory protein required for induction in the presence of nitrate. Mutation in *nit-4* makes *N. crassa* insensitive to nitrate induction (Fu *et al.,* 1989). *Nmr-1* gene is thought to be a negatively acting nitrogen regulatory gene. *Nmr-1* mutant strain has the ability to produce NR in presence of nitrogen metabolite repressor glutamine. So, it was contemplated that *Nmr*-1 gene product assists glutamine in mediating nitrogen metabolite repression. *Nmr-1* gene product in presence of co-repressor glutamine inactivates *nit-2* protein (Fu *et al.,* 1988).

RNA blot analysis with the help of a number of cDNA clones, reported from several plant species shows that the primary effect of NO^-_3 on NR induction is at the transcriptional level (Crawford, 1995; Sueyoshi *et al.,* 1995). A NR deficient mutant of *N plumbaginifolia* impaired in the production of NR transcript and protein has been transformed to restore the NR activity, with a chimaeric tobacco NR cDNA, composed of a CaMV 35 S promoter and a termination signal of tobacco NR gene (Vincentz and Caboche ,1991). The viable and fertile plants showed one fourth to three times the wild type NR activity in the leaves. In contrast to wild type plant which showed high level of NR mRNA in the presence of NO^-_3 and negligible amount in the presence of ammonium, the transformed plant showed fairly high amount of NR mRNA irrespective of the nitrogen source (Vincetz and Caboche, 1991). So NR gene transcription by 35 S promoter in transgenic plant is deregulated, which clearly indicates that nitrate regulates expression primarily at the transcriptional level. The study of Fernadez and Cardeans (1987) in the green algae revealed that depression of NR in cells grown in ammonium or methylammonium is dependent on mRNA synthesis since 6-methylpurine inhibited the depression process. However, in repressed cells incubated during 8-h in nitrogen free medium mRNA is formed. Warner *et al.* (1985) selected NR deficient mutants in barley using a qualitative *in vivo* assay to screen M_2 seedlings. About 35 mutants including three selected by resistance to chlorate have been partially characterized. Thus far, all mutants have low NADH-NR and high NiR activities. Characteristics of representative mutants are presented in (Table 1).

The mutants varied in NR associated activities. Mutant *nar 1h* has high $FMNH_{2,}$ but low NADH-NR activity. NR associated Cyt c reductase activities in the mutants vary from low (*nar 1a, nar 1h*) to intermediate (*nar 1b, nar 2a, nar 3b*) to high (*nar 1d*). These NR-deficient mutants are representative of four loci. In total, they obtained 18 alleles of the *nar 1* locus, one *nar 2* alleles, two *nar 3* alleles and one *nar 4* allele. Out of these, *nar 1* alleles are found to be co-dominant with the wild type allele while *nar 2* and *nar 3* alleles are recessive. However, all three exhibit segregation patterns characteristics of single Mendelian genes.

Table :1. Representative mutants of Genes controlling NR in Barley. (Warner *et.al.*, 1985).

Gene symbol	Selection Number	NR		Cyt.c Reductase	NiR	Xanthine Dehydrogenase
		NADH	$FMNH_2$			
nar 1a	Az 12	1	0	7	174	+
nar 1b	Az 13	1	0	61	174	+
nar 1d	Az 28	2	0	227	193	+
nar 1h	Az 32	2	167	10	156	+
nar 2a	Az 34	6	0	64	170	-
nar3b	X no 19	3	0	42	128	-
nar 4y	Az 72	1	0	-	135	-

Most barley mutants isolated so far are capable of at least limited NO^-_3 assimilation (Warner *et al.*, 1985). The *nar* 1 mutants (NR structural gene) are capable of growth with NO^-_3 as a sole source of N. When grown in the field, mutants *nar1a* and *nar1b* produced nearly as much total dry weight and reduced N as the wild type cultivar (Oh *et al.*, 1980). However, they could not find any direct relationship between NR deficiency and grain sterility. Characterization of NR present in *nar 1* mutants revealed that the mutants had greater activity with NADPH than NADH. The NR from mutants *nar 1a* has a pH optimum of 7.7 for both the NADH and NADPH and NADPH activities. Some other mutants exhibited optimum activities at a pH of 7.5 for the NADH and 6.0 for the NADPH (Dailey *et al.*, 1982). The lower pH optimum for NADPH activity is a typical of many NADH NRs and NAD-linked dehydrogenases (Beevers *et al*, 1964). Apparently, NADPH can bind to the enzyme when the pH favours ionization of the second hydrogen of the ribose 2- phosphate. Since the pH optimum for the NADPH activity of the *nar 1* enzyme is 7.7, it apparently has a NADPH binding site.

Molecular aspects of NR regulation in higher plants have been critically reviewed by Abdin *et al.* (1993). Accordingly the NADH: NR (EC 1.6.6.1) is the predominant form of NR in higher plants. NADH: NR has been cloned from corn leaves, scutella and roots of corn and analyzed NR induction by nitrate (Gowry and Campbell, 1989; Melzer *et al.*, 1989). In the leaves of wheat and barley the increase in steady state level of NR-mRNA through NO^-_3 induction has also been observed (Abdin *et al.*, 1992; Kumar *et al.*, 1993). The higher levels of NR mRNA in wheat under induced condition was correlated with the higher rates of *in vivo* NR protein synthesis (Fig. 5) Their findings implicit that nitrate induced increase in NR – mRNA is due to the activation of NR gene.

In corn it was reported that nitrate induction system is constitutive and the signal transduction must be mediated by modification of an existing gene regulatory protein (Abdin *et al.*, 1993). Accordingly NO^-_3 activates a putative nitrate sensor located at plasmalemma

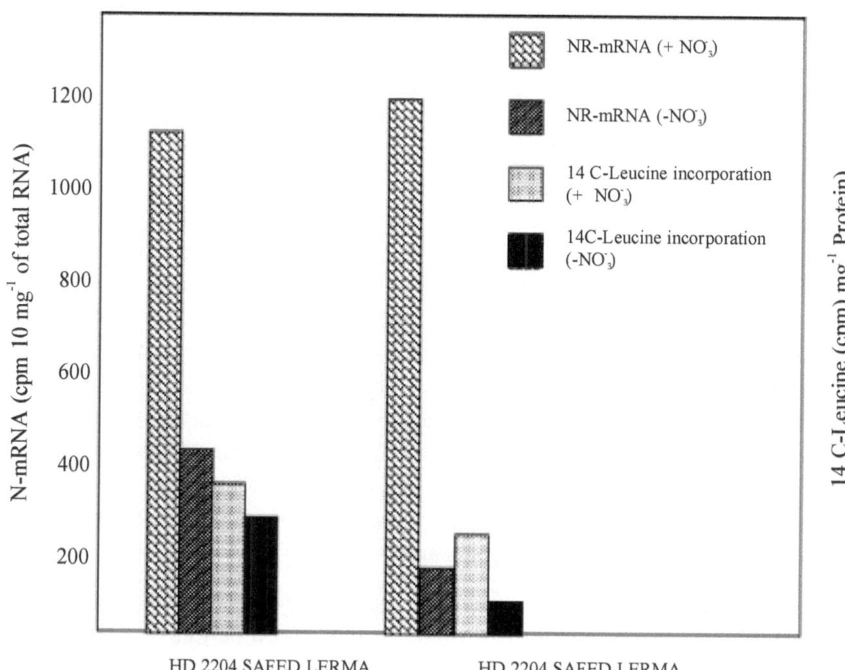

Fig.5. NR-mRNA levels and NR protein synthesis rates in two genotypes of wheat (Abdin *et.al.*, 1993).

after binding with it. This in turn activates a constitutive cytoplasmic regulatory protein by modifying it possibly through phosphorylation. This regulatory protein then migrate to the nucleus and binds to *cis*-acting sites of NR gene leading to activation of NR gene transcription *vis-a-vis* synthesis of mRNA (Redinbaugh and Campbell, 1991).

RNA analysis and transcription assays (Callaci and Smarelli, 1991) with isolated nuclei from barley and soybean indicate that nitrate induction of NR-mRNA is due to *de novo* synthesis of transcripts and not to activation of pre- mRNA or reduced m RNA degradation. The decline in NR- mRNA level after peak accumulation upon nitrate induction must result from decreased NR-mRNA production and /or increased NR-mRNA turnover (Hoff *et al.*, 1992).

NR under water stress

Several environmental factors including salinity, low temperature and moisture stress profoundly affects the various enzyme activities, specially enzymes of nitrogen metabolism and NR activities in particular (Pandey and Agrawal, 2002; Angrish *et al.*, 2001). For many years the causes have been sought for the metabolic changes occurring in plants exposed to dehydration. It has been proposed that low water availability may directly change the NR activity because of changes in free energy of water within the cells, changes in spatial relationships of membrane systems, volume changes and concentrating of enzymes resulting from losses of water, or decreases in the water of hydration surrounding macromolecules (Kramer and Boyer, 1995). In general nitrate reduction decreases during dehydration mostly because NO^-_3 is transported to the sites of nitrate reductase synthesis more slowly in the transpiration stream because of the decreased uptake of NO^-_3 by the roots. The decreased flux of NO^-_3 decreases the synthesis of the enzyme, and the natural degradation of the enzyme in the cell depletes the cell of reductase activity (Beevers and Hageman, 1983). Besides, decrease in enzyme activities in the cell in response to dehydration appear to result in part from the inhibition of NR protein synthesis followed by a decline in activity determined by the half-life (approx. 4 hours) of the enzyme in the cell (Mason *et al.,* 1988 a). Usually, the decrease in synthesis is detected as losses in the polyribosomal content of the tissue because most other methods of measuring protein synthesis require aqueous media that rehydrate the tissue. Because polyribosomes are complexes of messenger RNA and ribosomal RNA actually synthesizing the protein, their disappearance is evidence for a general slow down in protein synthesis. Reports also indicate that even 1hour of dehydration causes a substantial loss of poly ribosomes (Morilla *et al.,* 1973) in maize. Growing cells typically have large numbers of polyribosomes and undergo large losses during dehydration whereas mature tissues contain fewer polyribosomes and lose fewer (Bewley and Larson, 1982).

Bewley and Larson (1982) and Dasgupta and Bewley (1984) further found that the loss in polyribosome was not due to the loss in synthesis of a particular protein, although some synthetic differences were found. Dhindsa and Bewley (1976) were unable to attribute the polyribosome loss to increase in ribonuclease, and m RNA was conserved (Dhindsa and Bewley, 1978). This indicated that the losses in polyribosomes were not caused by a breakdown of mRNA. There is evidence that the polyribosomes translate some different mRNAs during dehydration (Creelman *et al.,* 1990; Oliver, 1991) and the changes in NR activities support this concept. Therefore, there is likelihood of occurrence of two phenomena during the dehydration; First, synthesis for certain proteins is more affected than for others and, second, the overall rate of protein synthesis tends to decrease . However, while there is some understanding of molecular regulation of individual protein synthesis such as nitrate reductase, the molecular control of the overall rate remains obscure.

Case studies of enzyme inactivation under water stress

Nitrate reductase (NR) has consistently shown a decrease in activity in corn seedlings when the leaf water potential (Fig. 6) decreased below – 2 bars, dropping to 25 % of the control at a potential of –13 bars (Morilla *et al.,* 1973). Similar results have been obtained with wheat (Rajgopal *et al.,* 1977), barley (Anikiev and Kuramagomedov, 1975), sorghum (Teare *et al.,* 1974). In leguminous crops drought stress reduced the activities of NR at all the growth stages *vis-a-vis* nitrogen acquisition of plants (Vyas *et al.,* 2001; Garg *et al.,* 1998). Vyas *et al.,* (1990) in sesame reported that there was nearly a loss of 40-60 % of NR activities under moderately water stressed conditions as compared to that of control plants. The decrease in NR is accompanied by an increase in free amino acids and a decline in protein synthesis as has been observed in wheat and sorghum (Sinha and Nicholas, 1981). It is suggested that under such conditions further NO^-_3 reduction is of little value to the plants.

The study in corn implicit that a part of the enzyme could be inactivated but a large amount would be degraded resulting in depletion of enzyme concentration in the plant tissue (Sinha *et al.,* 1982).

Although there are few critical studies on the relationship between growth and NR in plants experiencing water deficit, inhibition in growth appears to precede reaction in NR activity. Therefore, quicker loss in the activity of this enzyme during water stress and its activation on re -watering may be of adaptive importance.

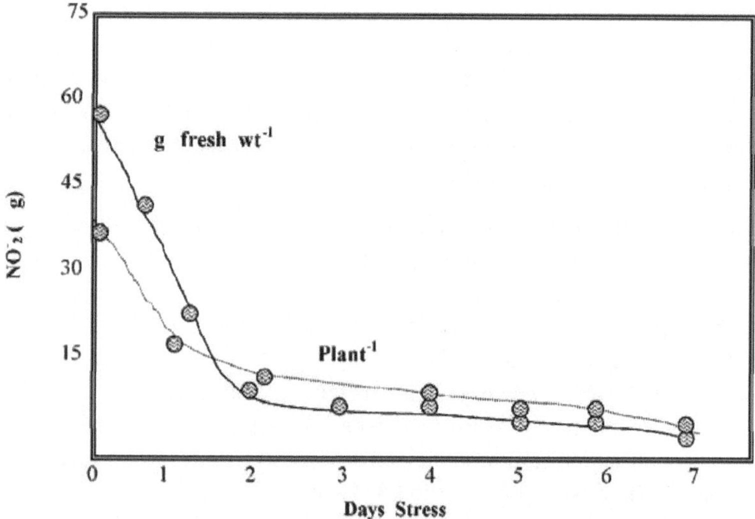

Fig.6. Changes of nitrate reductase activity of corn (Zea mays) during heat and moisture stress (Matlas and Pauli, 1965).

A comparative study was made by Huffaker *et al.* (1970) on activities of certain enzymes like NR, NiR, PEP carboxylase and Rubisco and nitrate content in barley in response to mild water stress ranging from – 0.2 to 0.4 MPa leaf water potential. Their study revealed that NR activity was affected more adversely than was nitrite reductase. Again the concentration of nitrate in the leaf tissue was less affected than the NR activity was. The activities of the two photosynthetic enzymes were also found to be considerably less sensitive to water stress than nitrate reductase. Moreover, the decrease in protein content was also relatively less than the loss of nitrate reductase. Consequent upon the decrease in activity of NR under water stress, nitrate accumulates in the cytoplasmic pool which is highly innocuous and causes the inhibition of the growth of cells, leaves and the whole plants (Sinha and Nicholas, 1981) in general.

Nitrite Reductase (NiR)

Structural- functional aspects of NiR

The enzyme nitrite reductase (NiR, EC 1.7.7.1) mediates the reduction of NO^-_2 to NH^+_4. It is a monomeric polypertide of about 60-63 kDa containing a siroheme prosthetic group. In contrast to NR which is localized in the cytoplasm, NiR is localized in the chloroplast in leaves and in the proplastids of roots and other non green tissue (Oak and Hirel, 1985). In green leaves, the electron donor is reduced ferredoxin, generated in the light by PSI. In the dark and particularly in the roots and other non green tissues a protein similar to ferredoxin may serve in this function (Solomonson and Barber, 1990) and energy for production of reducing equivalents is provided by glycolysis (Bowsher *et al.*, 1989).

According to Kamin and Privalle (1987) NiR is one of a small group of oxidative enzymes which catalyzes a multielectron reduction rather than the transfer of electrons one or two at a time. These multi electron reactions occupy crucial positions in metabolism: the four electron reduction of oxygen catalyzed by cytochrome oxidase, and the six-electron reductions of sulphite to sulphide, nitrite to ammonia, and dinitrogen to ammonia. One of the primary problems in reaching an understanding of the machinery and mechanism of these reactions is the nature of the process, whereby all of the electrons can be transferred without the release of free intermediates.

Spinach NiR is a siro-heme-containing enzyme containing a tetra nuclear iron-sulphur center (Lancaster *et al.,* 1979). The single polypeptide chain contained by NiR is very much analogous to hemoprotein subunit of the sulphite reductase of enterobacteria (Siegel *et al.,* 1973). Both sulphite reductase (SiR) and NiR can reduce the 'Cross' substrate, but in each case the apparent K_m is higher than for the normal substrate. Both NiR and SiR hemoprotein can be considered as the terminal enzyme in an electron transport chain for the reduction of nitrite and sulphite respectively. Both of the Fe_4–S_4- siroheme proteins appear to accept electrons from donors which can donate electrons one at a time. Two electrons equivalents from NADPH enter the flavoprotein at the FAD end, which serves as the electron acceptor. A reaction cycle involving fully reduced and semiquinone flavines cleaves the electron pair; the FMN serves as a one electron donor to the hemoportein (Kamin and Lambeth, 1982).

Earlier studies indicate that ferredoxin (Fd) is the physiological electron donor has not yet been fully successful. Studies of interactions between Fd and NiR are a necessary first step in studying the reductive phase of electron flow through the nitrite reduction system; most experimental studies have used reduced methyl viologen as the electron donor. Since the Fe_4-S_4 and siroheme groups of NiR are closely interactive, the question of whether electrons enter NiR 'in series' or 'in parallel' to the prosthetic groups become a matter of debate and the mechanism is not necessarily the same for Fd as for methyl viologen (Vega *et al.*,1980; Wilkerson *et al.,* 1983; Oak and Hirel, 1985).

Albeit, it now seems probable that electrons enter NiR one at a time from electron donors, it is by no means clear that NiR donates electrons to nitrite one at a time. Iron-sulphur groups are one- electron carriers, and so are hemoproteins. One might postulate a linear 'wire' model for NiR, delivering electrons one at a time to nitrite; intermediates would remain enzyme-bound until the six electrons have been delivered. This hypothesis is certainly plausible. But an alternate 'condenser' model should also be considered (Kamin and Privalle, 1987). It is suggested that only two ligated nitrogenous intermediates (other than nitrite itself) have been clearly identified as reaction intermediates: complexes formed by addition of NO and of hydroxylamine. The reduction of nitrite to NO requires one electron;

the reduction of NO to hydroxylamine requires three electrons, and the reduction of hydroxylamine to ammonia requires two electrons. Can the prosthetic group equipment of NiR give option which provides 'bursts' of one, two or three electrons? The X-ray crystallographic data of *E. coli* sulphite reductase hemoprotein have shown close proximity of the iron-sulphur center and the heme-iron and close proximity of the FeS iron to the edges one of the pyrrole rings of siroheme (McRee *et al.,* 1986). If this structure serves as an NiR model, and it the ring can itself serve as an electron carrier, then the prosthetic group (s) might carry up to the three electrons. Studies of the reductive steps which convert NO to its three-electron product hydroxylamine, would definitely add insight essential to the understanding of the process. However, existence of close contact between a cubane-S atom of the (4 Fe-4S) cluster and the edge of the siroheme macrocycle provides modes for rapid electron transfer in the enzyme through direct inner sphere mechanisms (Siegel *et al.*, 1987).

Since Hageman's successful demonstration of NiR in plants during 1920, it has been possible by now to understand mechanism of regulation to certain extent and has helped all of us to explore this knowledge in the crop plants from productivity point of view. However, the basic contention is that NO^-_3 enhances production of NiR at the transcriptional level whereas the light effect appears to be exerted at the translational level (Gupta and Beevers, 1985). The successful use of poly-A RNA to direct *in vitro* synthesis of NiR will ultimately provide a means for preparation of cDNA for NiR. The availability of such cDNA will allow for the quantification of mRNA for NiR in plants subjected to the various environmental perturbations. Further more, by using cDNA as a probe to quantify mRNA levels, it should be possible to assess to what extent genetic variability in NiR levels is controlled at the transcriptional level. Not withstanding with the advent of new techniques, genetic engineering can be well exploited on the molecular biology of NiR to generate information for crop improvement studies. Furthermore use of growth regulators like auxins and gibberellins on the activity of NiR (Devi, 1998) *vis-a-vis* nitrite reduction would improve the nitrogen use efficiency (NUE) of various crop plants.

Nitrite reduction in leaves and roots

It is an established fact that nitrite is reduced effectively and efficiently in leaves, which usually make a major contribution to NO^-_3 processing, nevertheless a substantial proportion, depending on species and developmental stage, may occur in roots. Despite the spatial compartmentation of NR and NiR (Fig. 3), as a rule, NO^-_2 rarely accumulates in intact plants under normal conditions. Root nodules of legumes are obviously exception to this rule of synchronisation of the activity of both enzymes (Marschner, 1995). In C_4 plants, mesophyll and bundle sheath cells differ in their functions not only in CO_2 assimilation but also in nitrate assimilation. Both, nitrate reductase and nitrite reductase are localized in the mesophyll cells and are absent in the bundle sheath cells (Vaughn and Campbell, 1988). The division of labour in C_4 plants, whereby mesophyll cells utilize light energy for nitrate reduction and assimilation and bundle sheath cells for CO_2 reduction, is most probably the cause for higher photosynthetic nitrogen use efficiency (NUE) in C_4 compared with C_3 plants (Moore and Black, 1979).

Several studies revealed that ferredoxin, being reduced by illuminated chloroplast materials or PSI, are able to pass electrons to leaf nitrite reductase for reduction of nitrite to ammonia (Hucklesby, 1987; Oaks and Hirel, 1985). In contrast, the root NiR which effectively functions with reduced viologens, the exact natural electron donating system of roots is still a matter of speculation. However, NiR from a non-green tissue, maize scutellum, was coupled to glucose-6-phosphate dehydrogenase via an NADPH- oxido-reductase, both obtained from the same tissue, giving a nitrite reducing system which required the supply of ferredoxin from leaf source (Hucklesby, 1987). Since the root NiR is plastid located as leaf NiR in chloroplast, so also possesses the similar properties as leaf NiR; the apparent connection of nitrite reduction with the pentose phosphate pathway of roots may not be ruled out. Further the presence of ferredoxin like protein reducible by NADPH supporting the nitrite reduction in the roots cannot be disregarded. Suzuki *et al* (1985) obtained a non-haem protein from maize roots which functions in the reactions of glutamine synthase and NiR and the protein so extracted from the roots strongly resembles that of leaf ferredoxin (Ninomiya and Sato, 1984) which mediates the reduction of nitrite using NADPH as electron donor.

During the past years, the root NiR has been vividly studied by different workers. The purified enzyme has an absorption spectrum with many characteristics in common with the leaf NiR (Nagaoka *et al.,* 1984). The spectrum appears to be that of a haem with an a-band at 573 nm and a soret band at 397 nm. The latter is of rather higher wavelength than is typical of leaf NiR, where 384-390 nm has been reported for spinach, *Cucurbita,* wheat and barley (Hucklesby, 1987). The overall conclusion is that root NiR may be a sirohaem-containing enzyme, based on its general similarity to the leaf enzyme. Neverthless, direct evidence of nature of prosthetic groups in NiR of roots is required and the degree of identity of the root and leaf enzymes needs elaborate description. The question here is how apparently similar enzymes cope with entirely different metabolic situations. Further, the relationship of the ferredoxin like cofactor of roots to leaf ferredoxin will be of great interest, particularly its kinetics and mode of functioning within the plastid necessitates detail study.

REFERENCES

Abdin, M. Z., Lakkineni, K.C. and Kumar, P.A. 1993. Molecular aspects of nitrate reductase regulation in higher Plants. *Proc. Indian Natl. Sci. Acad. B.,* **59**: 3 and 4: 219-226.

Abdin, M,Z., Kumar, P.A. and Abrol, Y P. 1992. Biochemical basis of variability in NR activity in wheat genotypes.*Plant Cell Physiol.,* **33**: 951-956.

Angrish, R. Kumar, B. and Datta, K. S. 2001. Effect of gibberellic acid and kinetin on nitrogen content and NR activity in wheat under saline conditions. *Plant Physiol.,* **6**: 172-177.

Anikiev, V.V. and Kuramagomedov, M.K. 1975. Effect of soil humidity on NR activity in barley shoot leaves. *Fiziol Rast.,* **22**: 354-358.

Aparicio PJ and Maldonado JM. 1979. Regulation of nitrate assimilation in photosynthetic organisms. In *Nitrogen Assimilation in Plants*: EJ Hewitt, CV culting, eds : Academic press, London, New York, 207-215.

Aryan AP, Batt RG and Wallace W. 1983. Reversible inactivation of NR by NADH and the occurrence of partically inactive enzyme in the wheat leaf. *Plant Physiol.,* **71**: 582-587.

Aslam M, Oaks A and Huffaker RC. 1976. Effect of light and glucose on induction of NR and on distribution of nitrate in etiolated leaves. *Plant Physiol.,* **58**: 588-591.

Barber MJ and Solomonson LP. 1986. Properties of the Mo domain of NR. *Polyhedron,* **5**: 577-580.

Beevers L and Hageman RH. 1969. Nitrate reduction in higher plants. *Annu Rev Plant Physiol.,* **20**: 495-522.

Beevers L and Hageman RH. 1983. Uptake and reduction of nitrate: Bacteria and higher plants. In *Inorganic Plant Nutrition* (Eds : Alauchli, RL Bieleski). *Encyclopedia of Plant Physiology,* **15**: 351-375.

Beevers L, Flesher D and Hageman RH. 1964. Studies on the pyridine nucleotide specificity of NR in higher plants and its relationship to sulphhydryl level. *Biochim Biophys Acta,* **89**: 453-464.

Bewley JD and Larson KM. 1982. Differences in the responses to water stress of growing and non-growing regions of maize mesocotls: Protein synthesis on total , free and membrane bound polyribosome fractions. *J Expt Bot.,* **33**: 406-415.

Bowman DC and Paul J.L. 1988. Uptake and assimilation of NO_{-3} and NH^{+}_{4} by nitrogen deficient perrenial ryegrass turf. *Plant Physiol.,* **88**: 1303-1309.

Bowsher CG, Long D M, Oaks A and Rothstein S J. 1991. The effect of light /dark cycles on expresdion of nitrate assimilatory genes in maize shoots and roots. *Plant Physiol.,* **95**: 281-285.

Bowsher CG. Hucklesby DP and Emes MJ. 1989. Nitrate reduction and carbohydrate metabolism in plastids purified from roots of pea. *Planta,* **177**: 359-366.

Brunelti N and Hageman R H. 1976. Comparison of in vivo and in vitro assays of NR in wheat seedlings. *Plant Physiol.,* **58**: 583-587.

Callaci JJ and Smarelli J. 1991. Regulation of inducible NR iso-form from soybeans. *Biochim Biophys Acta,* **1088**: 127-130.

Canvin DT and Atkins CA. 1974. Nitrate, nitrite and ammonia assimilation by leaves.: effect of light, carbon dioxide and oxygen. *Planta,* **116**: 207- 224.

Carlson DT 1980. *In The Biology of Crop Productivity,* Academic Press, New York, 413-453.

Champigny M L, Bismuth E, Talouizte A and Guiraud G. 1984. The role of the roots in nitrate reduction and metabolism of the carbohydrate product of photosynthesis for amino acid synthesis in wheat seedlings. In *Advances in Photosynthesis Research* (Ed:C. Sybesma) Vol.III. Martinus Nijhoff / Dr. W. Junk, The Hague, 875-878.

Chandok MR and Sopory SK. 1992. Phorbol 12 - Myristate replaces phytochrome mediated stimulation of NR in corn leaves. *Plant Physiol.,* **46**: 800-805.

Chatterjee SR and Naik MS 1993. Regulation of nitrate assimilation in higher plants under light dark conditions.*Proc Indn. Natl Sci Acad.,* **59**: 209-218.

Cheng CL, Adedo GN, Dewdney J, Goodman H M and Conkling MA. 1991. Differential expression of the two *Arabidopsis* NR genes. *Plant Physiol.,* **96**: 275-279.

Cheng CL, Dewney J, Nam H G, den Boer BGW and Goodman J M. 1988. A new locus (NIA – 1) in *Arabidopsis thaliana* encoding NR. *EMBO J.,* **7**: 3309 - 3314.

Cohen HJ and Fridovich I. 1971. Hepatic sulfite oxidase : the nature and functions of the heme prosthetic groups. *J Biol Chem*. **246**: 367-373.

Coruzzi G and Bush DR. 2001. Nitrogen and carbon nutrient and metabolite signaling in plants. *Plant Physiol.,* **125**: 61-64.

Cramer SP, Solomonson LP, Adams MWW and Mortenson LE. 1984. Molybdenum sites of *E. coli* and *Chlorella vulgaris* NR: a comparison by EXAFS. *J. Am. Chem. Soc.*, **106**: 1467-1471.

Crawford NM and Arst HNJ. 1993. The molecular genetics of nitrate assimilation in fungi and plants. *Annu Rev Genet.,* **27**: 115-146.

Crawford NM. 1995. Nitrate : Nutrient and Signal for Plant growth. *Plant Cell*, **7**: 859-868.

Crawford NM. Campbell WH and Davis RW. 1986. Nitrate reductase from squash : cDNA cloning and nitrate regulation. *Proc. Natl. Azad Sci.* USA, **83**: 8073 – 8076.

Creelman RA, Mason HS, Bensen RJ, Boyer JS and Mullet JE. 1990. Water deficit and ABA cause differential inhibition of shoot vs root growth in soybean seedlings. Analysis of growth, sugar accumulation and gene expression. *Plant Physiol*., **92**: 105-214.

Creelman RA, Mason HS, Bensen RJ, Boyer JS and Mullet JE. 1990. Water deficit and abscisic acid cause differential inhibition of shoot versus root growth in soybean seedlings. *Plant Physiol*., **92**: 205-214.

Dailey FA, Warner RL , Somers DA and Kleinhofs A. 1982. Characteristics of a NR in a barley mutant deficient in NADH nitrate reductase. *Plant Physiol*., **69**: 1200-1204.

Daniel – Vedel F, Dorbe MF, Caboche M and Rouze P. 1989. Cloning and analysis of the tomato NR encoding gene protein domain structure and aminoacid homologies in higher plants. *Gene*, **85**: 371-380.

Dasgupta J and bewley JD. 1984. Variation in protein synthesis in different regions of greening leaves of barley seedlings and effects of imposed water stress. *J Expt Bot*., **35**: 1450-1459.

Devi RK. 1998. Effect of IAA, GA_3 and kinetin on NR and nitrite reductase in the leaves of a tree legume, Parkia Javanica. *Indian J. Plant Physiol.,* **3**: 97-101.

Dhindsa R S and Bewley JD. 1978. Messenger RNA is conserved during drying of the drought tolerant moss. *Proc Natl. Acad. Sci*, USA, 842-846.

Dhindsa RS and Bewley JD. 1976. Plant desiccation: Polysome loss not due to ribonuclease. *Science*, **191**: 181-182.

Echevarria C, Mauriano SG and Maldonado JM. 1984. Reversible inactivation of maize leaf NR. *Phytochem*., **23**: 2155-2158.

Egmond F and Breteler H. 1972. Nitrate reductase activity and oxalate content of sugar beet leaves. *Neth J Agric Sci*., **20**: 193-198.

Fernandenz E and Cardenas J. 1987. Genetic control of NR in the green algae *Chalmydomonas reinhardtii*, In *Inorganic Nitrogen Metabolism*; WR Ullrich, P.J. Aparicio, PJ Syrett, F Castillo, eds; Springer- Verlag Berlin, 108-111.

Fernandeq E and Matagne R F. 1984. Genetic analysis of NR deficient mutants in *Chlamydomonas reinhardii. Curr Genet.,* **8**: 635-640.

Forde BG. 2002. Load and long range signaling pathways regulating plant response to nitrate. *Annu. Rev. Plant Biol.,* **53**: 203-224

Fritz BJ and Ninnemann H. 1985. Photoreactivation by triplet flavin and photoinactivation by singlet oxygen of NR of *Neurospora crassa. Photochem. Photobiol.,* **41**: 39-45.

Fu YH, Kneessi JY and Marzluf GA. 1989. Isolation of nit-4, the minor pathway regulatory gene which mediates nitrate induction in *Neurospora crassa. J Bacteriol.,* **171**: 4067-4070.

Fu YH, Young JL and Marzluf GA. 1988. Molecular cloning and characterization of a negative–acting nitrogen regulatory gene of *Neurospora crassa. Mol. Gen. Genet.,* **214**: 74-79.

Garg B K, Vyas S P, Kathju S and Lahiri A N. 1998. Influence of water deficit stress at various growth stages on some enzymes of nitrogen metabolism and yield in clusterbean genotypes. *Indian J. Plant Physiol.,* **3**: 214-218.

Gigon A and Rorison I.H. 1972. The response of some ecologically distinct plantspecies to nitrate and ammonium nitrogen. *J. Ecol*.**, 60** : 93-102.

Glass ADM and Siddiqi M Y. 1995. Nitrogen absorption by plant roots. In H.S. Srivastave and R.P. Singh (eds): *Nutrition in Higher Plants*. Associated Publishing Co, New Delhi, 21-56.

Gowry G and Campbell WH. 1989. cDNA clones for cornleaf NADH: NR and chloroplast NAD $(P)^{+}$glyceraldehydes-3-phosphate dehydrogenase. *Plant Physiol.,* **90**: 792-798.

Gupta SC and Beevers L. 1984. Synthesis and degradation of nitrate reductase in pea leaves. *Plant Physiol.,* **75**: 251-252.

Gupta SC and Beevers L. 1985. Regulation of nitite reduction. In JE Harper, LE Schrader RW Howell, eds : *Exploitation of Physiological and Genetic Variability to Enhance Crop Productivity*, American Society of Plant Physiologists, Rockville, Maryland, 1-11.

Hageman RH and Flesher D. 1969. NR activity in corn seedlings as affected by light and nitrate content of nutrient media. *Plant Physiol*., **35**: 700-708.

Hamat HB, Kleinofs A and Warner R L. 1989. Nitrate reductase induction and molecular characterization in rice (*Oryza sativa* L). *Mol. Gen. Genet.,* **218**: 93-98.

Hewitt EJ, Hicklesby DP and Notton BA. 1976. Nitrate metabolism. In *Plant Biochemistry* : J Bonner, JE Varner, eds. Academic Press, New York, 633 – 681.

Hirel B, Bertin Q, Bourdoncle W, Attagant C, Dellay C, Gouly A. Codiou S, Petailliau C, Falue M and Gallais A. 2001. Towards a better understanding of the genetic and physiological basis of nitrogen use efficiency in maize. *Plant Physiol.,* **125**: 1258-1270.

Hoff T, Truong HN and Caboche M. 1994. The use of mutants and transgenic plants to study nitrate assimilation. *Plant Cell Environ.,* **17**: 489-506.

Hoft T, Stummann BM and Henningsen KW. 1992. Structure, function and regulation of NR in higher plants. *Physiol Plant.,* **84**: 616-624.

Hopkins WG. 1999. In *Introduction to Plant Physiology*. 2nd Edn. John Wiley and Sons, Inc, New York.

House CM and Anderson JW. 1980. Light dependent reductionof nitrate by pea chloroplasts in the presence of NR and C_4- dicarboxylic acids. *Phytochem*., **19**: 1925 –1930.

Hucklesby DP. 1987. Nitrite reduction in leaf and root. *In* WR Ullrich, PJ Aparicio, PJ Syrett, F Castillo, eds; *Inorganic Nitrogen Metabolism*. Springer-Verlag, berlin, 123-125.

Huflaker RC, Radin T, Kleinkopf GE and Cox EL. 1970. Effects of mild water stress on enzymes of nitrate assimilation and of the carboxylative phase of photosynthesis in barley. *Crop Sci.,* **10**: 471-474.

Ingemarsson B, Oscarson P, Afugglas M and Larsson CM. 1987 . Nitrogen utilization in Lemna II. Studies of nitrate uptake using $^{13}NO^-_3$. *Plant Physiol.,* **85**: 860-864.

Jensen PE, Hoft T, Stummann BM and Hennigson KW. 1996. Functional analysis of two bean NR promoters in transgenic tobacco. *Physiol.Plant.,* **96**: 357-358.

Jhosi AK, Chanda SV and Singh YD. 1991. Nitrogen assimilation in pearl millet hybrids and parents. *Gujarat Univ. Res. J.* **20**: 49-57.

Johnxon JL, Hainline BE and Rajgopalan KV. 1980. Characterization of Mo cofactor of sulfite oxidase, xanthine oxidase and NR. *J. Biol. Chem.,* **255**: 1783-1786.

Kalita P and Nair TVR. 2001. Variability in kinetics of nitrate uptake in wheat genotypes. *Indian J. Plant Physiol.,* **6**: 411-413.

Kamin H and Lambeth JD. 1982. Role of flavins and iron-sulphur proteins in the reduction of cytochrome P- 450. *In* V Massey, CH Williams,eds; *Flavins and Flavoproteins*, VII. Elsevir, Amsterdam, New York. 655-666.

Kamin H and Privalle L S. 1987. Nitrate reductase. *In* WR Ullrich, P J Aparicio, P J Syrett, F Castillo, eds; *Inorganic Nitrogen Metabolism*, Springer- Verlag, Berlin, 112-117.

Kar M, Mishra BB and Patro BB. 1994. Changes in chlorophyll content, nitrate reductase and catalase activities in ragi seedlings in response to nitrogen nutrition. *Plant Physiol. Biochem*., **21**: 65-69.

Kashyap NN 1978. Inaugural Address. In *Nitrogen Assimilation and Crop Productivity* (Eds: S.P. Sen, Y.P. Abrol and S.K. Sinha). Published by Associated Publishing Company, New Delhi, 5-8.

Kelly JM, Graves W R and Aiello A. 2000. Nitrate uptake kinetics for rooted cuttings of Acer rubrum L. *Plant Soil,* **221**: 221 - 230.

Kenis J D, Silvente S T, Luna CM and Campbell WH. 1992. Induction of NR in detached corn leaves. The effect of the age of the leaves. *Physiol. Plant.,* **85**: 49-56.

Kramer PJ and Boyer J. S. 1995. In *Water Relations of Plants and Soils*. Academic Press, New York.

Kumar P A, Kumaran M L and Abrol Y P 1993. Hormonal reulation of NR gene expression in Hordeum vulgare. *Indian J Expt Biol*., **31**: 472-473.

Kumar SN and Singh C P. 2000. Nitrogen use efficiency in maize : Influence of nitrogen sources and doses and crop seasons. *J. Plant Biol.,* **27**: 161-170.

Kumar SN and Singh C.P. 2001. Nitrate reductase activity in maize leaves : Influence of N sources, doses and crop seasons. *Indian J. Plant Physiol.,* **6**: 152-157.

Lancaster JR, Vega JM, Kamin H, Orme N R, Orme WH, Krueger RJ and Siegel LM. 1979. Identification of the iron-sulphur center of spinach ferredox in – nitrite reductase, *In* M Edelman, R B Halleck, NH Chau, eds; *Methods in Chloroplast Molecular Biology*, Elsevier, Amsterdam, 723-734.

Li Z Z and Gresshoff PM. 1990. Developmental and biochemical regulation of constitutive NR activity in leaves of nodulating soybean. *J. Exp. Bot.,* **41**:1231-1238.

Lillo C. 1994. Light regultion of NR in green leaves of higher plants. *Physiol Plant.,* **90**: 616-620.

Lips SH and Roth-Bejerano N. 1969. Light and hormones: Interchangeability in induction of NR. *Science*, **166**: 109-110.

Losada M, Guerrero MG and Vega JM. 1981. The assimilatory reduction of nitrate. In *Biology of Inorganic Nitrogen and Sulphur* (H Bothe, A Trebst, eds), Springer, Berlin Heidel berg, New York, 30-63.

Lu J L, Ertl JR and Chen CM. 1992. Transcriptional regultion of nitrate reductase mRNA levels bycytokinin- ABA interactions in etiolated barley leaves. *Plant Physiol.,* **98**: 1255-1260.

Maldonado JM and Aparicio PJ 1987. Photoregulation of nitrate assimilation in Eukaryotic organisms. In *Inorganic Nitrogen Metabolism*, WR Ullrich, PJ Aparicio, PJ. Syrett, F Castillo, eds; Springer-Verlag, New York, 76-81.

Marschner H. 1995. In *Mineral Nutrition of Higher Plants,* Academic Press. New York.

Marwaha RS. 1998. Nitrate assimilation in potato cultivars during plant growth. *Indian J Plant Physiol.,* **3** : 147-151.

Mason HS, Mullet JE and Boyer JS. 1988a. Polysomes, messenger RNA and growth in soybean stems during development and water deficit. *Plant Physiol.,* **86**: 725-733.

Mattas RE and Pauli AW. 1965. Trends in nitrate reduction and nitrogen fractions in young corn plants during heat and moisture stress . *Crop Sci.,* **5**: 181-184.

Mc Ree DE, Richardson DC, Rihardson JS and Siegel LM. 1986. The heme and Fe_4 S_4 cluster in the crystallographic structure of *E. coli* sulphite reductase. *J. Biol. Chem.,* **261**: 10277-10281.

Mc. Clure PR, Kochian LV, Spanswick RM and Shaff JE 1990. Evidence for cotransport of nitrate and protons in maize roots. *Plant Physiol.,* **93**: 281-289.

Melzer JM, Kleinhofs A and Warner RL. 1989. Nitrate reductase regulation: effects of nitrate and light on NR mRNA accumulation. *Mol. Gen. Genet.,* **217**: 341-346.

Mitchell P. 1978. Promotive chemiosmotic mechanism in oxidative and photosynthetic phosphorylation. *Trends in Biochemical Sciences,* **3**: 58-61.

Moore R and Black jr CC. 1979. Nitrogen assimilation pathways in leaf mesophyll and bundlesheath cells of C_4-photosynthetic plants formulated from comparative studies with *Digitaria sanguinalis* L. Seap. *Plant Physiol.,* **64**: 309-313.

Morilla CA, Boyer JS and Hageman RH. 1973. Nitrate reductase activity and polysomal content of corn seedlings having low leaf water potentials. *Plant Physiol.,* **51**: 817-824.

Moses PB and Chua NH. 1988. Light switches for plant genes. *Scientific American,* **258**: 88-93.

Nagaoka S, Masakaza H, Fukushima K and Tamura G. 1984. Methyl viologen linked nitrite reductase from bean roots. *Agric. Biol. Chem.,* **48**: 1179 – 1186.

Nair TVR and Abrol YP. 1982. Nitrate reductase activity in flag leaf blade and its relationship to protein content and grain yield in wheat. *Indian J Plant Physiol.,* **25**: 110-121.

Nair TVR and Chatterjee SR. 1990. Nitrogen Metabolism in cereals, case studies in wheat, rice, maize and barley. In YP Abrol (ed), *Nitrogen in Higher Plants,* Research Studies Press, Tauton, England and John Wily and Sons, New York, 367 –426.

Nasrulhaq- Boyce A and Jones OGT. 1977. The light induced development of NR in etiolated barley shoots. An inhibitory effect of levulinic acid. *Planta,* **137**: 77-84.

Nelson – Schreiber BM. and Schweitzer LE. 1986. Limitation on leaf nitrate reducase activity during flowering and pod fill in soybean.*Plant Physiol.,* **80**: 454-458.

Ninomiya Y and Sato S. 1984. A ferredoxin like electron carrier from non-green cultured tobacco cells. *Plant Cell Physiol.,* **25**: 453-458.

Notton BA and Hewitt E J. 1979. Structure and P4operties or hither plant NR especially Spinacia oleracea. In *Nitrogen Assimilation in Plants* (EJ Hewitt, CV Cutting, ed), Academic Press, London, 227-244.

Oaks A and Hirel B. 1985.Nitrogen metabolism in roots. *Annu. Rev. Plant Physiol.,* **36**: 345-365.

Oaks A, Michell P, Good fellow VJ, Cass L A and Deising H, 1982. The role of nitrate and ammonium ions and light on the induction of NR in maize leaves. *Plant Physiol.,* **88**: 1067-1072.

Oaks A, Wallace W and Stevens D. 1972. Synthesis and turnover of NR in corn roots. *Plant Physiol.,* **50**: 649-654.

Oaks A. 1991. Nitrogen assimilation in roots : a reevaluation. *Bio Science*, **42**: 103-111.

Oh JY, Warner RL and Kleinhofs A. 1980. Effect of NR deficiency upon growth , yield and protein in barley. *Crop Sci.,* **20**: 487-490.

Okamoto PM, Fu YH and Marzluf GA. 1991. NIT-3, the structural gene of NR in *Neurospora crassa*; nucleotide sequence and regulation of mRNA synthesis and turn over. *Mol. Gen. Genet.,* **227**: 213-223.

Oliver MJ. 1991. Influence of protoplasmic water loss on the control of protein synthsis in the desiccation tolerant moss Tortula Ruralis. *Plant Physiol.*, **97**: 1501-1511.

Pandey V and Agrawal S. 2002. Aminotransferases and NR activity in *Cassia angustifolia* seedlings as affected by salinity stress. *Plant Physiol.,* 7: 179-182.

Pate J S. 1983. Patterns of Nitrogen metabloism in higher plants and their ecological significance. *In* Lee J A, Mc Neill S; Rorison I H (Eds). *Nitrogen as an Ecological Factor*. Blackwell, Oxford, 223-253.

Pattnayak D and Chatterjee S R. 1997. Molecular biology of the assimilatory NR gene. *Plant Physiol. Biochem.,* **24**: 1-9.

Rajagopal V, Balasubramanian V and Sinha S K. 1977. Diurnal fluctuations in relative water content, NR and proline content in water-stressed and non-stressed wheat. *Physiol Plant.,* **40**: 69-71.

Raveen J A and Smith F A. 1976. Nitrogen assimilation and transport invascular land plants in relation to intracellular pH regulation. *New Phytol.*, **76**: 415-431.

Raven J A. 1985. Regulation of pH & osmolarity generation in vascular land plants: costs and benefits in relation to efficiency of use of water, energy and nitrogen. *New Phytol.*, **101**: 25-77.

Redinbaugh M and Campbell W H. 1991. Higher plant responses to environmental nitrate. *Physiol Plant.*, **82**: 640-650.

Roldan J M and Butler WL. 1980. Photoactivation of NR from *Neurospora crassa. Photochem. Photobiol.*, **32**: 375-381.

Roth – Bejerano N and Lips SH. 1970. Hormonal regulation of NR in leaves. *New Phytol.*, **69**: 165-169.

Sagi M, Savidov NA, Lvov N pand Lips SH. 1997. NR and MoCofactor in annual ryegrass as affected by salinity and nitrogen suouce. *Physiol Plant.*, **99**: 546-553.

Salanoubat M and Dang H D B. 1993. Analysis of the petunia NR apoenzyme encoding gene: a first step for sequence modification analysis. *Gene,* **128**: 147-154.

Sawhney S K and Naik M S. 1972. Role of light in synthesis of NR and NiR in rice seedlings. *Biochem. J.*, **130**: 475-485.

Sawhney S.K, Naik M S and Nicholas DJD. 1978 b. Regulation of nitrate reduction by light, ATP and mitochondrial respiration in wheat leaves. *Nature,* **272**: 647-648.

Sawhney SK and Naik M S. 1988. Nitrate assimilation in Plants. *In Advances in Frontier Areas of Plant Biochemistry*, R Singh, S.K. Sawhney eds, Prentice Hall of India, Pvt. Ltd., New Delhi, 186-213.

Schnoor K M, Juricek M, Huang C, Culley D and Kleinhofs A. 1991. Analysis of barley NR cDNA and genomic coones. *Mol. Gen. Genet.*, **227**: 411-416.

Sehtiya H.L and Goyal S.S. 2002. Comparative uptake and assimilation of NO^-_3 by excised roots and leaves from seedling of C_3 (barley) and C_4 (Corn) plants: Effect of light, ambient CO_2 and exogenously supplied sucrose. *Indian J. Plant Physiol.*, **7**: 203-210.

Selvaraj K and Kumari EVN. 1999. Response of in vivo nitrate reduction in root nodules of legumes to nitrate, ammonia, molybdenum and tungsten. *Indian J. Plant Physiol.*, **4**: 271-276.

Shanker A K, Vijayalaksmi C, Senthil A and Pathmanabhan G. 2002. Heterosis for ammonia assimilating enzymes in sunflower. *J Plant Biol.*, **29**: 165-168.

Siegel L M, Murphy J M and Kanub H, 1973. The reduced NADP- sulphite reductase of enterobacteria : I. The *E. coli* hemoflavoproteins; molecular parameters and prosthetic groups. *J. Biol. Chem.*, **248**:251-264.

Siegel L M, Wilkerson J O and Janick P A. 1987. Structural studies on the siroheme (4 Fe-4s) cluster active centers of spinanch ferredoxin nitrite reductase and *E coli* sulphite reductase. *In* WR Ullrich, PJ Aparicio, PJ Syrett, F Castillo, eds; *Inorganic Nitrogen Metabolism*, Springer – Verlag, Berlin, 118-122.

Sinha SK and Nicholas JD. 1981. Nitrate reductase in relation to water stress. *In* L.G. Paleg and D Aspinall, eds, *Physiology of Biochemistry of Drought Resistance,* Academic Press, New York.

Sinha SK, Khanna-Chopra R, Aggarwal PK, Chaturvedi GS and Koundal KR. 1982. Effect of drought on shoot growth significance of metabolism to growth and yield. *In: Drought Resistance in crops in the Emphasis on Rice*. IRRI, Phillipines : 153-170.

Smiciklas KD and Below FE 1992. Role of cytokinin in enhanced productiviety of maize supplied with ammonium and nitrate. *Plant Soil,* **142**: 307-313.

Smirnoff N and Stewart GR. 1985. Nitrate assimilation and translocation by higher plants : comparative physiology and ecological consequences. *Physiol Plant.,* **64**: 133-140.

Smirnoff N, Todd P and Stewart GR. 1984. The occurrence of nitrate reduction in the leave of woody plants. *Ann. Bot.,* (London) **543**: 363-374.

Solomonson L P and Barber MJ. 1990. Assimilatory NR: Functional Properties and regulation. *Annu. Rev. Plant Physiol.Plant Mol. Biol.,* **41**: 225-253.

Solomonson LP and Barber M J. 1984. The flavin domain of assimilatory NADH : NR from *Chlorella vulgaris*. In *Flavins and Flavoproteins* (RC Bray, PC Engel, S G Mayhew, eds). De Gruyter, Berlin, New York, 247-250.

Solomonson L P and Barber M J . 1987. Structure- function relationships of assimilatory NR. In *Inorganic N Metabolism* (WR Ullrich, PJ Aparicio, PJ Syrett, F Castillo, eds), Springer-Verlag, Berlin, 71-75.

Solomonson LP, Barber M J Robbins A P and Oaks A. 1986. Functional domains of assimilatory NADH: NR from *Chlorella. J. Biol. Chem.,* **261**: 11290-11294.

Solomonson L P, Barber M J, Howard WD, Johnson JL and Rajagopalan KV. 1984a. Electron para magnetic resonance studies on the molybdenum center of assimilatory NADH : NR from *Chlorella vulgaris. J. Biol. Chem.,* **259**: 849-853.

Solomonson LP, Lorimer GH, Hall RL, Borchers R and Bailey JL. 1975. Reduced NAD : NR of *Chlorella vulgaris* : Purification, prosthetic groups and molecular properties. *J. Biol. Chem.***,** **250** : 4120-4127

Solomonson LP and Vennesland B. 1972. Properties of nitrate reductase of *Chlorella. Biochim. Biophys. Acta,* **267**: 544-557.

Somers DA, Kuo TM, Kleinhofs A, Warner L and Oaks A.1983. Synthesis and degradation of barley NR. *Plant Physiol.,* **72**: 949-952.

Sprent J I. 1980. Root nodule anatomy, type of export Product and evolutionary origin in some Leguminosae. *Plant Cell Environ.,* **3**: 35-43.

Srivastava HS. 1975. Distribution of nitrate reductase in ageing bean seedlings.*Plant Cell Physiol.,* **16**: 995-999

Stewart GR, Sumar N and Patel M. 1987. Comparative aspects of inorganic nitrogen assimilation in higher Plants : *In* WR Ullrich, P J Aparicio, P J Syrett, F. Castillo (Eds). *Inorganic Nitrogen Metabolism*. Springer Verlag. 39-44.

Stuber CW. 1997. Biochemistry, physiology and molecular biology of heterosis; In *The Genetics and Exploitation of Heterosis in Crops. Proceed. of an Int. Symposium* (CIMMYT, Mexico), 17-22 August.

Sueyoshi K, Kleinhofs A and Warner RL. 1995. Expression of NADH-specific and NADPH-bispecific NR genes in response to nitrate in barley. *Plant Physiol.*, **107**: 1303-1311.

Suzuki A, Oaks A, Jacquot J P, Vidal J andGadal P. 1985. An electron transport system in maize roots for reactions of glutamate synthase and nitrite reductase. *Plant Physiol.*, **78**: 374-378.

Syrett. P.J. 1987. Nitrogen assimilation by Eukaryotic Algae. In *Inorganic Nitrogen Metabolism* (Ed.: WR Ullrich, PJ Aparicio, P.J. Syrett, F (astillo), Springer-Verlag, Berlin, 25-31.

Taiz L and Zeiger E. 1991. In *Plant Physiology*. The Benjamin/Cummings Publishing Company, Inc. California.

Teare ID, Manam R and Kanemasu E T. 1974. Diurnal and seasonal trends in NR activity in field grown sorghum plants. *Agron. J.,* **66**: 733-736.

Thibaud JB and Grignon C. 1981. Mechanism of nitrate uptake in corn roots. *Plant Sci Lett.*, **22**: 279-289.

Travis RL and Key JL. 1971. Correlation between polyribosomes level and ability to induce NR indark grown corn seedlings. *Plant Physiol.*, **48**: 617-620.

Travis RL, Huffaker R C and Key J L. 1970. Light induced development of polyribosomes and induction of NR in corn leaves. *Plant Physiol.*, **46**: 800-805.

Ullrich W.R. 1983. Uptake and reduction of nitrate. Algae and Fungi. In*Inorganic Plant Nutrition* (Ed: A. Lauchli, RL Bieleski), *Encyclopedia of Plant Physiology*, **15**: 376-397.

Ullrih E. R. 1987. Nitrate and ammonium uptake in green algae and higher plants. In: *Inorganic Nitrogen Metabolism* (Eds : WR Ullrich, P.J. Aparicio, P.J. Syrett and F. Castillo), Published by Springer- Verlag, Berlin, 32-38.

Van Beusichem ML and Nelemans JA. 1990. Cation anion uptake balance and root nitrate reductase and nitrogenase activities in nodulated pea plants in relation to nitrate supply. *Proceedings of Intl. Congress of Plant Physiology* held at New Delhi from Feb 15 –20, 1988. 1054-1059.

Vaucheret H, Kronenberger J, Rouze P and Caboche M. 1989. Complete nuclcotide sequence of the two Romologous tobacco NR genes. *Plant Mol. Biol.,* **12**: 597-600.

Vaughn K C and Campbell W H. 1988. Immunogold localizationof nitrate reductase in maize leaves. *Plant Physiol.,* **88**: 1354-1357.

Vega JM, Cardenas J and Losada M. 1980. Ferredoxin nitrite reductase. *Meth Enzymol.,* **69**: 255-270.

Vincetz M and Caboche M. 1991. Constitutive expression of NR allows normal growth and development of *Nicotina plumbginifolia. EMBO J.* **10**: 1027-1035.

Vyas SP, Garg BK, Kathju S and Lahiri AN. 1990. Improvement of drought tolerance in sesame through early water stress. *Proceed of the Intl. Congress Plant Physiol* , **2**: 880-884., held at New Delhi Feb 15-20, 1988.

Vyas SP, Garg BK, Kathju S and Lahiri AN. 2001. Influence of potassium on water relations, photosynthesis, nitrogen metabolism and yield of cluster bean under soil moisture stress. *Indian J Plant Physiol.,* **6**: 30-40.

Wallace W and Patel JS. 1965. Nitrate reductase in the field pea. *Ann. Bot.,* (London) **24**: 655-671.

Wallace W. 1987. Regulation of nitrate utilization in higher plants. In *Inorganic N Metabolism* (W. R. Ulltrich, PJ Aparicio, P.J Syrett, F. Castillo, eds) Springer- Verlag, Berlin, 223-230.

Warner RL, Kleinhofs A and Narayanan KR. 1985. Genetics, biochemistry and physiology of NR-deficient mutants in barley. In *Exploitation of Physiological and Genetic Variability to enhance Crop Productivity*, J.E Harper, LE Schrader, RW Howell, eds; American Society of Plant Physiologists. Rockville, Maryland 23-30.

Wilkerson JO, Janick PA and Siegel LM. 1983. Electron paramagnetic resonance and optical spectroscopic evidence for interaction between siroheme and Fe_4-S_4 prosthetic groups in spinach ferredoxin nitrite reductase. *Biochem.*, **22**: 5048 –5054.

Witt HH and Jungk A. 1977. Beuteilung der Molybdanversorgung von pflanzen mit Hilfe der Mo-induzierbaren Nitratreduktase- Aktivitat. *Z Pflanzenernahr bodenk,* **140**: 209-222.

Section III: Environmental Stresses in Plants

Developments in Physiology, Biochemistry and Molecular Biology of Plants, 2005
Eds.: Bandana Bose and A. Hemantaranjan
Vol., 1, pp. 81-104, New India Publishing Agency, New Delhi
E-mail: spjain_niph@rediffmail.com web: www.bookfactoryindia.com

CHAPTER - 5

PLANT WATER RELATIONSHIPS AND STRATEGIES TO IMPROVE SALT TOLERANCE

VINOD GOYAL

INTRODUCTION

The purpose of writing this chapter is to develop a comprehensive understanding of the role of functions of water in plant growth and development. Water is essential to life on earth as the living organisms contain 80-90 per cent water. A typical crop or grassland will transpire approximately 500 ml of water per 10 g of dry matter produced. Water has a tremendous effect on the environment near the earth, which on regional basis we call 'climate'. Ecological importance of water is due to its physiological importance because water affects the plant growth through physiological processes. All metabolic activities in the plant cell or tissue depend on water status of the plant. For example, in a maturing seed, respiration rate is highest as it moves towards maturation stage, water content decreases and hence rate of respiration slow down. At the harvesting stage, when the water content is 10-20%, the respiration rates become negligible. Now in a stored seed if the water content will be high it will result into higher respiration, thus release energy, which will increase the temperature and will finally destroy the seed.

One of the biggest properties of water is that it is liquid at ordinary temperature, whereas substances with similar molecular weights, such as CH_4 and NH_3 become gases at temperature below zero (-161 and –33° C, respectively). Water is fairly good solvent for molecules that are biologically important including salts, sugars, amino acids and metabolic intermediates. The association of water molecules with polar groups on the surface of membranes and proteins contributes to their structures and biological activity and thus water participates in number of biochemical reactions.

Lack of an adequate supply of water throughout the life cycle is the single greatest limitation to realization of genetic yield potential of crop plants. By understanding the interactions of soil-plant atmosphere continuum (SPAC), as it affects water supply and demand, we can begin to develop both genetic and management strategies to increase productivity within the constraints of the water supply. By understanding the relative sensitivities of developmental and physiological systems to water stress, we can search for variations in stress tolerance mechanisms and genetically increase drought/salt tolerance, resulting in more stable productivity in water limiting environments.

An attempt is made to integrate the physical and chemical properties of water, water potential and factors affecting it, measurement of water potential, water movement from soil to atmosphere through complex plant structure and crop growth, development and yield under abiotic stresses especially salt stress.

PHYSICAL AND CHEMICAL PROPERTIES OF WATER

Chemically, water is hydrogen oxide with the molecular weight of 18. It consists of 2 hydrogen atoms attached to a single oxygen atom. As a result, one of the water molecules where hydrogen atoms are found attached is electropositive while the opposite side is electronegative. Such type of arrangement makes the water molecule dipolar in nature, which has differential distribution of electrical charges. Due to its dipolar nature, water act as an excellent solvent in which a number of solutes both electrolytes and nonelectrolytes can be dissolved in sufficient large quantity. Some of the important properties of water are described below:

***(a) Hydrogen bonding*:** Water molecules are present as individual molecules only in vaporous state. In the liquid form, molecules are joined together by hydrogen bonds. Each water molecule can join with a maximum of four more water molecules by hydrogen bonding. This happens when the temperature is below 0°C and water becomes ice. When the temperature rises, ice melts and hydrogen bonds are broken. Depending upon the temperature, each water molecules is attached to 1, 2 or 3 neighbouring molecule. Finally as the temperature rises, sufficiently water exists as individual free molecules when water is converted into vapour. This is why water is considered to become more viscous as temperature falls. It is one of the reasons for poor uptake of water by plants at low temperatures.

***(b) Specific heat*:** Water has a high specific heat. It is the energy required to raise the temperature of one gram of water by 1°C. It means water can absorb large quantity of heat without showing much rise in its temperature. This is of biological significance. Specific heat of water is 1 cal/g/degree at 25°C.

***(c) Heat of vaporization*:** Water has high heat of vaporization. It means to evaporate water large quantity of heat is required. This heat is absorbed from the evaporating surface such as the leaf as it happens during transpiration. As a result, the leaf becomes cooled. Heat required to evaporate 1 g of water at 100°C is 540 cal.

***(d) Heat of condensation*:** The opposite of vaporization is condensation, which is an exothermic reaction. On condensation of 1 g of water, 540 calorie of energy in the form of heat is liberated.

***(e) Dissociation*:** Water does not merely contain water molecules. Some of the molecules dissociate into H^+ and OH^- ions. During this one electron is transferred from hydrogen to hydroxyl group. Therefore, hydrogen becomes electropositive and hydroxyl ions as electronegative, however only a small fraction of pure water becomes dissociated. As the ions are electrically charged, water can conduct electric current.

***(f) Good transparent*:** Water is transparent for visible radiations and light will pass through water body. This property is very useful for photosynthesis in aquatic plants *i.e.* algae.

***(g) Long wave radiation absorber*:** Water act as an opaque body for long wave radiation, which cannot pass through water body. In this way water act as good heat absorber.

***(h) High surface tension and viscosity*:** Water has much higher surface tension and viscosity than most of other liquids because of high internal cohesive forces between molecules. This property provides the tensile strength and helps in upward movement of water in plants.

***(i) Density*:** Water has the high density among liquids. At 4°C density of water becomes greatest while its volume becomes least. Upon freezing, density of water decrease and volume increase and it can be evident from the common phenomenon of bursting of radiators and pipes at freezing point cold. Increase in volume in ice form of is useful as it results in floating. If water does not have this property then ice formed during cold days would have sunk into larger body of the water and would have destroyed the aquatic flora and fauna.

***(j) High dielectric constant*:** Due to this property, water is considered as almost a universal solvent. It has both positive and negative charges. Substances like electrolytes get dissolved in water. The water form cell is around the positive and negative ions. In case of nonelctrolytes like organic compounds water gets adsorbed on them so it is important for soil water relations.

(k) Colligative properties of water: Due to its physical and chemical properties, water also shows some of the colligative properties. In plants and soils, water never occurs in pure form but is in solution form. Dissolved solutes changes the various properties of water like vapor pressure, boiling point, freezing point, osmotic potential and chemical potential.

In a real sense, addition of solutes lowers the free energy of water by diluting it. It is also known as entropy effect. These properties are affected by the quantity and not by the quality of solutes added. Solutes reduce the vapor pressure of a solution, its freezing point and raise its boiling point. The decrease in freezing point of water due to addition of solutes is called freezing point depression. The decrease in freezing point and increase in boiling point can also be explained on the basis of change in vapor pressure (Fig. 1). The decrease of vapor pressure due to addition of solutes lowers the temperature at which solid, liquid and gas phases are in equilibrium. According to Rault's law, vapor pressure of non-dissociating substances is inversely proportional to its concentration.

CONCEPT OF WATER POTENTIAL AND ITS COMPONENTS

Osmotic relations of plant cells are based on the principles of thermodynamics, which is the science dealing with the change of energy in the system.

At a temperature above 0°C, the molecules possess some amount of free energy that keeps them in motion. This free energy is also called as Gibb's free energy or kinetic energy. The direction of movement of the molecules is always along concentration gradient of free energy. The ability of the molecules to move or the escaping tendency of molecules which in turn depends upon the kinetic energy is termed as chemical potential which is synonymous term for water potential. The water potential depends upon number of water molecules. Therefore, it should be expressed in terms of energy per mole. However, water potential of pure water is arbitrarily assigned zero value. This serves as a reference. It is denoted by the

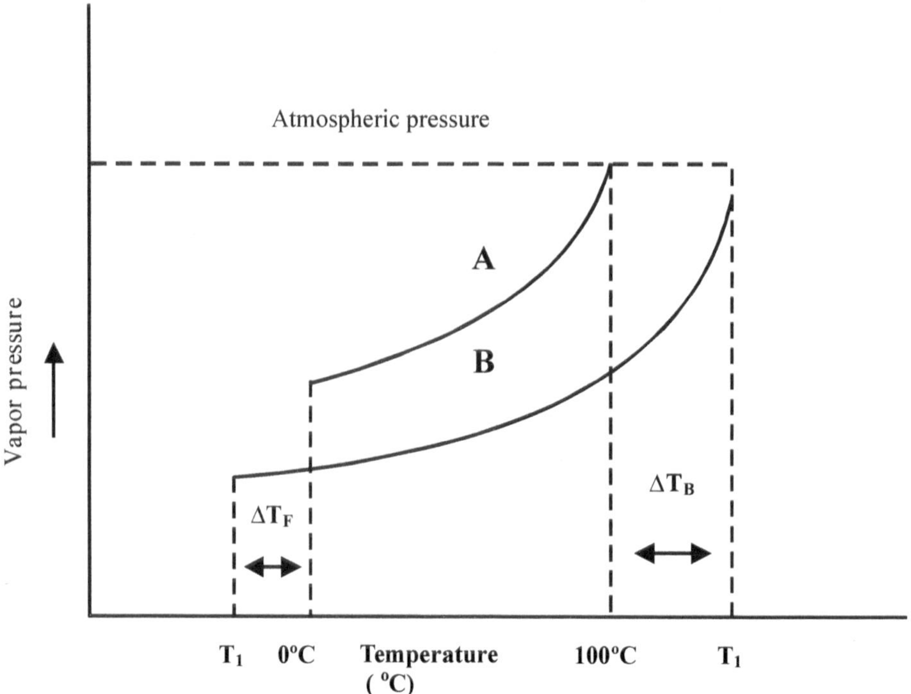

Fig. 1: Graph shows the change in colligative properties of water as depression in freezing point (ΔT_F) and elevation in boiling point (ΔT_B) on addition of solutes. A- pure solvent, B- solution and T_1- change in temperature on addition of solutes.

Greek letter psi (ψ) and is measured in atmosphere (atm), bars or Megapascals (MPa). The following relationships exist between the different units of water potential:

$$0.987 \text{ atm} = 1 \text{bar} = 0.1 \text{ Mpa}$$

Water potential is a measure of free energy status of water molecules. The factors that determine the water potential are:

a) Addition of solutes decreases the free energy of water. It is termed as osmotic potential (ψ_s).

b) Turgor potential (ψ_p) in plant cell. It is always positive inside the plant cells and increases free energy. Loss of turgor means the shrinkage of cell and ultimately will die (Fig. 2).

c) Surface of macromolecules *e.g.* Cellulose or the boundary walls exert pressure on water, which is negligible and is known as matric potential (ψ_m). It lowers the free energy of water.

So the water potential of plant can be expressed with the following equation:

$$\psi_w = \psi_s + \psi_p + \psi_m + \psi_g$$

Components of Water Potential

Chemical potential of water in a solution can be lowered by increasing the proportion of solutes to water molecules or rose by decreasing the proportion or by increasing the pressure. It is generally argued that the components include one ascribable to solutes present in the system which tend to lower the potential and other ascribable to the hydrostatic pressure which can exist in plant cells bounded by more or less rigid cell wall, which tend to raise it (Dale and Sutcliffe, 1986).

(a) Osmotic potential or solute potential: The osmotic potential of solution is most rigidly defined as the excess of hydrostatic pressure which must be applied to the solution in order to make the chemical potential of the solvent in solution equal to that of the pure solvent at the same temperature. For a dilute solution, we have

$$RT \ln a_w = -V_w \pi$$

$$\pi = (RT \ln a_w)/V_w$$

Where, R is gas constant, π is the osmotic pressure, T is the absolute temperature and a_W is the chemical activity of water. This Van't Hoff relationship applied to ideal dilute solutions and inclusion of a_W makes allowances for the interactions between solute molecules that occur in non-ideal solutions. In very dilute aqueous solutions, the presence of solutes tends to decrease the activity of water, the concentration of which is diluted. Since for pure water a_W is 1, and the presence of solutes reduces it, ln a_W becomes negative and osmotic pressure Π is positive. The symbol Π will be used henceforth only for osmotic pressure. To avoid the confusion, osmotic potential is given the symbol π, to indicate that it is component of water potential. Some writers use the symbols ψ_δ or ψ_s but single symbol π is preferred despite the possibility of confusion with Π. Numerically Π and π are identical but differ only in sign so that

$$-\Pi = \pi$$

Or in other words we can say that osmotic pressure is increased on addition of solutes whereas there is decrease in osmotic potential.

(b) Matric potential: There are some physical interactions between water and the solid matrix of the soil, which causes the chemical potential to be lowered. Matric forces can be considered to be those, which occur between liquid and solid phase in a soil or cell. They are short-range forces, involving hydrogen bonding and Vander Waal's forces. Matric potential is expressed in units of water potential. A dry colloid or hydrophilic surface such as filter paper, wood, soil, gelatin or the stripe of brown alga have a negative matric potential (as low as –300 MPa), while the same colloid in a large volume of pure water at atmospheric pressure has a zero matric potential. In general, when any colloid at atmospheric pressure is in equilibrium with its surroundings, the water molecules in the colloid have the same free energy as water molecules in the surroundings. So the matric potential of colloid is equal to the water potential of the surroundings. The matric potential is represented by ψ_m in the equation of water potential. Cell wall and membranes are the main sites at which solid-liquid interactions can occur and as such are the clear components of the matric potential.

The interactions between solid and liquid phases in the cells occurs only over a short distances (nm), it is reasonable to assume that the contribution of the matric term is negligibly small, at least when considering the mature vacuolated cells in which vacuolar volume is major component of total cell volume. Such assumption may unjustified for growing cells, although neglecting it is unlikely to lead to seriously erroneous conclusions.

***(c) Pressure potential*:** Application of pressure to an aqueous solution brings water molecules slightly closer together, making water molecules more concentrated and increase the chemical potential and also raise the water potential. For a plant cell, the difference in hydrostatic pressure between the inside of the cell and its surroundings is termed as turgor pressure (P) and it is defined as "The turgor pressure is outwardly directed excess hydrostatic pressure which is equal and opposite to the inwardly directed reaction due to the structured cell wall (wall pressure)" P is positive for a turgid cell and fall to zero of a plasmolysed cell. The measurement of pressure in plant is calculated after water potential and osmotic potential.

***(d) Gravitational potential*:** In most of the cases it is negligible within a root or leaf but it becomes important in comparing leaf water potential of terrestrial trees and in soil. Upper movement of water in a tree xylem must overcome a gravitational force of 100 Pa/m, whereas in soil drainage of water to the depths is the outcome of the gravitational pull or pressure.

CELL WATER RELATION

Water relation of plants is dominated by the cell water relations. In plants, majority of water is present in cells, chiefly in the vacuoles. Relative rigid walls and large vacuoles are the characteristics of plant cells, both of them play a key role in regulation of plant water relations. The cell wall exerts hydrostatic pressure on the water inside the cell, known as pressure potential. With its great tensile strength that limit extension of the protoplast, resulting in the development of turgor pressure. The pressure keeps the cell in maintaining its structure (Fig. 2). The cell wall is composed of cellulose, hemicellulose, pectin and cellulose micro fibrils. Water content (which is 5-40%) and elasticity of the cell walls are two important features of cell wall in relation to water relations. The volume of the cell changes continuously in the plants with concomitant change in the water potential and pressure potential with small change in solute potential. This can be better explained with helps of Hofler diagram (Fig. 3), which plots Water potential, pressure potential and solute potential as a function of relative volume for a hypothetical cell. In this example, 5% reduction in cell volume results in decrease of –1.8MPa in water potential. Pressure potential is mainly responsible for this decrease. The shape of the curve will be determined by the extent of rigidness of the cell wall. The degree of rigidity will be directly proportional to the steepness of the curve and large change in cell turgor will result with small change in cell volume. The tensile strength of the walls is measured as volumetric elastic modulus, symbolized by Greek letter (epsilon) and is given by

$$\varepsilon = \Delta \Psi_P / \Delta V/V$$

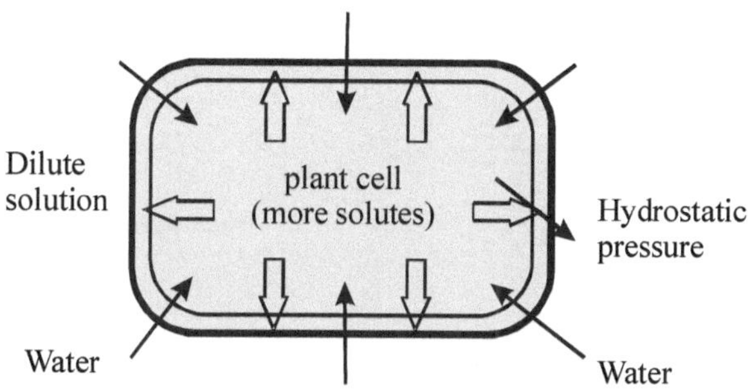

Fig. 2: Cell showing the entry of water as affected by the potential gradient across the cell membrane.

Characteristic features of the Hofler diagram:

(a) Turgor pressure approaches zero as cell lose mere 10-15% of their cell volume.

(b) Elastic modulus is not constant but changes with change in turgor and it is the slope of the Ψ_p curve.

(c) When the value of Ψ_p is less or small, than the changes in water potential are only due to changes in solute potential

So from the above, equation for the water potential for cell is simplified to

$$\Psi_w = \Psi_p + \Psi_s$$

The gravitational and matric potential is ignored because it is negligible when vertical distances are small. The entry of water in a cell is governed by the potential gradient and it always move from the higher to lower energy state. The self explanatory diagram illustrates the movement of water across the cell.

SOIL PLANT ATMOSPHERE CONTINUUM (SPAC)

The movement of water from soil to atmosphere takes place through a complex plant system. Water flows in the soil and xylem of plants by bulk flow according to pressure gradient whereas in vapor phase, water moves through diffusion. When water moves through membranes, the driving force is the potential difference of water across the membranes. Under all the circumstances, water moves towards region of low water potential or free energy. So as we move from root to leaves, the water potential goes on decreasing from – 0.3 MPa to –0.8 MPa. One major factor for the decreased water potential in leaves is due to synthesis of organic solutes, which lowers the osmotic potential and hence water potential (Kramer, 1983).

The movement of water through soil plant atmosphere system is an interrelated process. Various resistances that come across it from soil onwards to air hinder the movement of water. In general, rate of water uptake is affected by both, rate of water loss and rate at which it moves from soil to root.

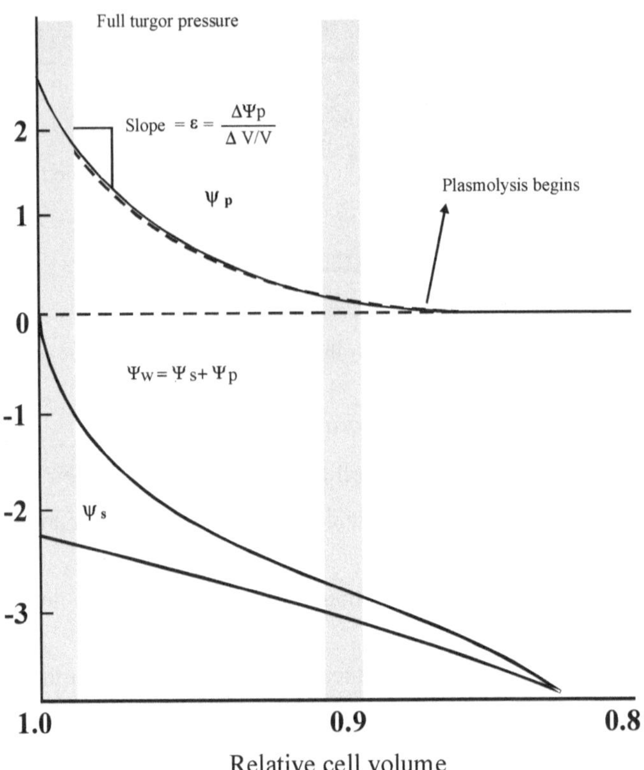

Fig. 3: Modified Hofler diagram representing the turgor pressure (Ψ_p) decreases steeply with the initial decrease in cell volume. In comparison Ψ_s changes very little (screened area at left). As cell volume decreases below 0.9 in the figure, the situation reverses. The change in Ψ_w is due to change in Ψ_s accompanied by relatively little change in turgor pressure (screened area at right). Slope of the curve will give the value of cell's volumetric elastic modulus (Taiz and Zeiger, 1998)

In general terms

$$\text{Flux} = \frac{\text{Driving forces}}{\text{Resistances}} \text{ or } \frac{\text{Differences in water potential}}{\text{Resistances (r)}}$$

This equation is very much similar to the equation of Ohm's law

$$V = I/R$$

Because of the similarity in both the equations, the water movement is often termed as Ohm's Law Analogy. It is very convenient concept to use in describing SPAC because factors affecting water movement can be discussed in terms of either these effects on driving forces ($\Delta\Psi_w$) or resistances (r). Explaining the various resistances coming across the water movement through SPAC on basis of Ohm's analogy is shown in the (Fig. 6).

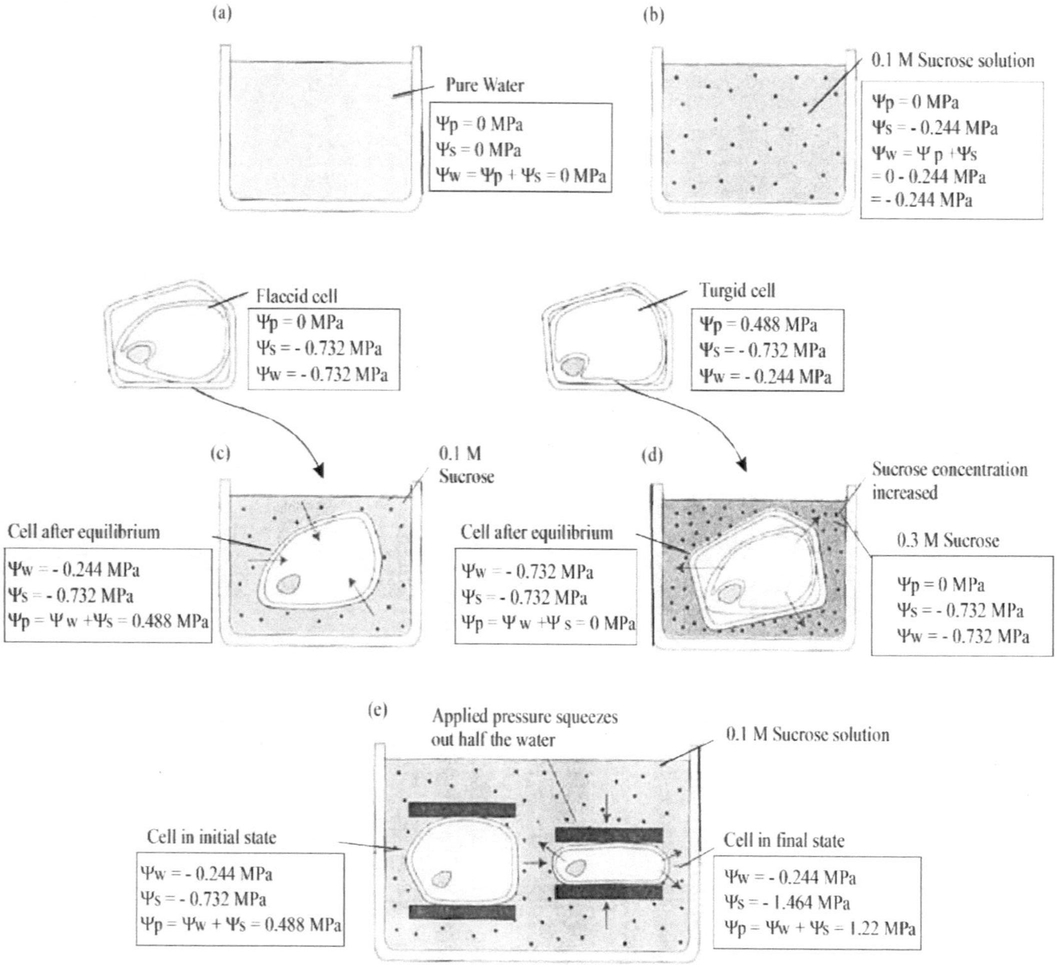

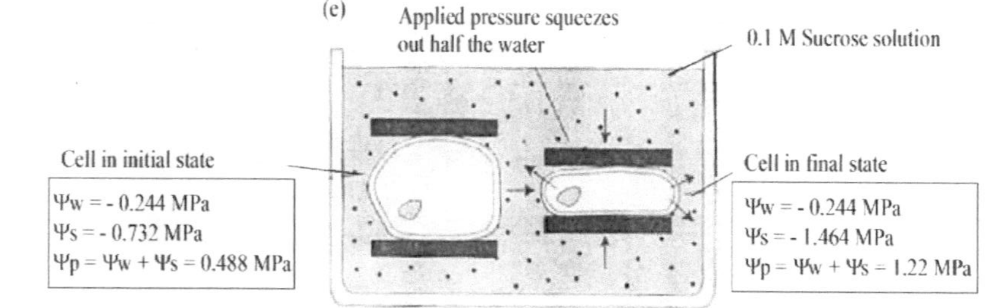

Fig. 4: Five examples illustrating the concept of water potential and its component.

(a) Pure water

(b) A solution containing 0.1 M sucrose.

(c) After a flaccid cell is dropped in the 0.1 M sucrose solution, because the starting water potential of the cell is less than the water potential of the solution, the cell takes up water. After equilibrium, the water potential of the cell rises to equal the water potential of the solution, and the result is a cell with a positive turgor pressure.

(d) Increasing the concentration of sucrose in the solution makes the cell lose water. The increased sucrose concentration lowers the solution water potential, draws water out from the cell, and thereby reduces the cell's turgor pressure.

(e) Another way to make the cell lose water is by slowly pressing it between two plates. In this case, half of the cell water is removed, so cell osmotic potential increases and turgor pressure increases correspondingly.

Total of resistances in series (R) = $r_{1} + r_{2} + r3$ ....

Total of resistances in parallel (1/R) = ($1/r_1 + 1/r_2 + 1/r_3$ ....)

So any factor affecting these resistances will affect the rate of water movement through SPAC (Fig. 5).

Boundary layer resistance: At the leaf surface, more air rest, this resists to water flow. During fast air or storms, this resistance is negligible. Boundary layer resistance increases with increase in wind speed. Presence of hairs on leaves also decreases the resistance. Size and shape of the sunken stomata decrease the boundary layer resistance.

Stomatal resistance: Any factor affecting opening or closing of stomata will affect the stomatal resistance. Closure is affected by less availability of water, high CO_2 concentration, thick cuticle etc.

Driving Forces

As is clear from the discussion of water potential concept, water will always move along with gradient of potential. All the factors like matric, pressure, gravitational and solute potential will affect the movement of water from soil to air.

$$\Psi_W = \Psi_m + \Psi_p + \Psi_s + \Psi_g$$

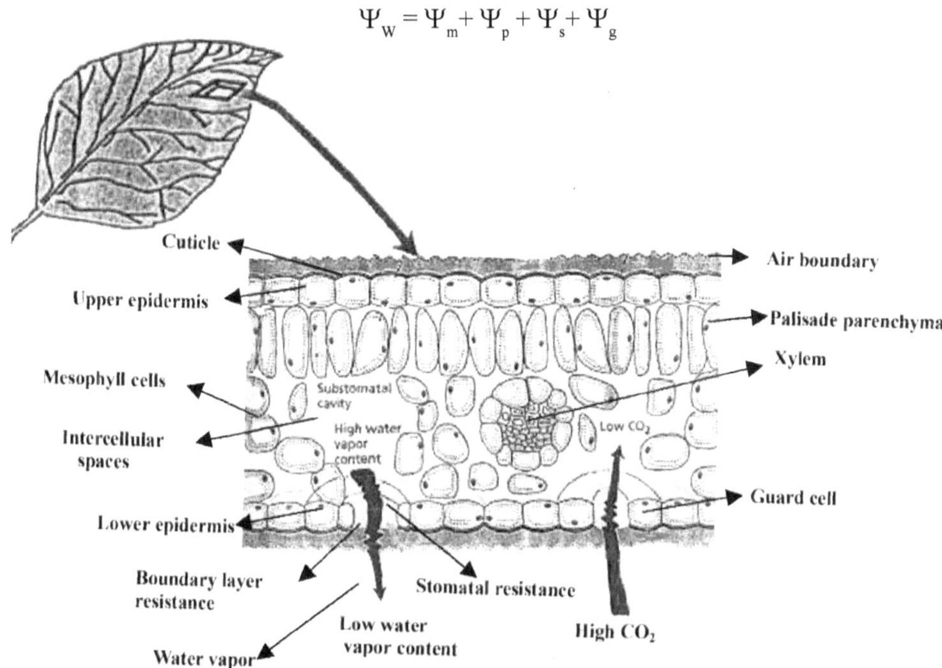

Fig. 5: Movement of water through leaf surface. Water is pulled from the xylem into the cell walls of mesophyll cells from where it evaporates into the air spaces within the leaf. By diffusion, water vapor then moves though the leaf air space through the stomatal pore and across the boundary layer of still air that adhere to the leaf surface. CO_2 diffuses into the opposite direction.

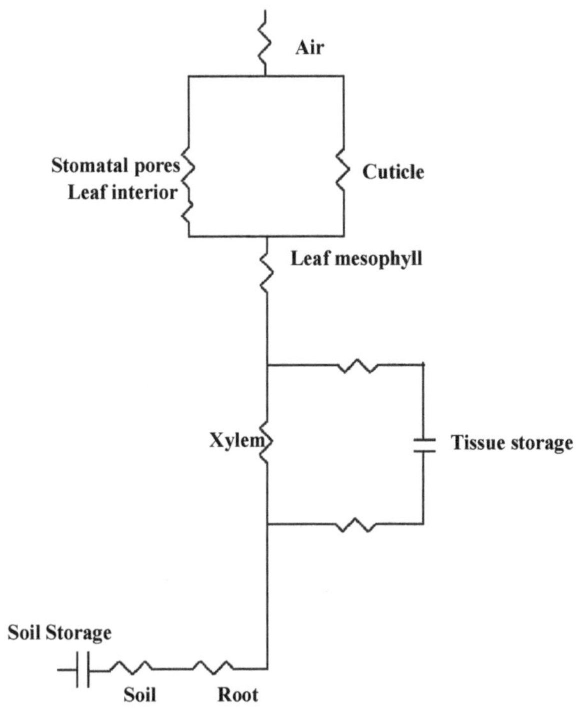

Fig. 6: Simplified diagram of water movement through the soil-air-atmosphere – continuum (SPAC). Kramer, 1983.

Now, as the evaporation increase in the upper parts of the plants, water potential of the leaf cells become more negative due to loss of water and hence due to this, osmotic potential also increase as solutes gets concentrated. Hence a gradient is developed.

Horizontal movement of water towards roots and upward movement toward the soil surface are also important. Upward movement of water occurs because of evaporation and absorption by plants, which results in decrease in the water content of the surface soil, thereby decreasing its water potential and causing upward movement against gravity. Assuming a constant rate of evaporation, the rate of upward movement to an exposed soil surface depends on the depth of the water table and the soil texture. It is estimated that upward flow to an evaporating surface from a water table at 60 cm depth would be only half as rapid in a coarse-textured soil as in a fine textured soil. The movement of water from soil to roots surface depends on the steepness of the water potential gradient and on the hydraulic conductivity of the soil. Both decrease rapidly as soil water content decreases. On the other hand, the root water potential also decreases as the rate of transpiration increase.

Water molecules remain joined to each other due to the presence of H-bonds between them. Although H-bond is very weak (containing about 5.0 Kcal energy) but when they are present in enormous numbers as in case of water, a very strong mutual force of attraction or cohesive force develops between water molecules and hence they remain in the form of a continuous water column in the xylem. Sometimes magnitude of this force is very high (upto 350 atm). Therefore, the continuous water column cannot be broken easily due to the force of gravity or other obstructions offered by the internal tissue in the upward movement of water.

When transpiration takes place in the leaves at upper parts of the plants, water evaporates from the intercellular spaces of the leaves to the outer atmosphere through the stomata. More water is released into the intercellular spaces from the mesophyll cells (Fig.5). In turn, the mesophyll cells draw water from the xylem of the leaf. Due to all this, a tension is created in water in the xylem elements of the leaves. This tension is transmitted downward to water in xylem elements of roots through the xylem of petiole and stem and water is pulled upward in the form of continuous water column to reach the transpiring surfaces up to the top of the plants.

Assumptions of SPAC: Following are the assumptions of SPAC

(a) Constant flow is necessary

(b) Resistances to be assumed constant but they vary with the rate of water movement

(c) Phase of water also varies, thus driving forces varies

However, comparison of water flow in SPAC to the flow of electricity in conducting system requires considerations of two other factors *i.e.* conductance and capacitance.

Conductance versus resistance: There are two means of expressing the hastening of water in conducting system with constant driving force, one is resistance (s cm^{-1}) and other is conductance (cm s^{-1}).

Resistance = Δ V/flow

Conductance = 1/resistance

It is easy to measure resistance in Ohm and calculate conductivity by reciprocate. Likewise resistance to water flow can be estimated at various points in the system and total flow.

Capacitance: Another property of some electrical circuits is the storage of electricity in condensers and release later in circuits. Storage of water in parenchyma cells can be regarded as analogues to the storage of current in condensers, as it is readily available to replace water lost by transpiration.

DEVELOPMENTAL TECHNIQUES

Different methodological approaches are used to measure the water content of leaves. Plant leaves generally have lower (more negative) water potential than pure water and hence they tend to osmotically absorb water and become turgid. A measure of this property is the relative water content (RWC), which expresses the leaf water content as percent of the turgid leaf water content. It is calculated by the formulae given below:

$$\text{RWC} = \frac{\text{Fresh weight} - \text{Dry weight}}{\text{Fully turgid weight} - \text{Dry weight}} \times 100$$

A similar parameter is water saturation deficit (WSD) that can be computed using following formulae:

$$\text{WSD} = \frac{\text{Fully turgid weight} - \text{Fresh weight}}{\text{Fully turgid weight} - \text{Dry weight}} \times 100$$

The water absorbed is to the existence of a water saturation deficit so that RWC + WSD = 100% or WSD = 100 – RWC (Kramer, 1983).

Basically three types of methods are available for determination of water potential *i.e.* (i) Compensation methods (ii) Psychometric methods (iii) Pressure equilibrium methods.

The compensation method was given by Chardakov's. In this, the concentration of the solution in which plant tissue is immersed changes (increase/decrease) depending upon the tissue and the solution. If the solution is hyper tonic (with less Ψ_w), the tissue will lose water and solution gets diluted. Whereas reverse occurs in case of hypotonic solutions i.e. tissue will gain water form solution and solution will become concentrated. The transfer of water is detected by measuring (a) Change in concentration of the test solution by refractometric or smear methods, (b) Change in size/volume of tissue sample and (c change in the volume of the test solution by potometer method. Changed potential of the test solution can be measured by Van't Hoff equation:

$$\Psi_s = \text{miRt}$$

Where Ψ_s is osmotic potential, m is molality of the test solution (moles of solute/100 g water), R is gas constant and t is absolute temperature (K) or (°C + 273) and I is constant that accounts for ionizations from perfect solutions. Thus if m, i and t are known, Ψ_s can be calculated. Thus the ψ_s of this solution gives the Ψ_s of the tissue (Salisbury and Ross, 1969).

Psychrometric techniques have now becomes the hands of research in plant water relations. The Ψ_w is measured by determining the 'wet bulb' depression in a closed gaseous system, which is in equilibrium with the sample. The wet bulb depression is measured in three ways. In the first type a thermister with attached miniature chamber, which can be filled with water, is used to measure wet and dry bulb temperature. In the second type wet loop or droplet Psychrometer the output of a thermocouple with the thermo junction permanently welted by a small drop of water is maintained. This type of psychrometer needs a long term thermal stability of the vessel or an additional similar thermocouple without a droplet included in every sample vessel and measured as a reference. Third type of junction is used alternatively as wet and dry bulb. The difference between two readings is taken as equivalent to the wet bulb depression and calibrated using known osmotic potential solutions.

For measurements, plant sample is placed on the sample holder and is then inserted into the chamber. The chamber is sealed by tightening the knob. When the tissue/sap attains the equilibrium with chamber atmosphere, the voltage is measured on the display panel. The Ψ_w or Ψ_s can be read from the calibrated curve. Now improved models of the psychrometer are available with auto calibration facility (Fig. 7).

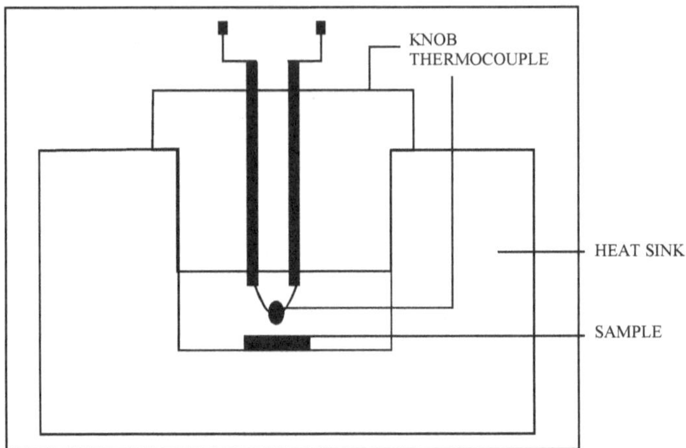

Fig. 7: Hypothetical sketch of a thermocouple psychrometer sample chamber

The pressure equilibration methods typically utilize the pressure chamber apparatus that measures negative hydrostatic pressure in xylem columns of leaves and shoots. This hydrostatic pressure is fairly equal to the average Ψ_w of whole the shoot because the Ψ_s of the xylem water per se is negligible and xylem water is in intimate contact with most of the cells of the plant shoot. When a plant shoot is cut, water is pulled up in the xylem capillaries due to negative hydrostatic pressure. An applied balance pressure that counteracts this hydrostatic pressure is therefore equal to the shoot Ψ_w (Scholander *et al.*, 1965) (Fig. 8).

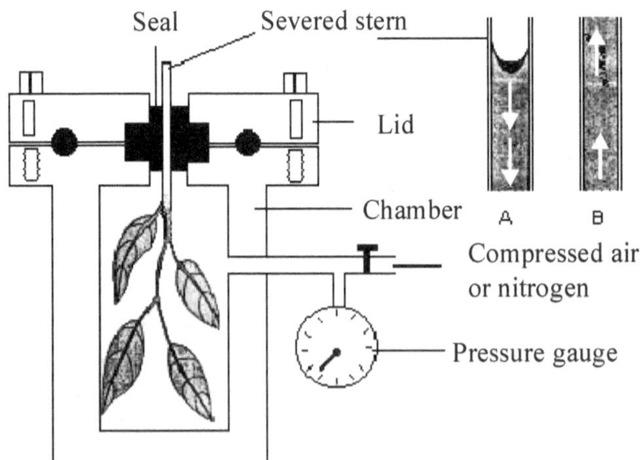

Fig. 8: Pressure chamber apparatus for the measurement of shoot water potential. (A) The position of the water column in the severed stem when shoot is just fixed in the chamber and (B) when the chamber is pressurized to the point of balance pressure and the severed stem appears wet.

ABIOTIC STRESSES AND CROP YIELD

In nature, plants are exposed to ever changing environmental conditions or stresses, which are broadly, put under two groups – biotic and abiotic stresses. Abiotic stresses include water deficit, water logging salinity and alkalinity, heavy metals, low and high temperature, ozone, UV radiations *etc.* To survive these challenges, plants have developed adaptive mechanisms that manifest themselves in morphological, physiological, developmental and molecular changes (Cheeseman, 1988). Among the environmental factors, salinity and water deficit are the two potential factors affecting the process of plant development and growth right from germination up to maturity in almost all the crops (Levitt, 1980). This statement was confirmed by the study done in USA in which crops like wheat, rice, maize, mustard and peanut gave only 10-20% yield under stress prone ecosystem. Since in India, the weather conditions are highly unpredictable, it is important to reduce the yield loses caused by various stresses. Sensitivity of the crop determines the extent of stress and yield losses.

Plants adapt to osmotic stress/salt stress by accumulating physiologically compatible solutes (osmolytes) like sugars, proline and glycinebetaine (Hare *et al.*, 1999) or by excluding Na^+ or K^+ ions from the roots at the expense of metabolic energy (Apse *et al.*, 2000). Due to similarities in the physiological mechanism of acclimation to these stresses, some common genes such as dehydrins, LEA, *Atbi*, RAB *etc.* are expressed under stress conditions (Choi *et al.*, 1999).

GENETIC ENGINEERING TO ABIOTIC STRESSES

Agricultural biotechnology will eventually touch the lives of virtually everywhere is some way. The agricultural sector, while only a subset of the entire universe of the industrial biotechnology will influence, contains a full complement of the opportunities created by the technology.

On broad terms, genetic engineering is the manipulation of the genetic material for the purpose of attaining useful products, crops that give higher yield and are resistant to pests and environmental stresses. It codifies a variety of modern techniques of molecular biology ranging from recombinant DNA technology to cell fusion, somaclonal variations and monoclonal antibody technology. Thus, although selective or traditional breeding might be considered genetic engineering in lateral sense, it is in fact imprecise and crude relative to the technique now emerging to improve agricultural production.

The key step in extending plant genetic engineering methods to any application appears to be the identification and availability of desired gene(s). So, the most critical input in this endeavor is to identify the genes, which could provide tolerance to biotic or abiotic stresses. Every plant has its own defense mechanisms or we can say that plant adapt to the changing environmental conditions by manifesting various metabolic and physiological changes associated with optimal metabolism of carbohydrates, lipids, nucleic acids and proteins. During conventional breeding, it was found that when crop was exposed to sub lethal levels of stress, it could induce stress tolerance to that particular type of stress. Increased tolerance was due to synthesis of some metabolites (sugars, proline and glycinebetaine, proteins *etc.*).

Transgenic approaches offer powerful means to gain valuable information towards a better understanding of mechanism that govern stress tolerance. However, it has paved the way for the new opportunities to improve tolerance by incorporating a gene involved in stress protection from any source into agriculturally important crops.

The steps in the successful genetic engineering of a plant have generally involved five distinct tasks using the most widely practical method of plant gene insertion. Each of them must be accomplished properly though necessarily not in sequence after the first two steps

(a) Identification and characterization of the gene to be inserted.

(b) Insertion of new gene into the plant cell (this is an imperfect process and only a small per centage of the cells subjected to the insertion of the DNA actually take up the DNA in a functional manner, thus necessitating the next step).

(c) Separation of the cells that have successfully incorporated the new gene from those that has not. This process, called selection, is intended to assure that only those cells that have taken up the new gene will reproduce to form the whole plant. This is essential if the seed ultimately attained is to exhibit the desired new genetic trait.

(d) Expression of intended trait by the new gene. Generally, genes code for proteins. When the new gene inserted into plant tissue results in production of the desired protein, the new gene is said to be expressed in the plant. Expression must occur in such a way that the new DNA is integrated into chromosomes of the cell and is stably passed on to future generations. Once a foreign gene has been inserted and stably expressed in plant tissue, the tissue is considered to be "transformed". Plant transformation is the term used to refer to this process.

(e) Regeneration of the cell into a whole plant that carries the new genes in all its cells and expresses them in some or all of them.

An important and sometimes overlooked part of this process is the seed increase that must be carried out to attain a volume of seed sufficient first for testing and then for commercial use. This single regenerated whole plant becomes the basis for production of seed through the traditional techniques requiring a number of generations of field production. At some appropriate stage, field evaluation of the new plant must be carried out. This typically requires two or more seasons to assure adequate experience under a number of fields and weather conditions and to allow for the assessment of performance of the new variety in farmer's fields. Fundamentally, the task of adding new genes into plants is same as adding new genes to microbes. The scientist must insert the new DNA into the plant cell in such a way that the rDNA get expressed. Practically there are more hindrances to genes transfer in plant system than bacterial transformations. Therefore, several hurdles are to be overcome by the plant genetic engineer to get success. Once the gene is inserted into the plant cell, it must be coaxed to develop into a whole plant capable of reproducing viable seed that can pass on the newly inherited trait and must not interfere/disrupt the normal functioning of the cell. The limited tolerance of the rDNA or new trait is related to two factors.

The genetic makeup of the plant comprises a finely balanced system of specialized cells programmed to carry out complex functions in response to developmental as well as situational conditions. Addition of new genes can disrupt these higher functions even though the critical functions necessary for cell metabolism may remain intact. Discovery of such a potential malfunction may not be evident to the research team at the outset. Rather, it may take field growing conditions to bring out the differences and field testing is not likely to be carried out until much later in the process.

Second, the plant will ultimately be required to survive in the field and produce a yield comparable to the best available conventional seed. This means that the genetically engineered plant must not only have its full functional complement of capabilities, but must also be capable of performing well under the wide variety of circumstances that typically accompany the field environment.

The transformation of genes into the plant system has been through *Agrobacterium tumifacians* a soil bacterium 'the gene ferry' in the hands of plant genetic engineers. In recent years, methods such as particle gun, electroporation, particle bombardment (biolistics) and microinjections have been cultivated for gene transfer. Further a host of DNA sequences called promoters (*e.g.* active promoter) are available for regulation of these genes and expression can be obtained either in constitutive (with out any external or internal factor) or the inducible (in response to inducers) mode.

SALT STRESS DETERMINANTS

Essential of targeted approach is the availability of some background information. It can be explained by taking the example of salt stress (osmotic stress) conditions in relation to maintenance of turgor by means of osmotic adjustment. Osmotic adjustment enhances the water potential of the cell sap (towards more negative) by synthesis of organic solutes, which help in maintaining cell turgor even at low water potentials. On this basis, it is inferred that accumulation of osmotically active compounds (osmolytes) in cells in response to stress is an important component for the induction of stress tolerance. Genetic engineering approach is used to activate/deactivate the enzymes responsible for the production of these metabolites/ osmolytes. A number of genes whose expression increases in response to stress are listed in (Table 1).

Response to abiotic stresses has always been complex at genetic level as the expression of large number of genes is altered at the same time. This statement has been proved by the molecular biologists, as the expression of more than 100 transcripts was affected, when the plants faced osmotic stress (Grover *et al.*, 1998). RNA analysis had revealed the existing physiological/biochemical knowledge of plant responses is scanty and several mechanisms are yet to be understood. The proteins, which are up regulated under stress, are known as "stress proteins" and have been listed in (Table 2).

Salt stress determinants are defined as the effectors molecules (metabolites, proteins or components of biochemical pathways) that lead to adaptation and as regulatory molecules

Table: 1. Stress reponses of transgenic plants over expressing various genes involved in stress tolerance.

Gene	Gene product (and function)	Transgenic plant	Performance of transgenic plant
bet A	Choline dehydrogenase (Glycine betaine biosynthesis)	Tobacco	Increased tolerance to salt.
cod A	Choline oxidase (glycine betaine biosynthesis)	*Arabidopsis, rice*	Seedlings were more tolerant to salt and showed increased germination under cold conditions.
ITM 1	Myo-inositol O-methyl transferase (D-ononitol biosynthesis)	Tobacco	Improved performance under drought and salt as shown by better photosynthetic rate.
mtl 1D	Mannitol-1-phosphate dehydrogenase (mannitol biosynthesis)	Tobacco	6-week old plants showed better growth in terms of percent change of height and fresh weight under high salinity.
		Arabidopsis, tobacco	Enhanced seed germination under high salinity.
		Tobacco	Increased tolerance to methyl viologenin-duced oxidative stress as documented by in creased retention of chlorophyll in transgenic plant leaves under stress.
p5cs	D1-pyrrolline-5-carboxylate synthetase (proline synthesis)	Tobacco, rice	Enhanced biomass and flower development development under salt stress.
odc	Ornithine decarboxylase	Carrot (cell line)	Transgenic cell lines could with stand high salt over a short period.
HVA 1	Group 3 LEA protein	Rice	Increased tolerance to water deficit and salt stress as shown by better growth of transgenic seedlings under 100 mM NaCl or 200 mM mannitol; 3-week old seedling per formed better under stress conditions in soil.
DREB1A	Transcription factor	*Arabidopsis*	Increased salt, drought and cold tolerance.
Nt 07	Glutathione-S-transferase	Tobacco	Transgenic seedling grew faster than controls when subjected to salt and cold stress.

Table : 2. Salt stress responsive genes/proteins in selected crop species

Plant species	Genes/proteins
Arabidopsis thaliana	At myb 2, Sal 1
Brassica napus	BnD 22
Citrus sinensis	Cit-SAP
Hordeum vulgare	HVA 1, 26 and 27 k Da proteins
Lycopersicon esculenutum	Osmotin, TSW 12, le 16
L. pimpinellifolium	14.5 kDa protein
Medicago sativa	psm 1409, pa 9
Nicotiana tabacum	Proteins of 30 and 43 kDa, osmotin
Oryza sativa	rab 21, sal T, Em gene, SAP 90, 104

(signal transduction pathway components) that control the amount and timing of these affecter molecules. Stress adaptation effectors are categorized as those that mediate ion homeostasis, osmolyte biosynthesis, toxic radical scavenging, water transport, and transducers of long-distance response coordination (Kjellborn *et al.*, 1999).

(a) Ion Homeostasis

A hypersaline environment, most commonly mediated by high NaCl, results in perturbation of ionic steady state not only for Na^+ and Cl^- but also for K^+ and Ca^{2+} (Hasegawa and Bressan, 2000). External Na^+ negatively impacts intercellular K^+ influxes, attenuating acquisition of this essential nutrient by cells. High NaCl causes cytosolic accumulation of Ca^{2+} and this; apparently, signals stress responses that are either adaptive or pathological. Ion homeostasis in saline environments is dependent on transmembrane transport proteins that mediate ion fluxes, including H^+ translocating ATPases and pyrophosphates, Ca^{2+}-ATPases, secondary active transporters (SAT) and channels. A role for ATP-binding cassette (ABC) transporters in plant salt tolerance has not been elucidated, but ABC transporters regulate cation homeostasis in yeast.

(b) Aquaporins

The hydrophobic nature of lipid bilayer presents a considerable barrier to the free movement of water into the cell and between intracellular compartments. However, plasma membranes and tonoplasts can be rendered more permeable to water by proteinaceous transmembrane water channels called aquaporins. Water movement through aquaporins can be modulated rapidly. Evidences suggest that these channels may facilitate water movement in drought stressed tissues and promote the rapid recovery of turgor on watering. The aquaporin RD28 is located in the plasma membrane. Genes encoding MIP-related proteins have been identified in *M. crystallinum* and Arabidopsis. The abundance of mRNA transcripts of RD28 genes correlates with the turgor changes in leaves of plants subjected to a 400 mM NaCl shock treatment. The amount of transcript first decreases after the initial shock, as does turgor. Transcript concentrations then increase as turgor is restored. Greater transcript concentrations, enhanced translocation, and activation of existing proteins each may constitute a mechanism for regulating aquaporin abundance and activity in response to water stress. Expression studies in *Xenopus oocytes* link ⇐-TIP phosphorylation with increased permeability of cell membrane to water.

(c) Osmoprotectants

Transgenic plants with improved tolerance to stress have been produced using various genes encoding enzymes that synthesize osmoprotectants. Principle behind the engineering is identifying the enzyme regulating the biosynthetic pathway of the metabolite and then putting the promoter to increase the expression of that particular protein. How these osmolytes protect the cell hydration layer is shown in (Fig. 9).

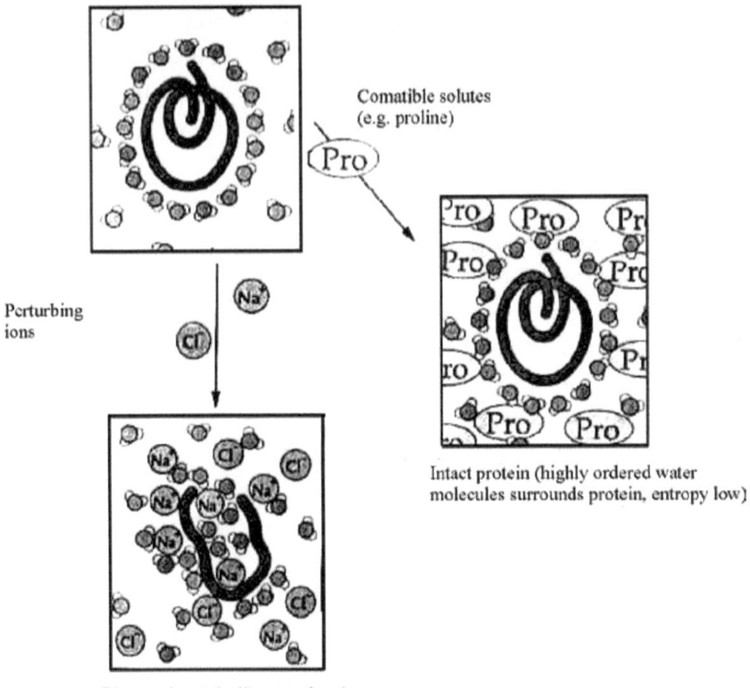

Fig. 9: The hydration shells of macromolecules are not disturbed by compatible solutes. Depicted is a protein with a hydration shell, which is surrounded by ordered water molecules. Ions such as sodium and chloride can penetrate the cells and interfere with the non-covalent interaction that maintains the structure of protein. Compatible solutes do not penetrate the hydration shell, so protein and solutes do not come into direct contact.

(a) Proline: The enzyme Ä-pyrrolline-5-carboxylate synthetase (P5CS) catalyzes the conversion of glutamate to Ä-pyrrolline-5-carboxylate, which is then reduced to proline. Over expression of a gene encoding for moth bean P5CS in transgenic tobacco plants resulted in accumulation of proline upto 10 to 18 fold over controls (Kavi Kishore *et al.*, 1995). This group also showed that transgenic plants demonstrated enhanced production and flower development under salt stress conditions, as determined by increased root length, root dry weight, capsule number and seed number per capsule. The same gene was introduced in rice under control of an ABA/stress inducible promoter (Zhu *et al.*, 1998) and it accumulated upto 2.5 fold more proline than control plants under stress conditions. Preliminary results showed that under stress, individual expression of P5CS transgene in second generation of transgenic rice plants showed an increase in biomass as reflected by higher fresh shoot weight under salt and water stress conditions compared with untransformed control plants. The extent of salt tolerance could be correlated with the levels of proline accumulated.

(b) Mannitol: Mannitol is the reduced form of the sugar mannose. This sugar alcohol is broadly distributed among plants. *In vitro* radiolabeling shows that mannitol concentration increase in response to osmotic stress. In sharp contrast to glycine betaine, mannitol accumulation appears to be regulated by inhibition of competitve pathways and by decreased rates of mannitol consumption and catabolism. In celery, salt stress inhibits sucrose synthesis, but does not seem to affect the enzymes that synthesize mannitol.

Mannitol as an osmoprotectant is basically found in microbes. Tobacco transformed with mtl1D gene encoding a mannitol-1-phosphate dehydrogenase accumulated mannitol (Tarczynski *et al*., 1993). These plants showed increased tolerance to high salinity. Arabidopsis was transformed with the same gene and it was found that transgenic seeds accumulating mannitol germinated in the presence of high salt, where as non-transgenic seeds did not germinate. It was also found that accumulation of mannitol made a negligible contribution towards osmotic adjustment in transgenic tobacco. Targeting mannitol biosynthesis to chloroplast in transgenic tobacco plants also resulted in accumulation of mannitol upto 10 mM in chloroplast and resulted in an increased tolerance to methyl viologen (MV)-induced oxidative stress, as shown by increased retention of chlorophyll in transgenic tissue upon MV treatment.

***(c) Glycine betaine*:** It is synthesized and accumulated by many algae and higher plants and its presence is cosmopolitan. Radiotracer evidences indicates that its accumulation in osmotically stressed plants results from the increased rates of its synthesis. The glycine betaine does not show the degradative pathway, but its concentration is controlled by its biosynthesis and its transport through phloem to growing tissues. The two key enzymes in the biosynthetic pathway of glycine betaine from choline in two step pathway are choline monooxygenase and betaine aldehyde dehydrogenase. Both the enzymes have been purified and their cDNA's are available. Activities of these enzymes have been increased several folds under osmotic stress.

Lilius *et al*. (1996) introduced bet A gene encoding for choline dehydrogenase from *E. coli* into tobacco. The transgenic plants were more tolerant to salt as measured by dry weight between transgenic and wild type plants at 300 mM NaCl. The bacterial choline oxidase (cod A) gene isolated from Arthobactor, which converts choline to glycine betaine, was introduced to Arabidopsis. The transgenic plants accumulated glycine betaine and showed enhanced tolerance to salt (Alia *et al*., 1999). The authors suggested that the possible mechanism for the glycine betaine action could be the stabilization of structure and function of proteins, maintain higher RWC, assimilated more carbon and maintained greater turgor than non accumulators.

***(d) D-ononitol*:** Expression of a cDNA encoding myo-inositol O-methyl transferase (IMT1) in tobacco during salt and drought stress resulted in the accumulation of methylated inositol (D-ononitol), which in turn conferred tolerance for both stresses (Sheveleve *et al*., 1997). Since the level of D-ononitol reached upto 600 mM in the cytosol and it did not enter the vacuole, its accumulation provided an osmolyte for osmotically balancing Na^+.

***(e) Polyamines*:** Polyamines are small, ubiquitous, nitrogenous cellular compounds that have been implicated in a variety of stress responses in plants. Polyamines accumulate under several abiotic stress conditions including salt and drought. Cultivars demonstrating a higher degree of salt tolerance contained higher levels of polyamines. Minocha and Sun

(1997) showed that transgenic carrot cell lines over producing mouse ornithine decarboxylase, which convert ornithine to diamine putrescine, can withstand salt and osmotic stress over a short period of 0 to 48 h. Further, tobacco plants over expressing the antisense construct of ornithine decarboxylase mRNA had high ADC activity and high tolerance to 200 mM NaCl. Since most of the osmolytes did not accumulate in amounts large enough to play a role in osmotic adjustments, it is not clear how they provide protection against dehydration stress. Alternatively, they may act as scavengers of reactive oxygen species. Mannitol has recently been shown to act in chloroplasts as a scavenger of hydroxyl radicals.

(f) Late embryogenesis abundant (LEA) proteins: The *LEA* genes were first identified as genes induced in seeds during maturation and desiccation. The amount of some lea gene products are now known to increase in the vegetative tissues of plants exposed to stresses that include water deficit component. Over expression of lea proteins in rice and yeast have been shown to enhance the resistance to specific water deficit stresses. The lea genes have been divided into five groups namely group 1,2,3,4 and 5 (Table 3).

Barley group 3 LEA proteins, HVA1 was found to be specifically produced in the aleurone layers and embryos during late seed development I rice, correlating with seed development (Xu *et al*., 1996). ABA and several stress conditions including dehydration and salt, rapidly induced expression of this gene in young seedlings. Transgenic rice plants expressing the barley HVA 1 gene, driven by a constitutive promoter from the rice actin gene, led to accumulation of HVA-1 protein in both leaves and roots of transgenic plants. The second-generation transgenic rice plants showed significantly increased tolerance to water deficit and salinity. A glyoxylate cDNA from *Brassica juncea* was ligated to the CaMV 35 S promoter and cloned into pB 1121. This plasmid was transformed into tobacco. Transgenic plant overproducing glyoxylate showed significant tolerance to high salt.

FUTURE STRATEGIES

Transgenic approaches for increasing plant tolerance to dehydration stresses are experimentally feasible. Preliminary results are encountered for enabling scientists to better understand the effects of single-gene transfers to plants. However, it should be emphasized that success to date represents only a beginning. Much more work is needed to gain a better understanding of the biochemical and physiological basis of stress tolerance. Once it is better understood how different single genes work, it is likely that several genes will need to be simultaneously transferred into plants to produce high levels of stress tolerance. This may more closely resemble what occurs in nature where stress tolerance is the cumulative effect of several genes. Recent advances in transformation of agronomically important crops and the development of better expression systems in terms of using stress-inducible promoters and adding MAR sequences hold much promise in this direction. These and other advances are needed to produce stress-tolerant transgenic crop plants that give significantly higher productivity under field conditions.

Table : 3. The five groups of LEA proteins.

Group	Representative proteins	Structural characteristics and shared motif	Properties	Proposed function (s)
Group 1 (D=19 family[a])	Em (early methionine labeled protein, wheat)	• Most (70%) protein conformation is random coil with some predicted short a-helices. • Rich in charged amino acids and glycine	• Most hydrated than most globular polypeptides.	• Binds water to minimize loss of cellular water content. • Over expression confers water-deficit tlerance on yeast cells.
Group 2 (D-11 family[a])	DHN 1 (maize) D-11 (cotton)	• Structure variable • Includes one or more conserved conserved lycine-rich regions that may form a-helices. The number of repeats per protein varies. • May or may not contain a poly (serine) region. • Contains regions of variable length rich in polar residues and either glycine or alanine and proline.	• Most members localize to cytoplasm and nucleus. Acidic members associated membranes.	• May stabilize macromolecules under conditions of reduced water content.
Group 3 (D-7 family[a])	HVA1 (Hordeum vulgare ABA-induced, barley) D-7 (cotton)	• Contains repeated motifs 11 amino acids long with the with the consensus sequence TAQAAKEKAXE. • Predicted to contain amphipathic a-helices. • Predicted to form dimers.	• D-7 is abundant in cotton embryos (0.25 mM).	• HVA1 promotes stress tolerance in transgenic plants. • Putative dimer of D-7 may bind as many as10 inorganic phosphates and their counter ions.
Group 4 (D-95 family[a])	D-95 (soybean)	• Hydropathy plots are unremarkable and slightly hydrophobic. • N-terminal region contains a possible amphipathic a-helix.	• A gene encoding a similar protein in tomato is expressed in response to nematode feeding.	
Group 5 (D-113 family[a])	LE 25 (tomato) D-113 (cotton)	• Shares sequence homology at the conserved N terminus, is predicted to form a-helix. • C-teminal domain is predicted to be a random coil of variable length and sequence. • Rich in alanine, glycine, and threonine.	• D-113 is abundant in cotton seeds (upto 0.3 mM).	• May bind membran-ces or proteins to main tain structural integrity. • May sequesterions to protect cytosolic metabolism. • LE 25 confers salt and freezing tolerance to yeast

* The protein families are named for seed proteins from cotton.

REFERENCES

Alia, K.Y., Sakamoto, A., Nonaka, H., Yagashi, H. 1999. Enhanced tolerance to light stress of transgenic *Arabidopsis* plants that express the cod A gene for a bacterial proline oxidase. *Plant Mol. Biol.* **40:** 270-288.

Apse, M.P., Aharon, G.S., Snedden, W.A., and Blumwald, E. 2000. Salt tolerance confirmed by overexpression of a vaculor Na^+/H^+ antiport in Arabidopsis. *Science*, **285:** 1256-1258.

Cheeseman, J. M. 1988. Mechanism of salinity tolerance in plants. *Plant Physiol.* **87**: 547-550.

Dale, J. E. and Sutcliffe, J. F. 1986. Water relations of plant cells. In: *Plant Physiology; A Treatise* (Ed. Steward, F. C.), Academic Press, Inc. 1-18.

Grover, A., Pareek, A., Singla, S. L., Minhas, D., Katiyar, S., Ghawana, S., Dubey, H., Aggarwal, M., Rao, G. U., Rathee, J. and Grover, A. 1998. Engineering crops for tolerance against abiotic stresses through gene manipulation. *Curr. Sci.*, **75**: 689-696.

Hare, P. D., Cress, W. A. and Staden, J. V. 1981. Proline synthesis and degradation-a model system for elucidating stress related signal transduction. *J. Exp. Bot.*, **50**: 413-434.

Hasegawa, P. M. and Bressan, R. A. 2000. Plant cellular and molecular responses to high salinity. *Ann. Rev. Plant Physiol. Plant Mol. Biol.*, **51**: 463-499.

Kavi Kishore, P. B., Honga, Z.A., Miao, G. H., Hu, C. A. and Verma, D. P. S. 1995. Overexpression of Ä pyrrolline–5-carboxylate synthetase increases proline production and confers osmotolerance in transgenic plants. *Plant Physiol.*, **108:** 1387-1394.

Kramer, P. J. 1983. In: *Water Relations of Plants*. (Ed. Kramer, P. J.), Academic Press, Inc. Orlando, 187-213.

Levitt, J. 1980. Salt stress. In: *Responses of Plant to Environmental Stresses*. Vol. II. Academic Press, New York.

Salisbury, F. B. and Ross, C. 1969. In: *Plant Physiology*, Prentice hall of India, Pvt. Ltd., New Delhi, 52-74.

Scholander, R. F., Hammel, H. T., Brodsteer, E. D. and Hemmingsen, E. A. 1965. Sap pressure in vascular plants. *Science,* **148**: 339-346.

Shevelava, E., Chmara, W., Bohnert, H. J. and Jensen, R. G. 1997. Increased salt and drought tolerance by D-ononitol production in transgenic *Nicotiana tabacum* L. *Plant Physiol.*, **115:** 1211-1219.

Taiz, L. and Zeiger, E. 1998. Water and plant cell. In: *Plant Physiology* (Eds. Taiz, L. and Zeiger, E.), Sinclair Associates, Inc. Sunderland. 61-80.

Tarczynski, M. C., Jensen, R. G. and Bohnert, H. J. 1993. Stress protection of transgenic potato by production of osmolyte mannitol. *Science,* **259**: 508-510.

Xu, D., Duan, X., Wang, B., Hong B., Ho, T. D. H. and Wo, R. 1996. Expression of the late embryogenesis abundant protein gene, HVA1, from barley confers tolerance to water deficit and salt stress in transgenic rice. *Plant Physiol.*, **110:** 249-257.

Zhu, B. C., Su, J., Chan, M. C., Verma, D. P. S., Fan, Y-L, and Wu, R. 1998. Overexpression of a Ä 1-pyrrolline–5-carboxylate synthetase gene and analysis of tolerance to water stress and salt stress in transgenic rice. *Plant Sci.,* **139:** 41-48.

Developments in Physiology, Biochemistry and Molecular Biology of Plants, 2005
Eds.: Bandana Bose and A. Hemantaranjan
Vol., 1, pp. 105-124, New India Publishing Agency, New Delhi
E-mail: spjain_niph@rediffmail.com web: www.bookfactoryindia.com

CHAPTER - 6

IMPACT OF HEAVY METAL POLLUTION IN PLANTS

A. K. BERA, ANITA BERA AND SAMADRITA BARMAN ROY

INTRODUCTION

One of the most dangerous ecological crises being faced by all living organism now-a-days is the pollution of the environment. It is easier to describe pollution rather than to define it. It is presumed that the environment in past was more or less pure. But due to various activities of man such as industrialization, urbanization, agricultural practices *etc,* composition of the environment gets changed which is called environmental pollution. Any substance that creates pollution is called 'pollutant'. There are different types of pollution among which pollution caused by toxic level of heavy metal pollutants is called heavy metal pollution.

Heavy metal pollution is a problematic field of research and is gaining immense importance now-a-days. We have 104 elements in the periodic classification of which 80 are metals. According to Passow *et al.* (1961), heavy metals are those having densities more than five (5 g cc^{-1}) and atomic weight twenty three. The term 'heavy metal' is imprecise but is widely used although others such as 'toxic metal', 'trace element' are possible alternatives. Sources of these heavy metals are varying; some of these are encountered in water, air effluent, solid waste and sewage. They are emitted out from various sources *e.g.* waste water from electroplating industry (Cr, Ni), air emission from fluorescent lamps (Cd, Be), paint pigment waste (Pb, Cr), auto-exhaust emissions (Pb, Te) *etc.*

All heavy metals are toxic to living organisms at excessive concentrations, but some are essential for normal healthy growth and reproduction by plants at low but critical concentrations. The essential heavy metals for plants include Co, Cu, Mn, Mo, Ni, and Zn. Other elements including Ag, As, Ba, Cd, Hg, Pb, Sb and Ti have no essential function and like the essential trace elements, cause toxicity above a certain tolerance level. The most important heavy metals with regard to potential hazards and occurrence in contaminated soils are: As, Cd, Cr, Hg, Pb and Zn. Thus fortunately, out of 80 metals, 18 metals are toxic and possess serious threat to plant system.

HEAVY METAL UPTAKE BY PLANTS

The uptakes of heavy metal are influenced significantly by other metal elements present in soil. Two most foremost reactions observed in the soil when two or more metals are present, are interaction and chelation. Interaction is known as a mutual or reciprocal action of one element upon another in relation to plant growth and also the different levels of a second element applied simultaneously. Thus, interaction produces an added effect that result modification of metallic regimes in the soil thereby affecting plant uptake and translocation. Chelation on the other hand is a special type of chemical reaction where a metallic ion remains firmly attracted with a molecule by means of multiple chemical bonds. The chelates have a tendency to hold tightly certain cations attached to them. Chelating compounds of various kinds act as a carrier of heavy metals in soils and plants. The chelators when added to the soil can increase solubility of metals many fold and promote metal convection and diffusion, hence potential uptake by plants (Lindsay, 1974).

Interaction among metals and their relative uptake by plants have been reported by many. Manganese interacts with many elements including heavy metals, alkaline earth metals and also either non-toxic element like phosphorus. Epstein and Stout (1951) suggested that, the antagonistic effect of high 'Mn' levels were mainly on 'Fe' translocation from roots to shoots. Ratner *et al.*,(1962) noted that 'Mo' counteracted the harmful effects of 'Al' on Alfalfa root tips. Similarly 'Al' toxicity in wheat could be completely overcome by increasing the concentrations of Ca, Mg, K and Na in the medium either individually or collectively (Ali, 1973). Bansal and Chahal (1990) studied Fe and Mn uptake in mungbean, grown in alkaline soils. There were 4 levels of Fe and Mn *viz.* 0, 25, 50 and 100 ppm. It was observed that an increase in the supply of Fe decreased the Mn content and its uptake.

Uptake and distribution of mercury and manganese during (Table 1) seedling growth of mungbean revealed that accumulation of 'Hg' in 6 days' old mungbean seedling occurred

Table : 1. Individual and combined effect of mercury and manganese (ppm) on distribution of heavy metal in different plant parts of 6 days old mungbean seedling (Data expressed in μg/g dry weight)

Combinations	Hg			Mn		
Hg:Mn (ppm)	Leaf	Stem	Root	Leaf	Stem	Root
0+0	0.000018	0.000016	0.000089	536.38	442.51	562.74
1+0	0.00028	0.00023	0.00538	459.88	247.32	736.16
10+0	0.00085	0.00041	0.02015	276.44	125.82	473.30
100+0	0.00878	0.00283	0.08173	-	-	-
1000+0	-	-	-	-	-	-
0+1	0.000041	0.000029	0.000155	709.12	656.11	847.56
0+10	0.000016	0.000014	0000045	1595.14	1471.62	11699.67
0+100	-	-	-	7582.42	7532.05	24153.99
0+1000	-	-	-	-	-	-
1+1	0.00019	0.00011	0.00156	259.88	147.32	610.00
10+1	0.00057	0.00025	0.01729	112.91	58.65	326.84
1+10	0.00012	0.00025	0.00371	1553.45	1364.69	8336.72
S. Em (±)	0.00	0.00	0.00	6.59	7.14	10.46
C.D. (P=0.05)	0.00	0.00	0.00	19.56	21.20	31.07

in the order root > stem > leaf when seedlings were grown in 100 ppm 'Hg' solution. However, supplementation of 10 ppm 'Mn' in the cultural solution reduces both uptake of 'Hg' by root system as well as its transport to other plant parts *i.e.* stem and leaf (Barman Roy and Bera, 2002).

PHYSIOLOGICAL AND BIOCHEMICAL EFFECTS OF HEAVY METALS IN PLANTS

The heavy metals are unique environmental and industrial pollutants. The growth and metabolism are adversely affected by excess supply of heavy metals. It has been observed that toxicity imposed by heavy metals involves an overall disruption in the synchronization of different metabolic processes occurring in the cells, the resultant effect being inhibition of growth, reproduction and crop yield (Nag *et al.*, 1981). A list of different abnormalities, noticed in various crop plants in response to heavy metal pollution is documented.

Germination and Seedling Growth

Mhatre and Chapekar (1982) studied the effect of three heavy metals *viz.* Mercury, lead and arsenic on seed germination and early seedling growth in 11 different plant species. It was noticed that mercury had the adverse effect on all the seedlings tested, while lead and arsenic induced variable responses. The relative susceptibility of the seeds to mercury was found to be as bajra > mustard > jowar > methi > alfalfa > cluster bean ≅ peas ≅ sunhemp ≅ mungbean ≅ radish ≅ lady's finger. The work of Sharma (1983), on the effect of mercury on germination and seedling growth of *Pisum sativum* cultivars revealed that pea germination as well as seedling growth were promoted by pre-soaking seeds in mercuric acetate solution a 4.98×10^{-5} M Hg but inhibited by 2.49×10^{-4} M Hg. The order or resistance to mercury was T-163 > Bonneville > Arkel. Wong *et al.* (1988) reported that germination and elongation of pea cv. Alaska radicles were significantly inhibited when cadmium was added to the culture solution. Cd also retarded shoot elongation and leaf development of young seedlings even at 1 μg/ml. Zhang *et al.* (1989) experienced the effect of mercury on the growth and physiological function of wheat seedlings. In a laboratory experiment, wheat seeds were cultured in Knop's solution containing 1, 5, 10, 20, 50 or 100 ppm $HgCl_2$. It was observed that Hg inhibited both germination and seedling growth.

While working on the effects of cadmium, Kalimuthu and Sivasubramanian (1990) found that soaking of seeds in 20, 50, 100 or 200 μg mercuric chloride or lead acetate/ml for 24 hours, decreased the germination and seedling growth of maize. Varshney (1990) studied cultivar specific growth of mungbean seeds to different concentrations (1×10^{-5} to 5×10^{-4} M Hg as mercuric acetate). Organ specific differences exist in cv. ML 384 and K-851, the former being more resistant than K-851. Mercury decreased the radicle length in cv. K-851 and increased it in cv. ML-384 (Upto 7.5×10^{-5} M Hg). Epicotyl length increased with low Hg concentration in K-851 and decreased with higher concentrations, whereas all Hg concentrations decreased epicotyl length in ML-384. Hg inhibited the hypocotyls length in both cultivars. Lima and Copeland (1990) worked on the effect of aluminium on the germination of wheat seeds. In green house experiments, the effect of 0-10 mM Al on wheat seed cv. 'Robin' (Al sensitive) and 'Carazinho' (Al tolerant) germination were examined. Germination was less sensitive to Al than seedling growth. Root length of 4 days old cv. Robin seedlings was reduced by 15, 30 and 50% in the presence of 1,5 and 10 mM Al while cv. Carazinho showd no inhibition of germination, root or shoot growth in the presence of Al concentration upto 10 mM.

Leena and Ramanujam (1992) worked on the effect of pre-treatment of groundnut seeds with nickel on germination and seedling growth. In their experiment, seeds of groundnut cultivar TMV-12 and VRI –2 were soaked in 10 or 50 ppm nickel-sulphate solution for about 24 hours. Seed germination was not affected by 10ppm Ni but it decreased from 97.0% to 80.8% in TMV-12 with 50 ppm Ni. Root length of 7 days old seedling was increased by 10 ppm Ni. Although shoot length was unaffected, root length was found to decrease compared to control. Influence of cadmium and lead on germination and growth of *Vigna unguiculata* L. Walp seedlings were examined (Battacharya and Mukherjee, 1994). It was observed that Cd at a concentration of 100 μM was absolute toxic in contrast to same concentration of lead. Increasing concentrations of both the metals decreased germination per cent and inhibition of root growth was more than shoot. Nandi and Bera (1995) noticed inhibition of blackgram seed germination by Hg and Mn. At 1000 mg./L Hg only 8% seed germination was observed while with same concentration of Mn, 96% germination was recorded indicating that Hg is more toxic than Mn.

Leaf Area, Stomatal Frequency and Stomatal Index

Different workers studied the effect of heavy metals on leaf area, stomatal frequency, photosynthetic rate *etc.* in various crops (Butler and Tibbits, 1979; Aggarwal *et al.*, 1990; Greger and Johansson, 1992). Butler and Tibbits (1979) studied stomatal opening mechanism determining genetic resistance of ozone in *Phaseolus vulgaris* L. In their experiment, the ozone sensitive bean cvs. Spurt and Blue Lake Stringless and the resistant cultivars, Black Turtle Soup and French's Horticultural were grown from seeds in a growth chamber. The resistant cultivars had 25% fewer stomata/mm^2 leaf area than the sensitive cultivars and exhibited partial stomatal closure following exposure to 134 p.p.h.m ozone for 1h, while the sensitive cultivars did not. Stomatal closure was determined to be more important than reduced stomatal frequency in providing resistance. Aggarwal *et al.* (1990) studied the effect of Cd and Ni on leaf area and photosynthesis of wheat and pigeon pea. It was observed that Cd is more toxic than Ni. Leaf area and photosynthesis were less affected in wheat than in pegion pea by both metals. In soybean cv. Gaurav, leaf area at the seedling, flowering and fruiting stages decreased with >5 μg Cd and >10 μg Zn (Avery and Sarkar, 1991). Similar reduction in leaf area was notice in *Raphanus sativus* L. cv. Pusha Rashmi treated with 2.5×10^{-4} M Cd (Lata, 1988).

In a separate experiment, sugar beet cv. Monohill seedlings grown hydroponically were subjected to 0-10 μM, $CdCl_2$ for 4 weeks by Greger and Johansson (1992). It was observed that transpiration rate increased with increasing Cd concentration, while the total stomatal aperture area/unit leaf area decreased. Cd increases number of defective and undeveloped stomata which were always closed or had small apertures which remain open and lacked a functional closing mechanism. Cd treatment induced closure of intact stomata and decreased the number of intact stomata/unit leaf area; total stomatal number increased slightly with increasing Cd concentration but this was due to higher numbers of small and defective stomata. Bindhu and Bera (2001) studied the effect of different concentrations of $CdSO_4$ (10^{-4} M, $5x10^{-4}$ M and 10^{-3} M) on leaf area, stomatal frequency and stomatal index in 6 days old mungbean seedling. It was observed that except at 10^{-4} M $CdSO_4$, leaf area decreased with an increase in the concentration of $CdSO_4$. Stomatal frequency on the leaf surface increased at $5x10^{-4}$ M and 10^{-3} M $CdSO_4$ in comparison to untreated control. However, stomatal index in both the leaf surfaces decreased with increased in the concentration of Cd (Table 2).

Table : 2. Effect of different concentrations of $CdSO_4$ on leaf area, stomatal frequency and stomatal index of 6 days old *Vigna radiata* cv. Pusa Baisakhi seedling.

Concentration of $CdSO_4$	Area of single Primary leaf (cm^2)	Stomatal frequency ($x10^3$)		Stomatal Index	
		Upper	Lower	Upper	lower
Control	0.99	46.209	88.642	22.38	36.65
10^{-4}M	1.435	44.794	56.577	22.02	30.15
$5x10^{-4}$M	0.705	47.090	68.840	21.83	28.35
10^{-3}M	0.635	53.479	75.162	18.75	26.17
S.Em ±	0.04	2.33	2.18	0.63	1.17
C.D (P=0.05)	0.13	-	6.70	1.93	3.60

Pigment

Heavy metal application cause yellowing of leaves which points towards a reduction in pigment status and consequently reduction in photosynthesis in higher plants. Nag *et al.*, (1981) studied the effect of toxic concentrations of $HgCl_2$ (0.4, 0.6 and 0.8 mM) $CuSO_4$ (0.5, 1.0 and 2.0 mM) and $ZnSO_4$ (10, 20, and 40 mM) solutions in rice seedlings. It was observed that chlorophyll development in the leaves of rice seedlings was reduced by heavy metal treatment with concomitant increase in chlorophyllase activity. Hill reaction activity of chloroplast isolated from rice leaves was also depressed. *Phaseolus vulgaris,* grown in nutrient culture solution containing 0, 10, 80, or 160 ppm Cd, showed reduction in chlorophyll and carotenoid pigments. Maximum reduction for both the pigments were noticed with 160 ppm Cd (Poschenrieder, *et al.*, 1983).

Stobart *et al.* (1985) worked on the effect of cadmium on the biosynthesis of chlorophyll in the leaves of barley and reported that Cd ions inhibited the production of chlorophyll in dark grown seedlings of barley cultivar 'Proctor' by affecting the synthesis of 5-amino leavulinic acid (ALA) and the photochlorophyllide reductase ternary complex with its substrates. The results were consistent with Cd inhibiting chlorophyll formation by reacting with essential thiol groups and enzymes involved in light dependent synthesis of

5-amino laevulinic acid. While studying the seedling of 'Okra' cultivars Nimbkar-5, Ankur-35 and Pusa Sawani, Shrivastava and Singh (1986) observed that cadmium chloride, cadmium acetate (each at 0-500 ppm) or a 1:1 mixture of both reduced the contents of chlorophyll 'a' and 'b'and carotenoids in all cultivars. The effect increased with concentration and cadmium acetate was found to be more harmful than cadmium chloride. Burzynski and Buczek (1989) studied the interaction between cadmium and molybdenum on chlorophyll content of *Cucumis sativus.* A strong antagonistic effect of Cd and Mo on chlorophyll content had been noticed. In another experiment, effect of cadmium treatment on level of chlorophyll and Hill activity of isolated chloroplast from, *Phaseolus aureus* cv.PS-16 was studied (Lata, 1989). Here, *Phaseolus aureus* seeds were imbibed in cadmium acetate solution containing $2.5x10^{-4}$ M Cd for 12 hours before sowing. It was observed that chlorophyll 'a' and 'b' contents of leaves and Hill activity in Cd treated plants were 60, 93 and 91% respectively over control. Soaking maize seeds in 20, 50, 100 or 200 μg mercuric chloride or lead acetate/ml for 24 hours significantly decreased chlorophyll contents but increased carotenoid contents of leaves (Kalimuthu and Sivasubramanian, 1990).

The effect of two heavy metals (Cd and Pb) on *Vigna unguiculata* was studied with reference to photosynthetic pigment (Bhattacharya and Mukherjee, 1994). Chlorophyll 'a' and 'b' and carotenoids showed a decreasing trend with increasing concentrations of the metals. While studying the effect of Ni and Co on growth performance of chick pea (*Cicer arietinum*), Khan *et al.* (1996) noticed significant reduction of leaf chlorophyll content. In general, carotenoids are less affected by heavy metals compared to chlorophyll in higher plants. Decrease in chlorophyll contents might be attributed due to senescing effect of heavy metals in plants (Clijsters and van Assche, 1985). Effect of different concentrations of $CdSO_4$ (10^{-4}M, $5x10^{-4}$ M and 10^{-3} M) on pigment content of 6 days old munbgean seedling was studied by Bindhu and Bera (2001). It was observed that chlorophyll-a, chlorophyll-b, total chlorophyll and chlorophyll a/b ratio increased at 10^{-4} M $CdSO_4$ and decreased there after with an increase in the concentration of Cd^{2+} (Table 3).

Table : 3. Effect of different concentrations of $CdSO_4$ on chlorophyll a, chlorophyll b, total chlorophyll and chlorophyll a/b ratio of 6 days old *Vigna radiata* cv. Pusa Baisakhi seedling (Data expressed as mg/g fresh weight).

Concentration of $CdSO_4$	mg/g fresh weight			
	Chlorophyll a	Chlorophyll b	Total Chlorophyll	Chlorophyll a/b ratio
Control	1.68	0.65	2.33	2.58
10^{-4} M	1.86	0.71	2.57	2.62
$5x10^{-4}$ M	1.54	0.59	2.13	2.61
10^{-3} M	0.91	0.42	1.33	2.17
S.Em ±	0.07	0.03	0.14	0.03
C.D. (p=0.05)	0.20	0.09	0.44	0.09

Carbohydrate Metabolism

In general, heavy metal pollution adversely affect carbohydrate synthesis, mobilization of storage reserves from cotyledon to developing seedling as well as partitioning of dry matter in various field crops. Naquib and Barakat (1989) experimented interaction of tin and strontium on the carbohydrate content of *Vicia faba* leaves. It was observed that leaf carbohydrate content decreased when seeds were soaked in 10^{-6} to 10^{-4} M stannous chloride before sowing in pots, while this effect was counteracted by the presence of strontium ions. Strontium increased leaf sucrose concentration without affecting the starch content. In a separate experiment, Greger and Bertell (1992) grew sugar beet cv. Monohill in nutrient solution with different combination of Ca^{2+} (0, 1, 5 or 20 μM) and analysed sucrose, fructose, glucose and starch contents in 5 weeks old plants. It was established that carbohydrate metabolism was affected by the presence of Ca or Cd. Cadmium reduced carbohydrate concentration while carbohydrate distribution between storage and growth processes was affected by Ca^{2+}. At low Ca^{2+} in the tissue, level of storage carbohydrates increased while at high Ca^{2+} the opposite was found. Both Cd^{2+} and Ca^{2+} decreased sucrose uptake in tap root.

Bhattacharya and Mukherjee (1994) treated seeds of *Vigna unguiculata* L. walp with different concentrations of Cd and Pb and reported that total soluble sugar content of plant at 7 days after treatment increases with increasing concentration of heavy metals but insoluble sugars decreased with reference to control. Nag *et al.* (1989) studied the deleterious effect of $HgCl_2$ and $ZnSO_4$ on growth and biochemical constituents of mungbean seedlings 5 days after treatment. In embryo, total soluble sugar was greatly reduced by $HgCl_2$ and $ZnSo_4$ treatment. At the highest concentration (0.8 mM) of $HgCl_2$, there was 80% decrease of total sugar content of embryo. However, at highest concentration (40mM) of $ZnSO_4$, the reduction was only 11%. The situation was reverse in case of cotyledons in which total sugar progressively increased under increasing Hg and Zn treatments.

Protein Metabolism

Similar to carbohydrate, protein synthesis, mobilization of protein from cotyledon to developing embryonic axis and its distribution in different plant parts is largely affected by heavy metals in various field crops. Stiborova *et al.* (1987) studied the effect of heavy metals on growth and biochemical characteristic of barley and maize seedlings and noticed that effect of Cu^{2+}, Zn^{2+}, Cd^{2+} and Pb^{2+} were stronger in barley than in maize seedlings. The total protein content of barley and maize roots declined with an increase in heavy metal ion concentration. The protein content of barley shoot was only slightly decreased by an increase in heavy metal concentration. Kalimuthu and Sivasubramanian (1990) investigated the effect of soaking maize seeds in 20, 30, 50, 100 or 200 μg mercuric chloride or lead acetate/ml for 24 hours and observed that protein contents were decreased in all seedlings at all the employed concentrations of Pb and Hg. The effect of different concentrations of cadmium on metabolism of protein in germinating rice seeds were analyzed by Mukherjee and Mukherjee (1990) and observed that protein content increased in endosperm and embryos on a fresh weight basis but decreased on seedling basis.

Mel' nichuk *et al.* (1991) noticed that soaking pea seeds in water or $2.5x10^{-5}$ M cadmium chloride for 7 hours or treating seeds with cadmium chloride for 7 days at the 10th hour of germination led quicker degradation of protein from cotyledon. Przymusinski *et al.*(1991) reported that increasing Pb' concentrations (175-1000 mg/liter) decreased the soluble protein content of roots in seedlings of *Lupinus lutens*. In other pot trials, wheat cv. WH-147 was given 100 or 200 mg Cd^{2+}/kg sand as cadmium chloride, 30 days after sowing by Malik *et al.*(1992) and observed that protein content decreased upto 4 days and then remained stable upto 10 days after treatment. *Sesamum indicum* cv. HT-1 seeds treated with 0.04-1.90 mM Pb^{2+} showed increase in total soluble proteins of roots and shoots/plant. It appears therefore that the increase in protein in the roots and shoots/plant may be a result of increased translocation of protein from the cotyledons to the roots and shoots/plant during early seedling growth in a Pb^{2+} enriched environment (Kumar *et al.*, 1983). Seven days old seedlings of *Vigna unguiculata* were treated with 10^{-5} M $PbCl_2$ or 10^{-5} M $CdCl_2$ by Bhattacharya (1995) and noticed that protein content decreased by heavy metals and the effects of Cd^{2+} were greater than those by Pb^{2+}. In a detailed study of metabolic disturbances due to nickel toxicity, Pillay *et al.* (1996) noticed the decrease in the concentration of soluble 'N' and protein in *Hyptis suaveolens* (L.) poit and *Helianthus annuus*. Hemlatha *et al.* (1997) reported that application of heavy metals like Cd^{2+}, Cu^{2+} and Hg^{2+} in 15 days old rice seedlings caused decrease in protein content at all concentrations (10,100 and 1000 μM).

Nucleic Acids

Different laboratory experiments have indicated that nucleic acid, the vital components of green plants, is highly sensitive to heavy metals, more so than any other component. Nag *et al.*,(1989) in their observation with different concentrations of Zn (10,20 or 40 mM $ZnSO_4$) or Hg (0.4, 0.6 or 0.8 mM $HgCl_2$) noticed that in mungbean (*Vegna radiata* L.), DNA and RNA content declined in the embryo and increased in cotyledons as Zn and Hg concentrations increased. In groundnut (*Arachis hypogea*), RNA content in root and cotyledon was found to increase compared with control, when seedlings were grown in $5x10^{-4}$ M $CdCl_2$ solution. Estimation of pure RNA in the presence of $CdCl_2$ indicated the possibility of Cd^{2+} binding to RNA (Satakopan and Rajendran, 1989). In another instance, Satakopan *et al.*, (1992) noticed changes in the level of RNA in chickpea seedling due to Al^{3+} toxicity. Chickpea, cv. Co-2 seedlings, grown at 0 or 1 mM $Al_2(SO_4)_3$, on 10th day of germination, showed decreases in RNA content of seedling. Nanda *et al.* (1993) reported that rice cultivar IR-36 seedlings on exposure to Hg contaminated waste soils, decreased nucleic acid concentration over time. The toxic effect of two heavy metals (Cd and Pb) on some biochemical parameters in *Vigna unguiculata* was studied by Bhattacharya and Mukherjee (1994). RNA content of seedling increased significantly after 24 hours of treatment but decreased later on. However, DNA content reduced form the beginning with increase in the concentration of both 'Cd' and 'Pb'. 'Cd' was found to be more toxic than 'Pb'.

Shah and Dubey (1995) investigated the effect of cadmium on RNA level in growing rice seedlings. Seedlings of rice cv. Ratna and Jaya were raised with increasing levels of cadmium nitrate (0-500 μM). It was observed that under 500 μM Cd^{2+}, RNA level was 0.28 to 1.90 times higher in roots and 0.20 to 1.47 times higher in shoots during 10-20 days growth. De and Mukherjee (1996) studied the effect of different concentrations (25 μM to 500 μM) of $HgCl_2$ in tomato seedling as well as tomato culture cell to compare biochemical make up of these two types of cells brought about by this heavy metal. It was observed that the amount of RNA increased upto 250 μM of $HgCl_2$ in both tomato seedling and culture cell and decreased in both with further increase in the concentration of $HgCl_2$. On the other hand, DNA steadily decreased with the increasing concentration of $HgCl_2$ in both types of cells.

Enzymes

A big group of ions such as Hg^{2+}, Pb^{2+}, Cd^{2+} Cu^{2+}, Zn^{2+} and Mn^{2+} are toxic to plants when present at an elevated level and affect plant metabolism in various ways. Nag *et al.* (1981) studied the effect of toxic concentrations of $HgCl_2$ (0.4, 0.6 and 0.8 mM), $CuSO_4$ (0.5, 1.0 and 2.0 mM) and $ZnSO_4$ (10, 20 and 40 mM) in germinating rice (*Oryza sativa* L.) seedlings. Gel electrophoretic analysis of buffer soluble proteins, basic proteins and peroxidase iso-enzymes shown an increase in the number of bands in metal treated samples. Whereas some of the isoperoxidase bands originally present in the water control disappeared in the treated sample. In an observation on the effect of fluoride toxicity on growth and metabolism of seedlings in 3 cultivars of chickpeas by Reddy (1985), it was observed that fluoride toxicity decreased catalase and amylase activities and increased peroxidase and polyphenol oxidase activities as determined on the 7th day of germination. Cadmium pre-treatment of *Phaseolus aureus* cv.T-44 seeds caused mark inhibition of seedling growth and activity of hydrolytic enzymes, α- amylase and protease. Since α –amylase and protease are responsible for hydrolysis of starch and protein respectively, this response of hydrolytic enzymes to Cd^{2+} is postulated to account for the inhibitory effects of Cd^{2+} on seedling growth (Lata, 1989).

Mittal and Sawhney (1990) noticed that activity of protease along with many other enzymes was reduced by Pb^{2+} when pea cv. Bonneville seeds were germinated in medium containing 0, 0.1, 0.5, or 1.0 mM Pb^{2+}. Similar decrease in α-amylase activity was noticed in germinating rice seeds, treated with different concentrations of Pb^{2+} (Mukherjee and Mukherjee, 1990). In an experiment to observe the differential response of *Phaseolus aureus* cultivars to mercury pre-treatment on protease activity, it was observed that the protease activity in seeds during germination was reduced (Varshney, 1990).The effect of $5x10^{-6}$ or $5x10^{-3}$ M solutions of Cd, Pb, Hg, Zn or Ni on activities of crude amylase, protease and peroxidase from imbibed *Cicer arietinum* cv.C-235 and wheat cb. HD-2250 seeds were studied *in vitro* by Kumar and Banerjee (1992). In C. *arietinum* peroxidase and protease activities decreased but amylase activity increased with heavy metal treatment, whereas in wheat, the activities of all the enzymes tested were decreased by heavy metals. Bishnoi *et al.* (1993) tested the effect of cadmium and nickel on activities of hydrolytic enzymes in germinating pigeonpea cv. UPAS- 120 seeds. The result revealed that cadmium depressed the activities of total amylase, protease and peroxidase in germinating seeds. However, the activities of these enzymes were stimulated at lower concentration of nickel and suppressed to higher ones.

The physiological and biochemical responses of *Vigna unguiculata* seedlings under the influence of cadmium and lead were investigated by Bhattacharjee and Mukherjee (1994). A decrease in protease activity was noticed with increasing concentrations of both the metals. Cd proved to be more toxic than Pb and the effect was maximum in roots and minimum in leaf. Pillay *et al.* (1996) studied the details of metabolic disturbances due to nickel (Ni) toxicities in *Hyptis suaveolens* and *Helianthus annuus* in terms of activities of certain enzymes and metabolites. While in both the plants, catalase activity decreased, the peroxidase polyphenol oxidase activities were found to increase. In case of sunflower, reduction in catalase activity was 5 times over control and the increase in peroxidase and polyphenol oxidase activities were 4 and 8 folds respectively. Individual and combined effects of mercury and manganese on amylase, protease and peroxidase enzyme activities in the seedlings of mungbean were studied by Barman Roy and Bera (2002). It was observed that activities of all the enzymes were stimulated at low concentrations of mercury and manganese but decreased thereafter as the concentration of metals increased. Stimulation in the activities of enzymes at low concentrations of Hg and Mn was considered as an adaptation to heavy metal stress and the depressive effect of heavy metals over the enzyme activities might be due to interference of Hg and Mn with SH system of the enzyme concerned (Table 4, 5 & 6).

Table: 4. Individual and combined effects of mercury and manganese (ppm) on total amylase activity in the cotyledons of 2, 4 and 6 days old mungbean seedling (total amylase activity expressed as µg starch degraded g^{-1} fresh $weight^{-1}$ min^{-1})

Combinations Hg (ppm)+Mn (ppm)	µg starch degraded g^{-1} fresh weight min^{-1}		
	2 days	4 days	6 days
0+0	630.00	869.55	1240.64
1+0	953.73	1056.42	1644.78
10+0	765.52	820.00	1025.58
100+0	432.48	645.25	735.36
1000+0	216.96	380.26	412.14
0+1	843.96	1075.12	1274.96
0+10	1126.94	1182.06	1480.75
0+100	626.42	824.85	1036.68
0+1000	412.36	560.02	622.42
1+1	864.16	916.16	1254.12
10+1	956.78	1080.00	1412.48
1+10	884.48	1140.38	1448.62
S.Em ±	60.96	69.58	67.07
C.D. (P=0.05)	177.94	203.10	195.77

Table : 5. Individual and combined effects of mercury and manganese (ppm) on protease activity in the cotyledons of 2, 4 and 6 days old mungbean seedling (protease activity expressed as µg amino acid release g^{-1} fresh weight min^{-1}).

Combinations Hg (ppm)+Mn (ppm)	µg amino acid released g^{-1} fresh weight min^{-1}		
	2 days	**4 days**	**6 days**
0+0	14.12	20.07	46.07
1+0	18.82	26.67	58.17
10+0	12.15	19.79	41.25
100+0	9.66	16.24	28.16
1000+0	4.58	10.61	16.22
0+1	16.67	24.52	68.60
0+10	21.96	37.58	72.82
0+100	12.85	21.79	45.64
0+1000	9.08	15.66	27.96
1+1	24.26	42.24	55.62
10+1	23.24	37.88	52.84
1+10	21.86	38.76	56.14
S.Em ±	1.25	1.10	1.13
C.D. (P=0.05)	3.63	3.20	3.30

Table : 6. Individual and combined effect of mercury and manganese (ppm) on peroxidase activity in the cotyledons of 2,4 and 6 days old mungbean seedling (Data expressed as unit of activity/g fresh tissue. A change in absorption (OD) by 0.01 per minute at 420 nm was accepted as a unit of activity).

Combinations Hg (ppm)+Mn (ppm)	Unit of activity g^{-1} fresh tissue		
	2 days	**4 days**	**6 days**
0+0	1620.00	3000.00	5272.00
1+0	3264.00	5325.00	7428.00
10+0	3582.00	6160.00	9434.00
100+0	1484.00	2142.00	5230.00
1000+0	840.00	1578.00	2222.00
0+1	2380.00	5142.00	7268.00
0+10	1548.00	2314.00	3333.00
0+100	1032.00	1764.00	2454.00
0+1000	660.00	935.00	1126.00
1+1	1868.00	3142.00	6500.00
10+1	2174.00	529.00	8274.00
1+10	1752.00	2281.00	6282.00
S.Em ±	156.43	141.35	115.90
C.D. (P=0.05)	456.60	414.35	338.31

Proline

Proline has been found to play an important role in stress tolerance mechanism in plants. A considerable amount of literature is available on the accumulation of proline under different stress situation. Satakopan and Rajendran (1989) reported changes in proline levels in germinating groundnut seeds under different stress conditions. Groundnut seeds were soaked water in $1x10^{-4}$ M $CdCl_2$ or $1x10^{-2}$ M $NaCl_2$ or CaCl solutions for 3 hours and germinated for 5 days in filter papers moistened with water or these solutions. Salt stress increased proline accumulation in cotyledons and roots of germinating seeds. The effect of nitrate salts of Cd, Co and Zn in B_5 medium and Pb in distilled water on accumulation of proline in the seedlings of *Cajanus cajan* cv.C-306 was tested under controlled sterile conditions (Saradhi and Saradhi, 1991). A considerable and proportionate increase in proline content was recorded with increase in concentration of heavy metals. Cd was the strongest inducer of proline accumulation and Zn was the weakest. Equimolar K concentration did not show any proportionate change in the level of proline when compared with control. It was therefore inferred that proline accumulation could be used as a marker to test the level of heavy metal pollution. Accumulation of large quantities of proline in seedlings raised in media with lead nitrate, which had a higher osmotic potential compared with those grown in B_5 medium (not supplemented with heavy metal) which had a lower osmotic potential (-0.125 M Pa) suggests that proline accumulation is not related to osmotic adjustment.

Alia *et al*, (1995) experimented the effect of zinc on free radicals and proline in *Brassica juncea* and *Cajanus cajan*. Seedlings were raised in modified B_5 medium supplemented with Zinc sulphate under controlled aseptic conditions and observed that level of proline were low in seedlings raised in concentrations upto 0.1 mM of Zn. However, at higher concentrations, Zn increased accumulation of proline. The shoots of *Cajanus cajan* exhibited higher levels of proline in comparison to *B. juncea* irrespective of the concentration of Zn-sulphate. A correlation between free radicals and accumulation of proline was noticed. When 7 days old seedlings of *Vigna unguiculata* were treated with 10^{-5} M $PbCl_2$ or 10^{-5} M $CdCl_2$, proline content was found to increase along with other parameters (Bhattacharya, 1995). Ameliorating effect of proline has also been reported. Khanna and Rai (1995) noticed that mercury induced inhibition of *Raphanus sativus* seedling growth could be reversed in presence of L-proline, L-histidine and 1-methionine. Tissue showed reduced levels of Hg when proline was supplemented exogenously. Thus effect of proline are related more to its inhibitory effects on Hg uptake rather than on Hg toxicity itself. Mungbean seedlings grown in different concentrations of $HgCl_2$ (1, 10, 100 and 1000 ppm) solutions accumulated proline in leaves and stems. The magnitude of accumulation correlated with concentration of metal (Barman Roy and Bera 2002) (Table 7).

HEAVY METAL POLLUTION AND CROP YIELD

The term yield is used in a very general way as an integrated expression of growth and development. Again, growth and development are complex factors interacting with the environment. In the previous discussions, it has been observed that heavy metals in excess of tolerance limit adversely affect many physiological and biochemical activities of plants.

Table : 7. Individual and combined effect of mercury and manganese (ppm) on proline content of leaf and stem during seedling development in mungbean (μg. g^{-1} fresh weight).

Combinations	2 days		4 days		6 days	
Hg +Mn (ppm)	Leaf	Stem	Leaf	Stem	Leaf	Stem
0+0	14.30	8.87	35.44	22.63	50.86	38.32
1+0	16.76	9.12	39.62	24.76	66.83	45.60
10+0	21.48	11.34	54.84	30.12	82.76	58.74
100+0	26.62	12.22	66.22	38.98	98.34	63.44
1000+0	-	-	-	-	-	-
0+1	14.42	9.66	38.85	22.80	54.35	39.16
0+10	15.96	10.85	41.14	25.62	55.40	40.00
0+100	20.35	11.41	48.36	26.16	65.56	52.38
0+1000	-	-	-	-	77.48	65.36
1+1	17.52	9.00	40.02	23.42	54.82	42.18
10+1	18.16	9.32	42.078	24.04	56.91	45.52
1+10	16.38	9.46	41.26	24.16	56.48	43.65
S.Em ±	1.05	0.97	1.28	1.18	1.13	1.14
C.D. (P=0.05)	3.11	2.86	3.78	3.46	3.34	3.36

Therefore, any attempt to curb the metabolic activities will reduce crop yield. Different heavy metals affect metabolic processes in various ways and there is no generalized pattern of inhibition. Baroccio and Dottori (1984) conducted a pot trial with 6 maize cultivars, 6 durum wheat cultivars and 6 bread wheat cultivars where soil was enriched with 50-400 ppm Zn, 25-200 ppm Cu and 25-100 ppm Cd per kg soil separately. It was observed that decreases in dry matter production, N-content and number of leaves were less for Zn than for the other elements. Differences in response among cultivars were noticed and a tolerance index was determined to detect the effects of metals. In another observation, wheat plants were grown in a field near mylochori, North Greece, part of which was abnormally high in soil Cu, Zn, Pb (207, 9780 and 6050 ppm respectively compared with 58, 135 and 88 ppm in the rest of the field). Plant height, ear length, number and weight of grains/ear, grain length and grain breadth were markedly reduced in the area with high heavy metal content (Babalonas and Karagiannakidou, 1987).

Cadmium, a well known heavy metal pollutant, affect yield of crops by various ways. Soaking *Vigna radiata* L. Wilczek seeds in 2.5×10^{-4} M Cd as cadmium acetate for 12 hours, decreased chlorophyll content, Hill activity and seed yield/plant compared with control (Lata, 1990). Sharma and Sharma (1993) investigated the effect of chromium on dry matter production and yield of maize in sand culture. It was observed that chromium application significantly decreased dry matter production and yield. The reduction being significant at higher levels of Cr^{2+} application especially at advanced growth stages of the crop. Seed formation was completely inhibited by 0.25, 0.5 and 1.5 mM chromium. In a pot experiment with sandy clay loam, 15 rice cultivars were given 0 or 25 mg Cd/kg soil. The grain yield was decreased by 2.7-13.2% and straw yield by 2.2-38.5%. There was also marked variation among cultivars in respect of Cd^{2+} content in the grain and straw (Sarkuna *et al.*, 1995).

Effect of foliar application of cadmium on growth, yield parameters and yield of mungbean was studied by Bindhu and Bera (2000). It was observed that foliar application of 3 $CdSO_4 \cdot 8H_2O$ at concentrations 10^{-4} M, 5×10^{-4} M and 10^{-3} M adversely affected growth, yield parameters and yield of mungbean compared to control. Although growth, yield parameters and yield were found to be influenced at 10^{-4} M $CdSO_4$, higher concentration (10^{-3} M) of $CdSO_4$ proved lethal (Table 8). Stimulatory effect at low concentration might be attributed due to higher metabolic activities of the plant to combat adverse situation and inhibitory effect at higher concentration is thought to be inactivation of enzymes as cadmium strongly binds with SH group of the enzymes.

Table 8. Effect of different concentration of $CdSO_4$ on some yield parameters of *Vigna radiata* cv. Pusa Baisakhi (Data were taken at harvest)

Concentration of $CdSO_4$	Total no. of nodes $plant^{-1}$	Total no. of pods $plant^{-1}$	Average length of pod (cm)	Average weight of pod (g)	Average no. of seeds pod^{-1}	Average weight of seeds pod^{-1} (g)	Yield $plant^{-1}$ (g)
Control	15	21	6.132	0.348	7	0.252	5.292
10-4 M	14	24	6.237	0.376	8	0.271	6.504
5x10-4 M	10	8	5.479	0.283	6	0.214	1.712
10-3 M	8	3	4.283	0.181	4	0.133	0.399
S.Em ±	0.93	0.77	0.19	0.01	0.56	0.00	0.21
C.D.(P=0.05)	2.85	2.37	0.59	0.04	1.72	0.01	0.64

BIO-MONITORIING OF HEAVY METAL POLLUTION

Biological approaches during the last decade for removal and accumulation of heavy metals from aquatic solutions have progressed to a great extent. Utilization of water hyacinth, an aquatic weed for removal of heavy metals like Hg, Cr, Cu, Ni, Zn, Mn, Ca etc. is well recognized. Buddhari *et al.* (1983) observed that aquatic weed *Eichornia crassipes* and *Lemna polyrhiza* has higher capacity to accumulate Pb than terrestrial plants like *Solanum lycopersicum* and *Oscimum sanctum* and more lead was accumulated in root than shoot. An effort has been made for the removal of chromium from tannery effluent using three aquatic weeds *viz*. Water hyacinth, pseudo-water hyacinth and *lemna* species. It was observed that water hyacinth a cumulated more chromium followed by Lemna and Pseudo-water hyacinth. In all the three species, the root accumulated higher amounts of chromium than foliage (Singaram, 1994).

Gaur *et al.* (1994) studied the efficiency of two aquatic vascular plants viz. *Spirodela polyrhiza* and *Azolla pinnata* in accumulating the heavy metals like Cd, Co, Cr, Cu, Ni, Pb and Zn. Both the plants accumulated Ni maximally and Cr was accumulated the least. In another observation Sen and Bhattacharya (1994) reported maximum removal of Ni (II) (20 $\mu g\ ml^{-1}$) from culture solution by *Salvinia natans*. This plant can survive better in a polluted environment caused by Ni (II) compared to other aquatic plants. Thus, *Salvinia* may be used for removal of Ni (II) form aquatic environment and act as an indicator of 'Ni' pollution. Bio-accumulation of Cadmium and chromium were examined using two aquatic macrophytes *viz*. *Hydrilla verticullata* and *Chara corallina* (Rai *et al*. 1995). Both the plants accumulated these heavy metals but more cadmium was accumulated than chromium.

Except aquatic weeds, the role of other bio-materials like leaves, lignocellulosic materials, agricultural wastes, *etc*. as bioabsorbents for heavy metals has not been exploited thoroughly. Such process can be used along with other biological or chemical processes and may prove commercially successful. Agricultural bi-products like spinach leaves, banana husk, saffola husk and onion skin have been found to be highly effective for binding heavy metals from waste water (Nigal and Jahagirdar, 1995). Various biotechnological methods have been developed employing different immobilized cells, microbial films or biomass for the separation of heavy metals. Biotechnological processes for separation of metals couple with subsequent recovery of the biomass and its use as food or feed supplement may prove as an economically and technically viable process.

CONCLUSIONS

Heavy metal pollution effects in plants are not always clearly understood, rather they may be the results of complex interactions between heavy metal in question with other essential or non-essential ions or environmental factors. Although excess of various metals may produce some common effects on plants but there are many specific cases of differential effects of individual metal on different plant species. Stimulation of physiological activities at low concentration and inhibition at higher concentration is a common feature in more or less all the heavy metals studied. Information on screening of corp cultivars for heavy metal tolerance and genetic control mechanism too is scanty which needs to be explored. Once the genetic control mechanism of metal tolerance is identified, it may be possible to combine metal tolerance with other desirable traits to produce plants that are better adopted to high metal soils. Biomonitoring of heavy metal pollution is less expensive but time consuming. Biotechnological approaches for qualitative and quantitative reduction of toxic metal pollutant in conjunction with public awareness and adoption of some safety measures might be advantageous to control toxic hazards of heavy metals. Such an attempt requires close collaboration among Soil Scientists, Plant Physiologists, Plant Biochemists, Plant Breeders, Government and Non-government organizations including Naturalists. Therefore, a multi-disciplinary approach seems to be essential if we really like to combat heavy metal pollution, rather to say any kind of pollution.

REFFERENCES

Aggarwal, N., Laura, J.S. and Sheoran, I.S. 1990. Effect of cadmium and nickel on germination, early seedling growth and photosynthesis of wheat and pigeonpea. *Intl. J.Tropical Agriculture*, **8**: 141-147.

Ali, S.M.E. 1973. Influence of cations on aluminium toxicity in wheat (*Triticum aestivum Vill Host*) Ph.D. Thesis, Oregon State Univ. Corvallis, Ore.

Alia Prasad, K.V.S.K. and Pardha Saradhi, P. 1995. Effect of zinc on free radicals and proline in *Brassica* and *Cajanus*. *Phytochemistry.*, **39**: 45-47.

Avery, N. C. and Sarkar, S. 1991. Studies on the effect of heavy metal on growth parameters of soyabean. *J. Environ. Biol.*, **12**: 15-24

Babalonas, D.and Karagiannakidou, V. 1987. Influence of soil containing heavy metals on cereal crops. *J. Agron. Crop Sci.*, **159**: 356-359.

Bansal, R.L. and Chahal, D.S. 1990. Interaction effect of 'Fe' and 'Mn' on growth and nutrient content of mung (*Phaseolus aureus* L.) *Tro. Agric.*, **18**: 126-132.

Barman Roy, Samadrita and Bera, A.K. 2002. Amelioration of mercurial toxicity by manganese 1. A case study in mungbean seedling. *J.Environmental Biology,* **23**: 321-323.

Barman Roy, Samadrita and Bera, A.K. 2002. Effects of mercury and manganese on amylase, protease and peroxidase enzyme activities in the germinated seedlings of mungbean. *Ad. Plant Sci.,* **15**: 201-206.

Barman Roy, Samadrita and Bera, A.K. 2002. Individual and combined effect of mercury and manganese on phenol and proline content in leaf and stem of mungbean seedling. *J.Environmental Biology,* **23**: 433-435.

Baroccio, A. and Dottori, A. 1984. Tolerance of cereal cultivars to heavy metals. In *Proceedings VIth International Colloquim for the Optimization of Plant Nutrition,* **1**: 35-44.

Bhattacharjee, S. and Mukherjee, A.K. 1994. Influence of cadmium and lead on physiological and biochemical responses of *Vigna unguiculata* (L.) Walp seedlings. I. Germination behaviour, total protein and proline content and protease activity. *Poll. Res.*, **13**: 269-27.

Bhattacharjee, S. and Mukherjee, A.K. 1994. Influence of cadmium and lead on physiological and biochemical responses of *Vigna unguiculata* (L.) Walp seedlings. II. Cell injury, pigment, sugar, nucleic acid content and peroxidase activity. *Poll. Res.*, **13**: 279-286.

Bhattacharya, M. 1995. Heavy metal (Pb^{2+} and Cd^{2+}) stress induced damages in *Vigna* seedlings and possible involvement of phytochelatin like stbstances in mitigation of heavy metal stress. *Indian J. Experimental Biol.*, **33**: 236-238.

Bindhu, S. J and Bera, A.K. 2001. Impact of cadmum toxicity on leaf area, stomatal frequency, stomatal index and pigment content in mungbean seedlings. *J.Environmental Biology,* **22**: 307-309

Bindhu, S.J. and Bera, A.K. 2000. Effect of foliar spray of cadmium on growth, yield parameters and yeld of mungbean. *Environ. Ecol.*, **18**: 969-971.

Bishnoi, N.R., Sheoran, I.S. and Singh, R. 1993. Effect of cadmium and nickel on mobilization of food reserves and activities of hydrolytic enzymes in germinating pigeonpea seeds. *Biol. Plant.*, **34**: 583-589.

Buddhari, W., Virabalin, R, and Aikamphon, K. 1983. *Proc. Intl. Conf. Water Hyacinth*, Feb. 5-11, P.379 (India).

Burzynski, M. and Buczek, J. 1989. Interaction between cadmium and molybdenum affecting the chlorophyll content and effect of some heavy metals in the second leaf of *Cucumis sativus* L. *Acta Physiol. Plantarum*, **11**: 137-145.

Butler, L.K. and Tibbits, T.W. 1979. Stomatal mechanism determining genetic resistance to ozone in *Phaseolus vulgaris* L. *J. Am. Soc.Hort. Sci.*, **104**: 213-216.

Clijsters, H. and Van Assche, F. 1985. Inhibition of photosynthesdis by heavy metals *Photosynth. Res.*, 7: 1-40.

De, B. and Mukherjee, A.K. 1996. Mercury induced metabolic changes in seedlings and cultured cells of tomato, *Geobiosis*, **23:** 83-88.

Epstein, E. and Stout, P.R. 1951. The micronutrient cations iron, manganese, zinc and copper, their uptake by plants from the adsorbed state. *Soil Sci*., **72**: 47-65.

Gaur, J.P., Noraho, N. and Chauhan, Y.S. 1994. Relationship between heavy metal accumulation and toxicity in *Spirodela polyrhiza* (L.) schleid and *Azolla pinnata* R. Br. *Aquat Bot.* , **49** (2-3) : 183-192.

Greger, M and Bertell, G. 1992. Effects of Ca^{2+} on the carbohydrate metabolism in sugarbeet (*Beta vulgaris*). *J.Experimental Botany*, **43**: 167-173.

Greger, M. and Johanson, M 1992. Cadmium effects on leaf transpiration of sugarbeet (*Beta vulgaris*). *Physiologia Plantarum*, **16**: 43-45.

Hemlatha, S., Anburaj, A. and Francis, K. 1997. Effect of heavy metals on certain biochemical constituents and nitrate reductase in *Oryza sativa* (L.) seedlings. *J. Environmental Biol.*, **18**: 313-319.

Kalimuthu, K. and Sivasubramanian, R. 1990. Physiological effects of heavy metals on *Zea mays* seedlings. *Indian J. Plant Physiol.*, **33**: 242-244.

Khan, M.R., Khan, M. W. and Singh, K. 1996. Growth performance of chickpea under the influence of Nickel and cobalt as soil pollutants. *J.Ind. Bot. Soc*., **75**: 193-196.

Khanna, S. and Rai, V.K. 1995, Amelioration of mercury toxicity in radish *Raphanus sativus* L. seedlings by :-proline and other amino acids. *Indian J. Experimental Biol.*, **33**: 766-770.

Kumar, G., Singh, R.P. and Sushila, S. 1993. Nitrate assimilation and biomass production in *Sesamum indicum* L. seedlings in a lead enriched environment. *Water, Air Soil Pollution*, **66** : 163-171.

Kumar, S. and Banerjee, D. 1992. Effect of some heavy metals on *in vitro* activities of certain enzymes. *Plant Physiol.Biochem.*, **19**: 33-35.

Lata, S. 1988. Studies on the pre-treatment effect of cadmium on growth, yield and levels of some biochemical components of *Raphanus sativus Linn. J. Of Sci. Res. in Plants and Medicine*, **9**: 7-12.

Lata, S. 1989. Effect of cadmium on seedling growth, mobilization of food reserves and activity of hydrolytic enzymes in *Phaseolus aureus* L. seeds *Acta Botanica Indica*, **17**: 290-293.

Lata, S. 1990. Effect of cadmium on growth, yield, level of chlorophyll and Hill activity of chloroplasts isolated from *Phaseolus aureus* cv. T44. *Comparative Physiol.Ecology*, **15**: 103-105.

Lata, S. 1991. Effect of cadmium treatment of seeds on growth, yield, level of chlorophyll and Hill of chloroplasts isolated form *Phaseolus aureus* cv. PS-16. *Advances in Plant Sciences*, **2**: 286-290.

Leena, G.D. and Ramanujam, M.P. 1992. Effect of pre-treatment of groundnut seeds with nickel on germination and seedling growth. *Advances in Plant Sciences*, **5**: 627-629.

Lima, M.L. De and Copeland, L. 1990. The effect of aluminium on the germination of wheat seeds. *J.Plant Nutr.*, **13**: 1489-1497.

Lindsay, W.L. 1974. Role of chelation in micro-nutrient availability. In *The Plant Root and its Environment* ed. E.W. Carson pp 525-564. Charlottesville Univ. Press Virginia.

Malik, D., Sheoran, I.S. and Singh, R. 1992. Carbon metabolism in leaves of cadmium treated wheat seedlings. *Plant Physiol.Biochem.*, **19**: 223-239.

Mel' nichuk, Yu, P., Lishki, A.K. and Sobolev, A.S. 1991. Effect of cadmium on degradation of storage substances in cotyledons of germinating pea seed. *Fiziologiya I Biokhimiya Kul' turnykh Rastenii*, **23**: 187-1991.

Mhatre, G.N. and Chapekar, S.B. 1982. Effect of heavy metals on seeds germination and early growth *J. Envir. Biol.*, **3**: 53-63.

Mittal, S. and Sawhney, S.K. 1990. Influence of lead on enzymes of nitrogen metabolism in germinating pea seeds. *Plant Physiology and Biochemistry*, **17**: 75-81.

Mukherjee, C. and S. 1990. Metabolism of germinating rice (*Oryza sativa* L.) seeds as influenced by toxic concentration of cadmium. *Indian Journal of Plant Physiology,* **33**: 190-196.

Nag, P., Nag, P., A. K. and Kukherjee, S. 1989. The effect of heavy metals, Zn and Hg on the growth and biochemical constituents of mungbean (*Vigna radiata*) seedlings. *Bot. Bull. Academia Sinica,* **30**: 241-250.

Nag, P., Paul, A.K. and Mukherjee, S. 1981. Heavy metal effects in plant tissue involving chlorophyll, chlorophyllase, Hill Reaction Activity and Gel electrophoretic patterns of soluble proteins. *Indian J. Exptl. Biol.,* **19**: 702-706.

Nanda, D.R., Mishra, B. B. and Mishra, B.N. 1993. Effect of soild waste from a chlor-alkali factory on rice plants: mercury acculation and change in biochemical variables. *International Journal of Environmental Studies*, **45**: 23-28.

Nandi, S. and Bera, A.K. 1995. Effect of mercury and manganese on seed germination and seedling growth in black gram. *Seed Research*, **23**: 125-128.

Naquib, M.I. and Barakat, N.M. 1989. Interaction of tin and strontium on the carbohydrate and nitrogen component of *Vicia faba leaves Egyptian J.Bat.*, **32**: 45-52.

Nigal, J.N. and Jahagirdar, D.V.J. 1995. Removal of heavy metal ions from water and waste water by using agricultural bi-products. *J.Aqua Biol.*, **10**: 34-36.

Passow, N., Rothstien, A. and Clarkson, T.W. 1961. Toxic metal cadmium, phytotoxicity and tolerance in plants. In *Advances in Environmental Science and Technology* (Vol. 1) by R.K. Trivedy, Ashish Publishing House, New Delhi-110002 : 225-256.

Pillay, S.V., Rao, and Rao, K.V.N. 1996. Effect of nickel toxicity in *Hyptus suaveolens* (L.) *Poit and Helianthus annus* L. *Indian J.Plant Physiol.,* **39**: 153-156.

Poschenrieder, C., Cobot, C. and Barcelo, J. 1983. Influence of high concentrations of cadmum on the growth, development and photosynthetic pigment of *Phaeolus vulgaris*. *Anales de Edafologia Y Agrobiologia*, **42**: 315-327.

Przymusinski, R., Spychala, M. and Gwozdz, E. 1991. Inorganic lead changes growth and polypeptide pattern of lupin roots, *Biochmieund Physiologieder Pflanzen*, **187**: 51-57.

Rai, U.N., Tripathy, R.D., Sinha, S. and Chandra, P. 1995. Chromium and Cadmium bioaccumulation and toxicity in *Hydrilla verticillata* (l.f.) Royle and *Chara corallina* wildenow. *J. Envir. Sci. Hlth.*, **30**: 537-551.

Ratner, E.I., Smirnoy, A.M. and Kuan, K. H. 1962. Importance of molybdenum for growth of isolated roots of Lucerne depending on the acidity of the medium and its content. *Al.Friziol. Rast.*, **9**: 279-288.

Reddy, K.J. 1985. Effect of fluoride toxicity on germination, growth and metabolism of seedlings in 3 cultivars of chckpea. *Indian Botanical Reporter,* **4**: 115-117.

Saradhi, A. and Saradhi, P.P. 1991. Proline accumulation under heavy metal stress. *J. Pl. Physiol,* **138**: 554-558

Sarkunan, V., Misra, A.K. and Mahapatra, A.R. 1995. Effect of cadmium on yield and uptake of cadmium by different rice varieties. *J. Ind. Soc. Soil Sci.*, **43**: 298-300.

Satakopan, V.N. and Bhaskaran, G. and Shankar, M. 1992. Changes in levels of RNA, proline, phenol and ascorbic acid in Al^{3+} toxicity in chickpea seedlings. *Indian J. Plant Physiol.*, **35**: 272-274.

Satakopan, V.N. and Rajendran, L. 1989. Cd. Interaction with RNA in germinating *Arachis hypogaea. Indian J. Plant Physiol.*, **32**: 129-132.

Sen, A.K. and Bhattacharya, M. 1994. Studies of uptake and toxic effect of Ni (II) on *Salvinia natans. Water Air Soil Pollut.* **78**: 141-152.

Shah, K. and Dubey, R.S. 1995. Effect of cadmium on RNA level as well as activity and molecular forms of ribonuclease in growing rice seedlings. *Plant Physiol. Biochem.*, **33**: 577-584.

Sharma, D.C. and Sharma, C.P. 1993. Effect of chromium on growth and biological yield of maize (*Zea mays* L.) cv. Ganga-5. *Indian J. Plant Physiol.*, **36**: 61-64.

Sharma, S.S. 1983. Effect of mercury on germination and seedling growth of *Pisum sativum* cultivars. *Ind. J. Ecol.*, **10**: 78-82.

Shrivastava, G.K. and Singh, V.P. 1986. Effect of cadmium on chlorophyll and carotenoid contents in three varieties of *Abelmoschus esculentus* (L.). Moench. *Plant Physiol. Biochem.*, **131**: 20-24.

Singaram, P. 1994. Removal of chromium from tannery effluent by using water weeds. *Indian J. Envir. Hlth.,* **36**: 197-199.

Stiborova, M., Ditrichova, M. and Brezinova, A. 1987. Effect of heavy metal ions on growth and biochemical characteristics of photosynthesis of barley and maize seedlings. *Biologia Plantarum,,* **29**: 293-298.

Stobart, A.K., Griffiths. W.J., Ameen Bukhari, I and Sehrwood, R.P. 1985. The effect of Cd^{2+} on the biosynthesis of chlorophyll in leaves of barley. *Physiologia Plantarum,,* **63**: 293-298.

Varshney, A.K. 1990. Differential response of *Phaseolus aureus* cultivars to mercury pretreatment on mobilization on 'N' and protease activity during germination and seedling growth. *Geobios,* **17**: 39-42.

Wong, Y.S., Lam, H.M., Dhillone, E., Tam, N.F.Y. and Leung; W.N. 1988. Physiological effects and uptake of cadmium in *Pisum sativum. Environment International*, **14**: 535-543.

Zhang, Z.J., Lu, Q.F. and Tang, F. 1989. Effect of mercury on the growth and physiological function of wheat seedlings. *Chinese J. Environmental Science*, **10**: 10-16.

Section IV: Biological Nitrogen Fixation

Developments in Physiology, Biochemistry and Molecular Biology of Plants, 2005
Eds.: Bandana Bose and A. Hemantaranjan
Vol., 1, pp. 125-158, New India Publishing Agency, New Delhi
E-mail: spjain_niph@rediffmail.com web: www.bookfactoryindia.com

CHAPTER - 7

BIOLOGICAL NITROGEN FIXATION IN PULSES AND CEREALS

J.D.S. PANWAR AND VIJAY LAXMI

INTRODUCTION

After water, nitrogen is the most critical nutritive element for crop productivity and human efforts to produce food and energy have greatly changed the nitrogen cycle of the earth. Sustainable cropping systems throughout history and across the world have relied on the combination of a cereal with an N-fixing legume. In recent history, cereals have dominated global agriculture, while legume areas and productivity have stagnated or even declined. An atmosphere around us contains nearly 78% nitrogen, which is in free form. Hence it is not available to plants. At present, India produces around 206 million metric tones of foodgrains for its growing population, increased foodgrain cannot be produced unless we carefully make use of biological nitrogen. Biological nitrogen fixation is the key to sustain agricultural productivity through the application of biofertilizers in the field. However, there is an urgent need to transfer this technology on the field to farmers and industry by producing these fertilizers on large scale.

Today, global agriculture is at crossroads as a consequence of climate change, increased population pressure and detrimental environmental impacts. New mechanisms must be found for ensuring food security through sustainable crop production systems that supply adequate human and animal nutrition, without harming the agro-ecosystems. Crop diversification is a key component for sustainable cropping systems particularly in resource poor environments and legumes should play a pivotal role in these systems. Thanks to their unique ability to fix atmospheric nitrogen, legumes are able to provide a multipurpose crop commodity, with a myriad of potential benefits on agro-ecosystems and , especially on poor farmer's livelihood. By the year 2050, world population is expected to double from its current level of more than 5 billion. It is reasonable to expect that the need for fixed nitrogen for crop production will also at least double. If this is supplied by industrial sought synthetic fertilizer nitrogen use will increase to about 160 million tons of nitrogen per year, about equal to produce biological process. Nitrogen is an essential plant nutrient. It is the nutrient that is most commonly deficient, contributing to reduced agriculture yields throughout the world. Molecular nitrogen or dinitrogen (N_2) makes up four-fifths of the atmosphere but is metabolically unavailable directly to higher plants or animais. It is available to some species of microorganism through Biological Nitrogen Fixation (BNF) in which atmospheric nitrogen is converted to ammonia by the enzyme nitrogenase, Microorganisms that fix nitrogen diazotrophs. The pulses have the capacity to fix the atmospheric nitrogen by forming the nodulation on roots that protects the nitrogenase enzyme activity. The bacteria infect the roots and nodules formed. The nodulation can be classified as determinate and indeterminate type depending on the longevity and continuity of nodule infection. The determinate types had the non-persistent meristem having finite life span and fixed growth period and the senescence of nodules occurs almost at the same time. Whereas, in the indeterminate types had the persistence meristem the root growth continues and hence provide the new infected zone . Perennial nodules always had indeterminate habit and senescence of nodules was noted at different times.

Yield of pulse and legume crops got benefit from rhizobial inoculation. The most common type of symbiosis occurs between members of the plant family *Leguminosae* and soil bacteria of the genera *Rhizobium, Bradyrhizobium, Azorhizobium, Sinorhizobium* and *Photorhizobium* (collectively called rhizobia). In addition, certain amount of nitrogen is left over in soil which is taken by other plants also. It has been reported that there is more yield of subsequent crops in *Rhizobium*-inoculated fields than in un-inoculated control. Response to Rhizobium inoculation has been amply demonstrated with most of the legumes like pigeonpea, urdbean, mungbean, gram, soybean *etc*. Besides, legume cultivation also leaves behind a naturally nitrogen enriched soil for subsequent cultivars. Scientific data analysis indicates that with *Rhizobium* inoculant usage, fair sum of money can be saved by marginal farmers of India, provided quality tested inoculants are used by them. In this regard, it is recommended that small or large-scale industries for *Rhizobium* inoculants be established for each agro-climatic region of India under the guidance of scientific agencies to ensure quality control. Rhizobia have the ability to induce the formation of nodules on the leguminous plants. In these nodules the bacteria are able to fix nitrogen and provide the host plant with ammonia for its growth, individual bacterial species and strains nodulate a particular set of host plants. The process of nodule formation display striking developmental similarities in different legumes.

Bacteria/Crop	**N Fixed/ha/year**
Legumes	35 t
Free living bacteria	15 kg
Cynobacteria	7-80 kg
Associative bacteria	36 kg
Azolla/Anabaena	45-450 kg
Frankia	2-262 kg
Rhizobium/legumes	24-584 kg

ROLE OF BNF IN PULSE PRODUCTION

India is fully sufficient in terms of the technology of obtaining efficient rhizobial and other cultures from the native populations and also to produce inoculants on large scale to meet the country's requirements and also minimizing the dependence on chemical fertilizers by generating relevant information on BNF both in legumes and cereals through symbiotic, associative and non-symbiotic nitrogen fixing microorganisms and exploiting such knowledge in Indian agriculture. It builds and stabilizes the population of efficient nitrogen fixers in field for improved soil fertility and crop productivity in different cropping systems. Evolve a strategy for improving the nitrogen fixing potential of productive nitrogen fixing strains and systems.

BNF ecominimize the economy of nitrogen fertilizer *vis-à-vis* nitrogen fixers along with incorporation of farmyard manure, legumes and green manure crops biomass. Also minimize the changes in physical, chemical and microbiological properties of soil as a result of continuous application of biofertilizers. There are varieties of nitrogen fixing microorganisms present in the nature. These are broadly divided into three categories *viz.*

(i) Symbiotic microorganism *e.g. Legume-Rhizobium-Symbiosis*;
(ii) Asymbiotic or free living *e.g. Azotobactor, Blue green algae*;
(iii) Associative Symbiosis, *e.g. Azospirillium*

Symbiotic Microorganism e.g. legume-Rhizobium-Symbiosis

Rhizobium : Rhizobia were first classified into seven cross inoculation groups according to their host range. The use of modern methods of bacterial classification such as numerical taxonomy.

Associations between host plants and rhizobia

Plant host	Rhizobial Symbiont
Parasponia (a non-legume, formly called Trema	*Bradyrhizobium* spp.
Soybean (*Glycine max*) (slow growing)	*Bradyrhizobium japonicum*
	Sinorhizobium fredi (fast growing type)
Alfalfa (*Medicago sativa*)	*Synorhizobium meliloti*
Sesbania (aquatic)	*Azorhizobium* (forms both root and stem nodules; the stem have adventitious roots)
Beans (*Phaseolus*)	*Rhizobium leguminosarum* by. Phaseoli; *Rhizobium tropicii*; Rhizobium etli.
Clovers (*Trifolium*)	*Rhizobium leguminosarum* bv. Viciae
Aeschenomene (aquatic)	Photorhizobium (Photosynthetically active Rhizobia)

Nucleic acid hybridization and sequencing through polymerase chain reaction (PCR) has led to the definition of two major groups, commonly known as the fast and slow growing *Rhizobium* and *Bradyrhizobium,* respectively (Elkan, 1992.) Both type almost exclusively infect and form nodules growing *Bradyrhizobium* species can infect a broad range of diverse legume hosts, that form stem nodules, probably associated with adventitious roots, Rhizobia with hydrogenase enzyme (which can split the H_2 formed and generate electrons for N fixation) can improve the efficiency of N fixation . Formation of effectivity of nitrogen fixing nodule is one of the key factors in harnessing the full potential of BNF in pulses. The efficient function of BNF is dependent on rhizobia, host, environment and their interactions.

Infection and nodule formation in legume - rhizobium symbiosis

Nodulation begins when rhizobia attach themselves to epidermal cells. Epidermal cells with immature or as yet unformed root hairs are the usual sites for bacterial penetration/ prior to attachment, communication between the two symbiotic partners is required and a certain minimum period of contact is needed. Infected hairs are always shorter than mature intact hairs, due to marked curling upon infection. At the point of infection, the root hair wall forms a depression that invaginates deeply, forming an infection thread lined by a continuation of the root hair cell wall and membrane (Fig.1).

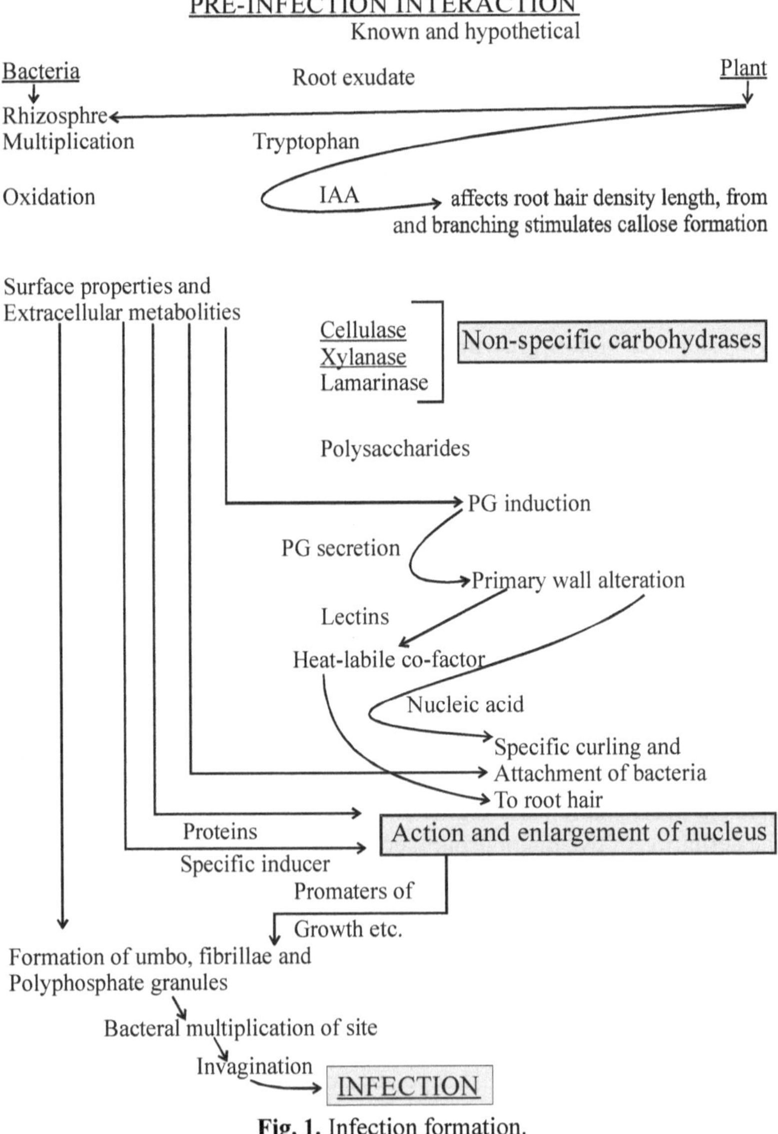

Fig. 1. Infection formation.

Infection threads may branch within a root hair. The infection threads, with its included dividing bacteria, grow 60 to 70 μm of base to the root hair cell. The cortex adjacent to infected root hairs becomes meristematic and produces a wedge-shaped area of dividing cells even before any infection threads enter. These mitosis increase cell number in the cortical layer, which then becomes the main area of infected cells. The combination of multiple threads and branching of threads in the cortex results in penetration of many, but not all, of these cells. The peripheral uninfected area becomes the nodule córtex, which includes a scleroid layer and several vascular bundles. As sometime during following mitotic activity, rhizobia are released into cortical cells through their areas on the tips of the infection threads. *Bradyrhizobia* are called bacteroids after their release into host cell. Their cell walls have been considerably modified or, in the case of peanut entirely removed. Mitosis in infected cortical cells ceases about 14 days after infection. Subsequent increases in the volume of infection tissue are due entirely to cell enlargement. A series of very finely tunned interactions results in the formation of nodules. Specific flavonoid compounds secreted by host plant roots activate the bacterial nodulation (nod) genes. These nod genes are responsible for the synthesis of nod factors that function as signaling molecules which bigger nodulation in host plant. The development of nodule starts with the attachment of rhizobia to the root hair of host plant, which causes deformation and curling of root hair, followed by the formation of infection thread in the curled hair. The root nodule bacteria enters the root through these threads or intercellular spaces. Concomitantly with the infection process, the nod factors induce the cell division in the root cortex, which leads to primordium function. The infected threads grows towards the centre of mitotic activity and enters primordium cells, and bacteria are released into the plant cells to try an endocytotic process. Then the nodule differentiates into mature nodule, which provide the proper environment for the bacteria to fix nitrogen. Successive steps in nodule formation are marked by the expression of nodule specific plant genes that code for plant proteins called nodulins. Depending upon their growth pattern, indeterminate and determinate nodules are formed. As the nodule matures, oxygen binding leghaemoglobin develops gradually in the host tissue and the nodule becomes pink, remaining so until it begins to senesce. As leghaemoglobin forms, bacteria ceases dividing, and dinitrogen fixation commences.

Bacterial attachment to root hairs occurs within minutes of inoculation and is followed within 12 hours marked curling of root hairs. Infection threads, first visible within 24 hours infection, reach the base of the root hair by 48 hours after inoculation. Anticlinal division of the adjacent cortical cells has already occurred, giving rise to nodule primordia. Infection thread penetration of this extensively dividing meristem not observed until 48 hours after inoculation. Anticlinal divisions of the adjacent cortical cells has already occurred, giving rise to nodule primordia. Infection thread penetration of this extensively dividing meristem is not observed until 48 to 96 hours after inoculation. Bacteria are released from the infection thread to form bacteroids within 7 to 10 days after inoculation, A spherical mass of cytoplasmically rich cells, which have been invaded by infection threads, divide and differentiate into the central zone of N_2- fixing cells with 12 to 18 days after inoculation. During nodule development unique proteins are produced, nitrogenase, an enzyme responsible for reduction of free nitrogen (N_2) to ammonia (NH_3) and leghaemoglobin that function in transport of oxygen utilized in ATP production (Fig. 2.)

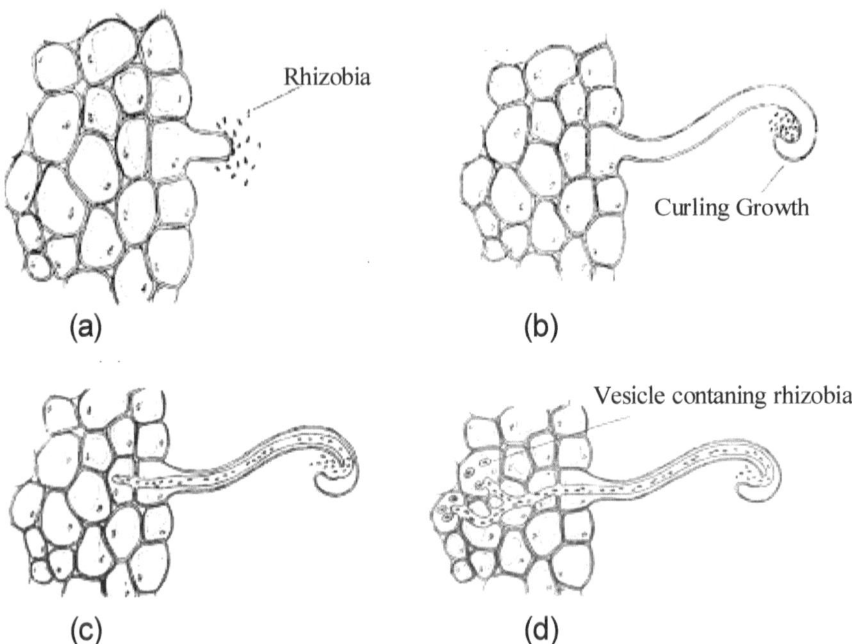

Fig. 2. The infection process in pulse root hairs depicting : (a) Rhizobia being attracted by the emerging root hairs. (b) Abnormal curling growth of root hairs and rhizobia cells proliferate within the coil. (c) After degradation of root hair wall the infection thread reaches in the interior of cell and form a channel. (d) The infection threads extends and reaches target cells and vesicle containing bacterial cells are released into the cytosol.

BNF requires energy, these remind that nitrogen independent of other organisms are called free living. The free diazotophes require a chemical energy source of nonphotosynthetic, whereas the photosynthetic diazotrophs utilize light energy. The free-living diazotrophs contribute little fixed nitrogen to agricultural crops. Associative nitrogen-fixing microorganisms are those diazotrophs that live in close proximity to plant roots *i.e.,* in the rhizosphere or within plants and can obtain energy materials from the plants. They may make a modest contribution of fixed nitrogen to agriculture forestry, but quantification of their potential has not been established. The symbiotic relationship between diazotrophs called rhizobia and legumes (For example, clover and soybean) can provide large amounts of nitrogen to the plant and have a significant impact on agriculture. The symbiosis between legumes and the nitrogen-fixing *rhizobium* occurs within nodules mainly on the root and few cases on the stem.At present the terrestrial input of nitrogen from biological N_2 fixation is held to be in the range of 139-170x10^6tN/year, as compared to 65x10^6tN/year provided by fertilizer nitrogen. There are varieties of nitrogen-fixing microorganisms present in nature.About 16 ATPs are required to reduce one N_2 molecule and evolve one NH_3 and one H_2 molecule in an accompanying reaction. In mungbean, urdbean and other *Vigna* species, the product for assimilation in stems, leaves and other plant parts is ureide-N (glutamine and asparagine).

Leghaemoglobin

Leghaemoglobin constitutes up to 30% of the total soluble host plant protein in nodule and confers the red color. Leghaemoglobins are monomeric proteins, which will reversibly bind oxygen, and are similar in many ways to myglobin, the red oxygen carrying protein bound in mammalian muscle. They have one haem group and can bind one oxygen molecule. They have a very high affinity for oxygen and are half saturated at oxygen concentrations of around 10-20nM. The synthesis of leghaemoglobin is interesting from the point of view of the formation of the functional symbiosis. The synthesis of the leghaem group was rather more difficult to establish. This is because both the bacteria and the plant require the haem group for other proteins such as cytochromes, so that the specific synthesis of haem for leghaemoglobin must be decided on a quantitative rather than an all-or none basis.

Two enzymes that initiate the synthesis of the tetrapyrrole ring, the core of the haem group, are a-aminolevulinic acid synthase (ALAS).

$$\text{Succinyl-CoA+glycine} \Rightarrow \delta \text{ aminolevulinic acid} + CO_2 + H_2O \quad (1)$$

And δ-aminolevulinic acid dehydrase (ALAD);

$$2\delta\text{-aminolevulinic acid} \Rightarrow \text{Parphobilinogen} + H_2O \quad (2)$$

Investigation of the activity of these two enzymes in both bacteroid and plant fractions of the nodule showed that ALAS could be detected only in the bacteroid fraction and that its activity increased as the nodule developed. The second enzyme, ALAD, is found in both plant and bacteroid fractions, but the activity in the plant fraction declines with nodule development while the activity in the bacteroid fraction increases. The bacteroids that synthesize the haem is that ALAD is found only in the bacteroids of nodules where leghaemoglobin is produced and is absent from the bacteroids of ineffective nodules that lack leghaemoglobin. The haem group then must be exported to the plant cytoplasm, where it is united with the globin make the complete molecule.

The haem synthesis is induced by low oxygen tension, after which haem synthesis was increased ten fold. The gene for the synthesis of leghaemoglobin is expressed only in the nodule so that it must require some signal to switch it on. The nature of this is not known.

Regulation of nitrogenase by combined nitrogen

Under natural conditions, ammonium and nitrate are the forms of mineral nitrogen which are most likely to be available to support microbial and plant growth. With the notable exception of *Rhizobium,* uptake of either of these ions can repress both the synthesis and activity of nitrogenase. This is done by some what different routes depending on the ion which is accumulated. Studies with bacteria such as *Klebsiella* or *Azotobacter* have shown that the effect of ammonium on nitrogenase synthesis cannot be explained by a mechanism involving direct repression of the nitrogen fixing *(nif)* genes. For ammonium to act as a repressor the cell must be capable of ammonium accumulation. If this is prevented by treating cultures with methionine sulphoximine, a potent inhibitor of the primary ammonium-assimilating enzyme, glutamine synthetase-then nitrogenase synthesis is not repressed by ammonium. Nitrogenase synthesis in mutants which lack the ability to produce glutamine synthetase is also unaffected by ammonium. Although it was initially thought that glutamine synthetase itself might be directly implicated as a regulatory element for nitrogenase synthesis, further genetic studies now suggest that it is some other, as yet uncharacterized, products of the gene system coding for glutamine synthetase production which, by interacting with the regulatory genes in the *nif* gene cluster, switch *nif* gene expression on or off. In addition to this regulation of the transcription of the *nif* genes, ammonium has also been shown to destabilize *Klebsiella* mRNA, thus diminishing the coding capacity of the cell for nitrogenase (Eady *et al.,* 1981.)Nitrate also represses nitrogenase synthesis, not as might be expected by a direct effect on the *nif* genes or by its reduction to ammonium, but through its reduction to the intermediate product nitrite by the action of nitrate reductase. Evidence for this comes first from observations of the failure of nitrate to repress chlorate-resistant mutants of diazotrophs, which do not show nitrate reductase activity. Nitrogenase synthesis is, however, repressed by nitric in these mutants. Secondly, (Hom *et al.,* 1980) monitored the rate of nitrogenase synthesis in *Klebsiella* by following the appearance of radioactivity in nitrogenase polypeptides after pulse feeding ^{14}C-labelled amino acids, and found that nitrogenase synthesis was not, repressed by nitrate immediately, but that repression coincided with the appearance of nitrite in the growth medium. Repression by nitrate clearly does not involve the same *nif* regulator genes as ammonium repression, but seems to be linked to a redox change in the cell which results from nitrate reduction. The mechanism of nitrate repression of nitrogenase synthesis may be similar to the regulatory effects which oxygen has on the *nif* gene cluster, since nitrate can act as an alternative terminal electron acceptor to oxygen (Eady *et al.,* 1981).

The activity of nitrogenase is not affected directly by nitrate and those effects, which have been observed, can be ascribed to damage of nitrogenase by nitrite. On the other hand, the addítion of ammonium often inhibits the nitrogenase activity of many diazotrophs, the degree of inhibition being dependent upon growth conditions such as oxygen supply and pH.

Location: The location of leghaemoglobin within the nodule cell has been a matter of some controversy. If it is to function as an aid to oxygen diffusion then it should be present throughout the nodule cell. Claims that it is present either only within the peribacteroid membrane or only within the cytoplasm were based on evidence that looked equally convincing.

Function: It has been for a long time that the rate of nítrogen fixation within the nodule is closely correlated within the concentration of leghaemoglobin. This close linkage of leghaemoglobin with nitrogen fixation even led at one time to the suggestion that leghaemoglobin was the nitrogen binding agent. Leghaemoglobin reversibly binds oxygen and its function will be related to this. There are three possibílities, any of which could be functions of leghaemoglobin.

(i) To facilitate oxygen diffusion.

(ii) To act as an oxygen buffer.

(iii) To act as oxygen carrier to specific sites on the bacteroid surface.

NITROGEN ASSIMILATION

Ammonium, the first free product of nitrogenase action, is a toxic compound with a range of effects on metabolism, such as uncoupling oxidative phosphorylation or competing with the cationic cofactors of certain enzymes. The accumulation of ammonium in the microbial cell during nitrogen fixation must therefore be prevented. This is achieved in free-living nitrogen fixers by rapid utilization for growth and in symbiotic organisms by excretion into the host plant cell cytoplasm where it is assímilated into organic form. It is in the form of amino acids, amides or ureides that fixed nitrogen is made available to the host plant for its growth.

Amide

Glutamine synthetase activity is negligible in *Rhizobium* bacteroids and in *Frankia* vesicles, and the ammonia excreted during nitrogen fixation is assimilated into organic form by host plant enzyme. Glutamine, the first product of ammonium assimilation, is usually present in xylem sap collected from amide transporting plants but is not the major nitrogenous constituent. The role is assumed by asparagine, which is more soluble and less metabolically active than glutamine. It has a more favourable C:N ratio of 4:2 (Compared with 5:2 in glutamine) which effectively reduces the energy cost of nitrogen transport. Asparagine synthesis is catalysed by asparagine synthetase, an enzyme for which the K_m for glutamine is at least two orders lower than for ammonia. Consequently it utilizes the amide group of glutamine as the nitrogen donor in a reaction that requires ATP:

glutamine +aspartate+ATP => asparagine +glutamate+AMP +Ppi

Aspartate is generated in a transamination reaction between glutamate and oxaloacetate. It can be seen from the flow chart for asparagine biosynthesis that the equivalent of seven ATPs are required for the synthesis of one aspargine molecule. An additional ATP is required to drive the anaplerotic reaction that incorporates CO_2 to replenish the citric acid cycle. Each carbon atom in aspargine represents a potential loss of six ATPs which could be obtained from its respiration by the citric acid cycle. The total cost of asparagine synthesis is therefore 32 ATP per asparagine molecule, or 16 ATP per ammonia assimilated. There is an additional, unquantified, expenditure of energy for the biosynthesis of the enzymes involved in amide synthesis and also in transport costs (Fig. 3).

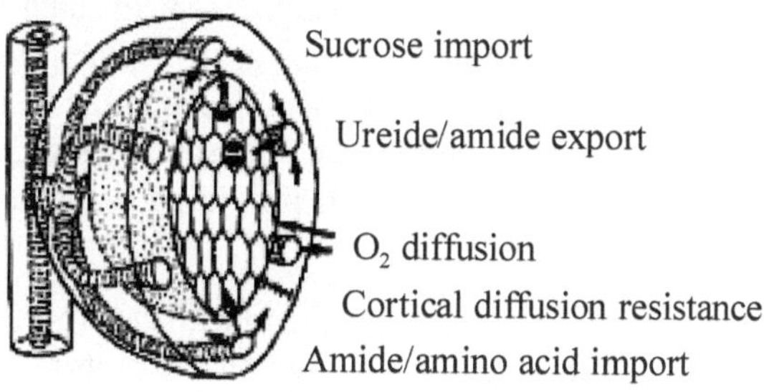

Fig. 3. Plant Root Nodule

Ureides

The major derivatives of urea which have been found in nodulated plants are allantoin, allantoic acid and citrulline. The formulae of these compounds are shown in (Fig.4). The ratio of C: N in the first two compounds is 1.0 which provides a clear advantage over the amides in terms of carbon economy for nitrogen transport. The amides are the more common transport compounds in temperate legumes such as *Pisum, Vicia* or *Lupinus* and in most actinorhizal species. Allantoin and allantoic acid, the hydrolysis product of allantoin, are the favoured transport compounds of tropical and subtropical legumes such as *Glycine, Phaseolus* and *Vigna.* Citrulline is found as a major nitrogenous constituent of a smaller number of nodulated woody species. For example, it is present together with allantoin and allantoic acid in *Albizzia* xylem sap and it is a major amino acid in the nodules of *Alnus.*

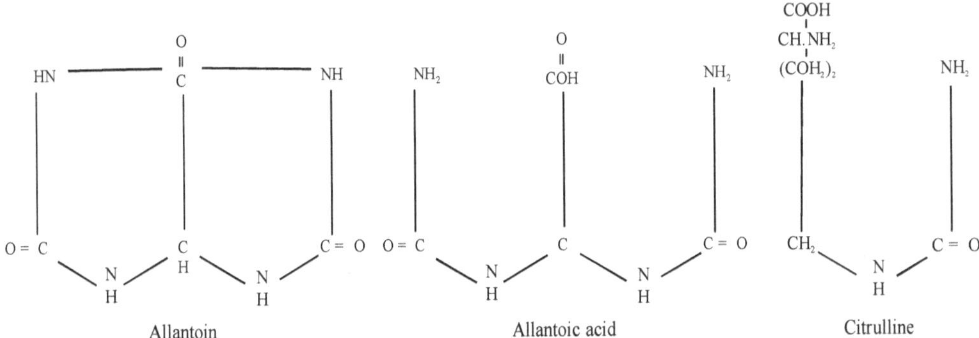

Fig. 4. Structures of allantoin, allantoic acid and citrulline.

Ureide metabolism in tropical legumes

NH_3 is the first stable product of N_2 fixation in legume nodules. It is excreted into the host cell cytosol, where it is assimilated and used in the synthesis of organic nitrogen for transport. Based on the products used for transport, N_2 fixing legumes have been classified as amide exporters or ureide exporters. The amide exporters transport asparagine, glutamine or 4-methylene glutamine and generally belong to the tribes *Vicieae, Genistcae,* and *Trífoliacae* while ureide exporters transport either allantoin (ALN) and allantoic acid (ALA) or citrulline and are members of the tribe phaseolae, Allantoin and allantoic acid account for about 60-90% of the total N in the xylem sap of many tropical legumes including soybean, cowpea, gardenbean, pigeonpea and other legumes grown symbiotically. These legumes contain very high levels of ureide in stems, leaves and developing fruits, where deposition occurs after translocation from xylem. Seeds also contains ureides and in some species their level increases sharply following germination, providing a transportable and readily metabolized nitrogen source for seedling growth.

Synthesis of ureides in nodule is closely associated with the process of nitrogen fixation, as indicated by their rapid labelling an exposure of nodulated roots to labelled N_2. Moreover nodulated legumes contain higher concentration of ureides than non-nodulated. Similarly the presence of nitrate or ammonium in the rooting medium decreases the content of ureides in xylem sap, which further correlates with the decrease in nodule mass and N_2 fixing activity indicating that ureides synthesized in nodules are the primary products of recent N_2fixation in legumes of tropical region.

Physiological significance of ureide

The ureides is more efficient form of N transport for carbon used and the energetics of synthesis. In ureides the ratio of C:N is one where as in amides it is two. The ureides produced are more efficient in use of photoassimilates compared to amides. Cowpea (Ureide producer) used 5.5g C/g N fixed while lupin (amide producer) consume 6.9gC/gN fixed, In terms of ATP consumption (6ATPs/N compared to 9ATP/N for Arginine 13ATPs/N for citrulline) also, ureides are least expensive. The ureide producing plants have the ability to recapture much of the energy costs associated with purine synthesis by coupling purine oxidation and other associated reaction to the reduction of NAD(P). The carbon skeleton associated with these ureides is highly oxidized and thus the cost of these compounds are quite low. Ureides are less soluble and are better suited for storage than amides. Ureide storage may have an advantage in special situations where photosynthetic and/or sink activity is affected hence, storing N in the form of ureides may help in coordinating the rate of protein synthesis in relation to the availability of carbon skeletons from photosynthesis. The molar concentration of amides in the xylem is several fold higher than the corresponding concentration of ALN and ALA in the tropical legumes and as a result, the water use efficiency for N export is higher for amide exporters. Ureides before assimilation are degraded to NH_4 and CO_2. The CO_2 released may be refined by RuBP carboxylase and / or PEP carboxylase . Thus assimilation of ureides may help in raising the internal, level of CO_2 which in turn may optimise photosynthesis rate. The PEP carboxylase activity is much higher in amide exporters than in ureides exporters.

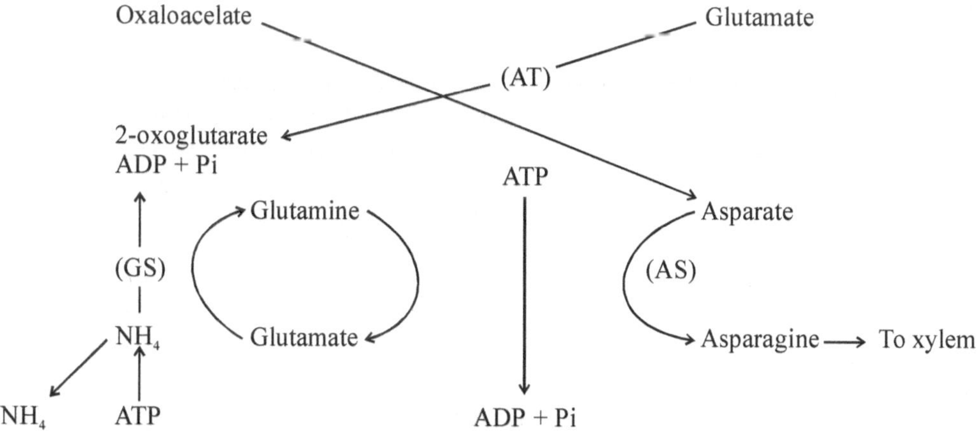

Fig.5. Metabolic flow chart for aspargine biosynthesis. Enzyme are indicated in parentheses : AT, aminotransferase; GS, glutamine synthetase; As, asparagine synthetase.

Combined inoculation

Effect of PSB and PGPR on *Rhizobium* growth: Growth pattern of *Rhizobium* indicated that *Bacillus megaterium* (PSB-1) was found to inhibit the urdbean. *Rhizobium* growth showing clear zone around filter paper discs. Profuse and vigorous growth pattern of *Rhizobium* was observed around discs of PSB-1. The cell free extract of these organisms also gave similar effect on *Rhizobium* growth suggesting secretion of soluble compounds by organisms in the medium as reported by Sarwar and Kremer, (1995).

Plant growth promoting rhizobacteria (PGPR) for increasing crop production

PGPR's were shown to produce auxins and cytokinins by plant bioassay. They also solubilized considerable amount of phosphorous from insoluble sources and increased rice grain yields even in the presence upto 75% and 100% recommended dose of N & P. A new strain of Bacillus (SB-1) more effective in solubilizing P than reference strains of *Pseudomonas striata* or *Bacillus megaterium* var. *phosphaticum* was isolated from the rhizosphere of soybean. Dual inoculation of *Azospirillium lipoferum* and Bacillus SBI significantly improved the grain yield and nutrient content of irrigated rice in a clay loam.

FACTORS AFFECTING BIOLOGICAL NITROGEN FIXATION

Biological N_2 fixation is the largest natural pathway for the introduction of gaseous N into the biosphere. Because of their capacity to symbiotically fix N_2, grain and forage legumes are of special agronomic significance in most cropping systems. It is anticipated that their importance will continue to increase with the increasing development of sustainable agricultural practices. However, symbiotic N_2 fixation by legumes is particularly sensitive to environmental stresses, with salinity and soil drying being two of the most important.

Salinity

The symbiotic nitrogen fixation in legume nodules has been shown to be extremely sensitive to water deficits and salinity stress, which could result in decreasing N accumulation and yield of legume crops under these conditions. The effects of salts stress have been investigated on several legume-*rhizobium* symbiosis. Nitrogenase activity was substantially inhibited by sodium chloride (NaCl), and this inhibition was associated with a significant decrease in plant growth and N content. The effect of salinity on nitrogenase activity has been associated with changes of the nodule permeability to oxygen diffusion. A large genetic variation has been found among legume species and cultivars in the N_2 fixation sensitivity to salinity. Drought sensitivity of N_2 fixation has been compared among several grain legume species. Those species that transport ureides from the nodules were found to be much more drought sensitive than those that transport amides. It was concluded that a feedback mechanism involving ureide level may control N_2 fixation under drought. Consistent with this observation, the drought tolerance of N_2 fixation in soybean was associated with low concentrations of ureides in plant tissues. Regardless of the physiological mechanisms for the inhibition of N_2 fixation, there is now strong evidence that legumes species and cultivars present a large spectrum of genetic variability and can thus, be selected for decreased sensitivity of N_2 fixation to salinity and drought stress. Study of salinity effects on four grain legumes including broadbean, chickpea, lentil and soybean confirmed the effects of soil salinity on crop yield, total nitrogen uptake and nitrogen fixation (VanHoorn *et al.*, 2001).

The existence of inter-and intra-specific-variability in the sensitivity of N_2 fixation to salinity has been reported in legumes (Serraj *et al.*, 1998a; 2001). Overall, the existence of genetic variability in the sensitivity of N_2 fixation to salt among legume species and cultivars, may be useful to further elucidate the NaCl inhibition of symbiotic N_2 fixation and to select optimal *rhizobium*-legume symbiosis for agricultural production in soils subjected to salinity. Salt tolerant *Bradyrhizobium japonicum* strain SRAU-1 for soybean was at par with the most effective reference strain ABR-1 in a saline soil. Similarly performance of salt tolerant *Azospirillum* STA-1 was similar to reference strain Sp-7.

Drought

Symbiotic nitrogen fixation is highly sensitive to drought, which results in decreased N accumulation and yield of legume crops. The effects of drought stress on N_2 fixation usually nave been perceived as a consequence of straight forward physiological responses acting on nitrogenase activity and involving exclusively one of three mechanisms; carbon shortage, oxygen limitation, or feedback regulation by nitrogen accumulation. The sensitivity of the nodule water economy to the volumetric flow rate of the phloem into the nodule offers a common framework to understand each of these mechanisms. As these processes are sensitive to volumetric phloem flow into the nodules, variations in phloem flow as a results of changes in turgor pressure in the leaves are likely to cause rapid changes in nodule activity. This could explain the special sensitivity of N_2 fixation to drying soils. It seems likely that N feedback may be especially important in explaining the response mechanism in nodules. A number of studies have indicated that a nitrogenous signal (S), associated with N accumulation in the shoot and nodule, exists in legume plants so that N_2 fixation is inhibited early in soil drying. The existence of genetic variation in N_2 fixation response to water deficits among legume cultivars opens the possibility for enhancing N_2 fixation tolerance to drought through selection and breeding. The effects of drought stress on N_2 fixation were perceived during the 1970s and 1980s as a straight forward physiological process focused only on carbon metabolism and nodule permeability to O_2. Since that time, nitrogen feedback has also been hypothesized as being involved in N_2 fixation regulation. A common factor that links these three factors is the sensitivity of phloem flow is decreased early in the development of soil water deficit. Due to the high sensitivity of nodules to phloem volumetric flow into the nodules, there are a number of possible consequences resulting in a high sensitivity of N_2 fixation to soil drying. Decreased phloem flow would probably decreased the import of carbon to the nodule, although photosynthate supply during water deficits has not been shown to be a major limitation to nodule activity. A lower rate of water flow in the nodule has the potential directly to influence nodule P_0 because of alterations in the configuration of the cortical cells and airspaces. The dependence of nodule export on the water flow into the nodule through the phloem also has the potential to have major influence on the accumulation of N export products in the noduie. In particular, an accumulation of N products in the nodules opens the possibility of an N feedback on nodule activity.

A clear distinction has been found among legume species between amide and ureide exporters in terms of N_2-fixation response to drought. Tropical legumes, which transport ureides appear to be more sensitive to soil drying. This includes some species such as cowpea and pigeonpea, which have the reputation of being generally tolerant to drought in terms of plant survival is likely that N may be the major limiting factor for plant growth and yield capacity under drought. Because genetic variation has now been documented among legume cultivars, there exists the opportunity genetically to alter the sensitivity of N_2 fixation to drought in commercial cultivars. Inoculation of mung and black gram with drought tolerant strains of rhizobia gave additional grain production of 69 and 45 kg/ha respectively in dryland conditions. *Rhizobium* inoculation saved the starter dose of 20 kgN/ha.

Temperature

High temperature

High root zone temperature have been shown to strongly affect bacterial infection and N_2 fixation in several legume species including soybean (Munevar and Wollum, 1982), guar (Arayangkoon *et al,* 1990). Peanut (Kizhinevsky *et al.,*1992) and beans (Hungria and Franco, 1993; Pina and Munns, 1987 and Panwar *et al.,* 1988). Elevated temperatures may delay nodule initiation and development and interfere with nodule structure and functioning in temperate legumes, whereas in tropical legumes N_2 fixation efficiency is mainly affected. The effect of high temperature depends on plant species, cultivar and on strain of *Rhizobium*. Hernandez-Armenta *et al.,* (1989) found that transferring nodulated bean plants from 26 to 35°C daily temperature markedly inhibited N_2 fixation. Soybean plants appear somewhat more tolerant, nitrogen fixation only being severely inhibited by day time temperature greater than 41°C and enhanced at 36°C (La Favre and Eaglesham, 1987). Root temperatures above 30°C have been shown to reduce nitrogenase activity in Peanut (Nambiar and Dart, 1983). Peanut *bradyrhizobium* symbiosis was completely inhibited by nodulation and inability of the nodule function at this temperature even they were formed. It is well known that respiration is increased with higher temperature, and this may imply that less carbon will be available for the symbiosis. Temperature is one of the chief factors affecting the survival of rhizobia. In many tropical soils, the surface temperature reaches and occasionally exceeds 40°C. Under these conditions, the numbers often decline readily.

Low Temperature

Low temperature delay root hair infection and decrease nodulation and nitrogenase activity (Waughman, 1977). Low temperature was also shown adversely to affect nodule development in *Trifolium* spp. *Stylosanthes* spp. and *Vicia faba* while not altering the sequence of anatomical changes (Fyson and Sprent, 1982). Effective nodule development and nitrogenase activity were completely inhibited at 8°C root zone temperature conditions decreased N_2 fixation activity by directly decreasing the activity of the nitrogenase enzyme complex (Layzell *et al.,* 1984) and by suppression and for delaying root infection and nodulation (Walsh and Layzell, 1986). The greater degree of sensitivity to low root zone temperature below 17.5°C was due to events related to or occurring before infection thread penetration to the base of the root hair (Xhang and Smith, 1994). Lie (1974) and Fyson and Sprent (1982) concluded that low temperatures delayed bacteriod tissue formation. The increased individual nodule mass and reduced specific nodule activity observed at low temperatures have been attributed to inhibition of nodule cell differentiation and bacteroid tissue development (Fyson and Sprent, 1982; Pankhurst and Layzell, 1984). In soybean-*Bradyrhizobium japonicum* symbiosis suboptimal root zone temperatures slowed down nodule development by disruption of signal exchange between the host plant and *Bradyrhizobium* (Zhang and Smith, 1997). The prior incubation of *Bradyrhizobium* with genistein has been shown to increase nodule number at lower root zone temperatures and the effect of genistein application on nodule number and N_2 fixation decreased with increasing root zone temperature (Zhang and Smith, 1995). Lower temperature resulted marked reduction in nodule efficiency and ARA per nodule dry weight in soybean. More of the N fixed at lower root temperature accumulated in the root, presumably due to decreased N translocation (Lynch and Smith, 1993). Cultivar varieties in low temperature response is also evident in the results of Rennie and Kemp (1981) who showed that the bean variety Kentwood failed to nodulate at 10°C and took 32 days to initial nodulation at 12°C, respectively. Strain differences in nodulating ability at low temperatures have been shown between arctic and temperate zone rhizobia (Ek-jander and Fahraeus, 1971; Schulman *et al.,* 1988).

Soil acidity

Soil acidity affects all aspects of nodulation and nitrogen fixation from survival and multiplication of the rhizobia in the soil, through infection and nodulation, to nitrogen fixation. The failure of legumes to nodulate under acid soil condition is common especially in soils with pH less than 5.0. The ínability of some rhizobia to persist under such conditions is one of the cause of nodulation failure (Granam *et al.,* 1982; Lowendorf and Alexander, 1983). (Pijnenborg *et al.,* 1991) suggested that the major factor contributing to the nodulation failure and the stunted growth of Lucerne in soils with pH values below 5.0 was the poor survival of R. *meliloti.* However, chickpea root nodule bacteria appeared to survive quite well under acidity even down to a soil pH of 4.2 and this was further shown by the enhanced lateral nodulation but, the limiting factor might be the inability of the host plant to tolerate acid soil (Scott and Howieson, 2000). Species of *Rhizobium* were known to differ in their tolerance to soil acidity; the slow growing *Bradyrhizobium* strains being generally more acid tolerant than fast growing especially *R.meliloti* (Munns and Keyser, 1981). (Rossum *et al.,* 1994) observed that the soil acidity enhanced nodulation in groundnut particularly when nodulation was caused by the indigenous *Bradyrhizobium* population. However, the enhancement was affected by groundnut cultivars. The acid tolerant and sensitive cultivars differed in their content and concentration of root exudates (Ramaq, 1992).

Elevated atmospheric CO_2

Elevated CO_2 is likely to affect carbon cycling and crop productivity by stimulating photosynthesis. Hartwig *et al.,* 1996 postulated that legumes, which can fix atmospheric N_2, would have an advantage over other plants. Legumes could respond to a potential C:N imbalance caused by elevated CO_2 by introducing N through symbiotic N_2 fixation one obvious hypothesis in predicting the response of symbiotic N_2 fixation to elevated atmospheric CO_2 would be that as the rate of photosynthesis increases more photosynthates would be delivered to the nodules and thus nitrogenase activity increases (Hardy and Havelka, 1976). The long-term increase of total nitrogenase activity would be the nitrogenase activity per unit nodule dry weight was not greater than that in the control (Finn and Brum, 1982; Ryle *et al.,* 1992). The CO_2 enhanced plant growth may result in an increased N-sink strength (Ingestad, 1982). As a result nitrogenase activity would be unregulated to meet the increased N demand, a concept by several workers (Hartwig and Nosberger, 1994; Hartwig *et al.,* 1994; Oti-Boateng and Salisbury, 1993; Parson *et al.,* 1993).

Low rhizosphere CO_2 concentration (below 100 ppm) result in significant decline in nitrogenase activity of legume nodules (Bethenod *et al.,* 1984). However, increasing the partial pressure of CO_2 (pCO_2) in the rhizosphere above ambient did not stimulate the nitrogenase activity of soybean. Therefore, direct affects of altered rhizosphere pCO_2 on nitrogenase activity are unlikely. Nitrogenase activity is rather oxygen than carbon limited (Hunt and Layzell, 1993).

Agrochemicals

The use of pesticides has become an essential part of agriculture. Pesticides have different effects on symbiotic systems. For example, the fungicides carbondazim has been found to have no significant effect on number of nodules in cowpea but increased the fresh nodule weight and total N content (Singh *et al.,* 1986), Yet, the same compound decreased the number of *Bradyhizobium japonicum* surviving on soybean seeds and decreased nodulation and yield in the field. But, when associated with oxine copper, carbondazim had no detrimental effect on *B. japonicum* or soybean growth (Catrooux and Arnand, 1991). Golebiowska *et al.,* (1967) reported that *Rhizobium* strains displayed different sensitivities to the fungicide thiram. When associated with carbondazim, thiram had no negative or positive effects on chickpea nodulation (Welty *et al.,* 1988).

The insecticide dimethoate (1%) reduced nodule number by 63% black gram and 64% in pea. Leghaemoglobin concentration, nitrogenase activity and size of the bacterial region were also reduced by higher rates of dimethoate (Soam and Agarwal, 1989). At lower concentration, insecticide phorate enhanced the activity of nitrogenase.

NODULATION IN CEREAL

Cereals crops have constantly increased during the 20th Century to reach remarkable levels in many areas. This increase in yield has been primarily facilitated by the introduction of genetically improved cultivars and use of chemical pesticides and fertilizers. Meanwhile, productivity of low input agriculture on marginal lands has stagnated or even declined. A critical challenge for the future of world food security is to ensure dramatic increases in crop yields to crop with the massive increase in global world population. Throughout history and across the world, civilization has depended on cropping systems that combined a cereal with a legume. In recent history cereals productivity has dramatically increased while legume production has neared a plateau, stagnated or even reduced, thus, there is clearly an urgent need to enhance legume yields for the direct benefit of protein production and the indirect effects on the productivity and sustainability of agriculture as a whole. Improvements in cereal grain production culminating in the 'Green Revolution' resulted from the selection of nitrogen-responsive crop varieties with large reproductive structures to accommodate more grain and increased nitrogen (N) through provision of fertilizer. Nitrogen is the soil nutrient element needed in greatest quantity by crops. Although abdundantly available in the air (78% of the atmosphere), plants cannot directly utilize elemental gaseous N. Biological N^2-fixation occurs naturally in legumes and certain non-legume woody species that form a symbiotic association with nirogen fixing bacteria. Total fertilizer use increased 10-fold between 1950 and 1990, per capita fertilizer use quintupled, whereas global crop land per capita decreased by nearly 50% over the same time interval. Cereal crops like rice, wheat and millets are the staple food for most of the countries in the world and they provide the major source of carbohydrates. The nitrogen requirements for these crops are the highest particularly after the introduction of high N responsive varieties and only wheat and rice received any substantial amount of these fertilizers in most of the developing countries. In 1978 the fertilizers use on arable land was 115 kg/ha in developed and 39 kg/ha in developing countries (FAO Year Book Fertilizers Development Corporation, 1978). The nitrogen requirement is being increasing every year and inspite of the increased output as chemical fertilizers, the gap between the demand and supply is increasing. Previous to that, nitrogen fixation by legumes was important in sustaining soil nitrogen balance, but in many traditional systems, particularly in tropics, legumes contributed little to soil nitrogen economy.The grain crops often have low leveis of nitrogen fixation and may even deplete soil N (Dart, 1986) left by legumes. Field data indicated that associative nitrogen fixation could potentially contribute agronomically significant amounts of nitrogen (more than 30-40 kg N/ha/year) to the N nutrition of plants in tropical agriculture including sugarcane and forage grasses (Chalk, 1991). The induction of the nodules/paranodules in cereals is an exiting field and many laboratories in different corners of the world are engaged in inducing the nodulation in cereals including wheat, maize, rice *etc*. Though the present literature is meager in this field, however an attempt has been made to cite the literature in the light of development made during the last 10 years in the field of nodule development in cereals. The role of associative symbiosis in cereais and the new technology developed in inducing nodulation and its effect on crop growth and development has been reviewed.

Nitrogen fixation is the process of making the unavailable dinitrogen to available form of nitrogen to the plants, *viz.,* nitrate and ammonia. This can be achieved either through chemical or biological processes. Biological nitrogen fixation is carried out by symbiotic enzymic processes involving nitrogenase enzyme or through beneficial effects of associative nitrogen fixation. The symbiotic process is enzymic process involving fixation of N_2 by nitrogenase enzyme in the nodules of host plants by the bacteria. Only prokaryotes are able to fix molecular nitrogen and can grow on atmospheric nitrogen. Though the nitrogen fixed symbiotically is the most important from agricultural point of view, still most studies were carried out with free living bacteria because of economy of space, time and labour. The associative fixation also envisages a fair amount of fixed nitrogen and enables the plant in uptake of various mineral elements and better physiological growth. In addition to their N_2 fixing capacity, legumes are extremely important in human and animal diets. They supply 33% of global human protein intake. Legume such as groundnuts and soybeans are important sources of oil, and they are also valuable sources of unique phytochemicals, which seem to promote heart health. Recent evidence from Africa is suggesting that the intensification of grain legumes within the farming systems resulted in significant increase in household health standards.

PARANODULATION

Nodule morphology and anatomy

The exogenous application of the synthetic auxin 2,4-D to rice and wheat and Triticale induces modified root outgrowth (MROs). These are probably modified lateral roots (Elanchezian and Panwar, 1999) rather than specialized structures, with carbon reserves similar to those found in the cortex or roots. They results from the induction of meristems initiated at or close to the root pericycle. The initiation of multiple meristems in condensed zones or the root may be a result of the plant initiating sink of auxin in order to reduce auxin levels. However, at certain specific concentration of externally applied 2,4-D, the initiated meristems are found to be so close together that they often fuse and may result from the interlateral root spacing being reduced to zero (Rolfe *et al.,* 1994).

Nodule-like structures showed distinct zoning of tissue and internal development of proto-vascular elements, which had a clearly defined differentiated tissue of the typical pattern of vascular bundle. Several kinds of phenomenon were observed in rice roots with the addition of 2,4-D (Ridge *et al.,* 1993) such as (i) Fusion of lateral roots emerging from the root in close proximity to each other; (ii) Formation of structures that outwardly resembles the nodules of legumes, termed nodule like structures; and (iii) Structures that grow from the root surface but had a callus like surface rather than an epidermis. Internal examination of the latter two structures by light microscopy showed that the nodule like structures were out growths with a number of different cell and tissue types present and that structures with a callus like surface have only slightly differing cell types within the body of the structure.

The ultimate aim of establishing endophytic interaction between diazotrophs and non-legumes is that the diazotrophs should fix nitrogen and transfer this fixed nitrogen to the plant. The establishment of endophytic nitrogen fixation in cereals would be one of the most significant contribution of agriculture (Cocking *et al,* 1994). Previously low levels of nitrogenase activity upto 112 n moles of ethylene accmulated over 24 hours , have been reported in experiments using wheat *cv.* Wembley inoculated with *A. caulinodans* strain ORS 571 (Cocking *et al.,* 1995). Recent findings have shown that when wheat plants are grown in pots in growth chambers, high levels of acetylene reduction activity can be detected

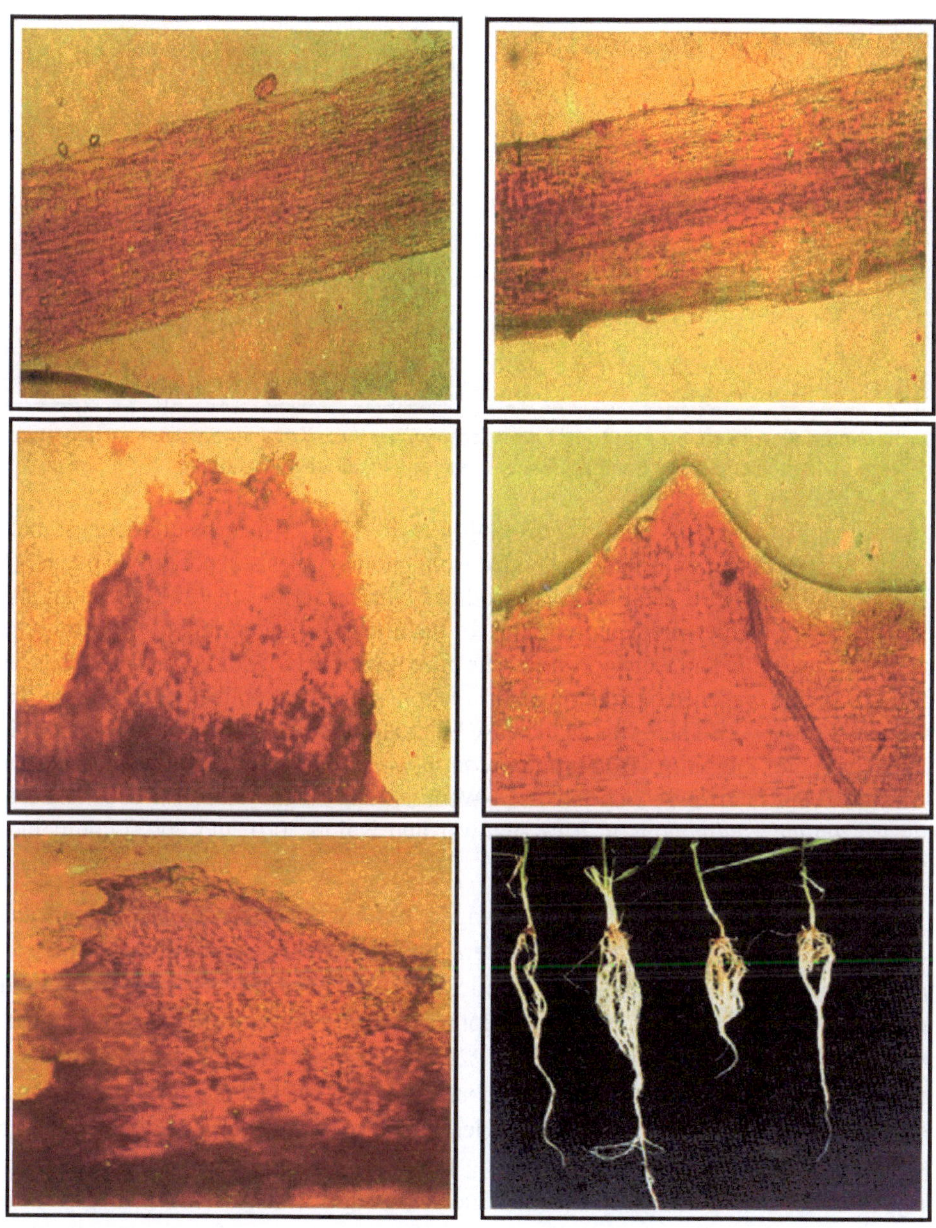

Fig. 6a. Transverse section of roots as affected by 2, 4-D and *Azospirillum brasilense* treatments (200 x magnification) showing untreated control (top left); roots inoculated with *Azospirillum brasilense* (top right); roots treated with 2, 4-D showing nodular outgrowth (middle left); nodule initation in 2, 4-D and *Azospirillum brasilense* treated roots (middle right); well developed nodule with 2, 4-D and *Azospirillum brasilense* inoculation (bottom left); and comparative root growth of different treatments, *viz.*, uninoculated control, *Azospirillum* inoculation, 2, 4-D treatment and *Azospirillum* combined treatment respectively from left to right (bottom right)

Fig. 6b. Nodulation in rice using 2, 4-D and *Azospirillum brasilense* in nitrogen for medium (b) nodules showing the crack entry and presence of bacteria inside the paranodule.

in plants inoculated with *A. caulinodans* (Sabry *et al.,* 1997). These results suggest that an adequate supply of carbon is being provided from photosynthesis in pot grown plants. One of the possibilities for N_2 fixation in paranodules with bacteria is that the deformations of roots may provide an O_2-free environment for the efficient functioning of the nitrogenase enzyme (Elanchezian and Panwar, 1999). The nitrogenase activity of *A. brasilense* has been reported with the presence of bacteria inside the root steler portion in the pearlmillet (Tilak and Subba Rao, 1987). There was an increase N_2 fixation in rice at low oxygen tension (1.5-3.0 kpa) by tumor inhibiting *Azorhizobium caulinodans* (Christiansen-Weniger, 1996), and by *Azospirillum brasilense* (Christiansen-Weniger, 1977) *Azospirullum brasilense* and *Arhizorhizobium caulinodans* in Rice (Panwar and Naidu 2001 and 2003) compared to untreated control.

The scientist all over the world are engaged in studying the possibilities of enhancing nodulation and nitrogen fixation process in cereals using different strains of N fixing bacteria and various hormonal studies. The Genetics of nodulation in cereals is not yet fully understood. There is a need to go into depth and possibilities are being highlighted in achieving the success in this direction. *Rhizobium* inoculation of legumes helps in increasing yields by 15-30% with residual benefits of about 30-40 kg N/ha. Inoculation with *Azotobacter* and *Azospirillum* in cereals gave 10-15% yield increase with benefits equivalent to about 10-15 kg N/ha primarily through promoting plant growth. Co-inoculation of *Rhizobium Azospirillum,* VAM and PSB has been found to be significantly better than their single inoculation. A rich germplasm collection of effective strains of agriculturally useful microorgasnisms are now available which are currently being commercially exploited. A range of stress (temp-, salinity) tolerant microorganisms and strains efficient in fixing N in the presence of recommended dose of N fertilizer have been developed. The local isolates are more effective than the best of the introduced isolates showing the importance of the ecological adaptation of the strains in a given habitat and the difficulty of introducing an organism in an alien environment.

Azospirillum

Nitrogen fixing bacteria of the genus *Azospirillum* have been isolated from the roots of many grasses including important crops such as rice, wheat and maize (Tarrand *et al.*, 1978). In the case of rice. *A lipoferum* and *A. brasilense* have both been isolated from the roots and stems (Baldani and Dobereiner, 1980, Ladha *et al.,* 1982) and *A. amazonense* has been isolated from the roots (Pereira *et al.,* 1988). Due to their ability to colonize root surface and interior of many cereals and forage grasses, the first report showing the presence of *Azospirillum* in cells of cortex, in intercellular spaces between the cortex and endodermis and in the xylem cells of maize roots by applying the tetrazolium reduction staining technique (Patriquin and Dobereiner, 1978) was accepted. (Baldani *et al.,* 1993) and Vermeiren *et al.,* (1998) have presented microscopical evidence as to the endophytic nature of *Azospirillum* in rice. (Christiansen-Weniger 1997) has reported the colonization of 2,4-D induced *Para nodules* by an ammonium excreting mutant of *A. brasilense.* They may have initially entered the cracks that had formed when the tumors emerged from the epidermis. Incorporation of biologically fixed N and enhancement of crop yields of cereals by inoculation with nitrogen fixing bacteria have been observed in many field experiments.

Plant growth promotion by *Azospirillum* has been demonstrated in field (Baldani *et al.,* 1987) and in greenhouse experiments (Gaskins *et al.,* 1977) and attributed to several mechanisms including nitrogen fixation (Van Bulow and Dobereiner, 1975: Smith *et al.,* 1976) and auxin production (Tien *et al.,* 1979). Wheat plants inoculated with *A. brasilense* showed better root development (Panwar and Sirohi, 1989: Panwar *et al.,* 1990). Christiansen-Weniger and VanVeen (1991) showed that an ammonium excreting mutant of *A. brasilense* (Wa 3) promoted better growth of wheat plants as compared with the wild type. They also concluded that another strain (C3) was able to transfer the nitrogen fixed directly to the maize plant and that the amount transferred was increased in para-nodulating plants obtained by treatment with 2,4-D (Christiansen-Weniger, 1994). Ability of the *Azospirillum* bacterial nitrate reductase to help in the incorporation of the nitrogen assimilated from soil by the plant has been demonstrated (Boddey *et al.,* 1986; Ferreira *et al.,* 1987). Enhanced photosynthetic rate, NRA, chlorophyll content and grain yield were reported in wheat inoculated with VAM and *Azospirillum* (Panwar, 1991, 1992) even under stress conditions (Panwar, 1993). Despite these different mechanisms exerted by facultative endophytic diazotrophs in association with graminaceous plants, increases in yield in the range of 5-30 per cent have been observed in several inoculation experiments with *Azospirillum* (Baldani *et al.,* 1983; Okon and Gozalez, 1994). Nodulation perse a physiological process and were induced using plant growth regulators (2, 4-D, NAA and IBA) in nitrogen free growth in rice *cv.* Pusa 834(Panwar and Naidu, 2003). The Bacterial cultures *Azospirillum brasilense* (SP7) and *Azorhizobium caulinodans* (ORS 571).

Nodule induction were used for 2, 4-D (0.5 ppm) was better than NAA (8ppm) and IBA (8 ppm) each. The nitrgenase activity and leghaemoglobin content was found in the 2, 4-D+ *A. brasilense* in nitrogen free growth medium. The bacterial entry was through the crack between the nodules and root . The vascular connectivity with central system in the root was observed . The colonization of nodule by *A. culinodans* was more than *A. brasilense.* After transplanting to pots the treated plants showed higher chlorophyll content, Ps rate and stomatal conductance and better Harvest Index. 2-4D + Ab/Ac showed maximum growth than with IBA and NAA. The enhanced N content in grain and straw due to inoculation confirm the N fixation and its translocation to different plant parts.

The treatment of roots of rice seedlings with synthetic auxins such as 2, 4-D (0.5 ppm),NAA (8ppm) and IBA (8ppm) resulted into the formation of nodule like structures *(para-nodules)* Naidu and Panwar (2003). These *Para-nodules* were colonized by *Azorhizobium caulinodanse* which penetrated the *Para-nodule* by migrating in between loosely arranged cells that covered their surface or by penetrating the space at the junction of root and *Para-nodule* (crack entry). Nitrogenase activity was enhanced in *para-nodulated* roots inculated with *Azorhizobium caulinodans* due to the increased internal bacterial colonization within the *para-nodules,* which provided the niche that protected the bacteria from the higher concentration of oxygen on the root surface. Bacteria of genus *Azospirillum* are free-living nitrogen (N_2) - fixing rhizobacteria that are found in close association with plant roots and are able to promote plant growth and increase yield. The objective of this work, over the last years, was to identify maize-endophyte associations with increased plant productivity compared with uninoculated controls. In maize, root colonization by *Azospirillum* takes place mainly in the root elongation zone. Inoculation increases the density and length of the root hairs, as well as appearance and elongation rates of lateral roots, thus increasing the root surface area. Enhanced nitrogenase activity and legheamoglobin were observed in maize roots inoculated with *Azospirillum,* compared to uninoculated control. This is due to the efficiency of the host to encapsulate the intracellularly colonizing bacterial cells with a membrane layer, resembling the peribacteriod membrane of the legume nodule. The photosynthetic rate, stomatal conductance, chlorophyll content, NR activity, grain yield and N content in grain and stover were also found to be higher in inoculated plants. It is assumed that the bacteria affect plant growth mainly by the production of plant growth promoting substances, which lead to an improvement in root development and an increase in the rate of water and mineral uptake, thus increasing the yield. Thus , the gramineous plants are potentially able to create a symbiosis with diazotropic bacteria, promoting a higher level of N_2- fixation for the better growth and deveiopment of the plants(Sakia *et al.,* Research over the last few years has shown that inoculation with nitrogen-fixing bacteria of the genus *Azospirillum* and *Azorhizobium* presents an alternative for (or supplement to) chemical fertilization, mainly due to the bacteria's capability of producing plant growth promoting hormones. The strain of *Azospirillum brasilense* Sp7 and *Azorhizobium caulinodans* ORS 571 were shown to have different abilities to colonize the root interior of an Indian maize cultivar. Whereas *Azorhizobium caulinodans* strain ORS 571 was only detectable on the root surface, *Azospirillum brasilense* strain Sp7 was able to penetrate to outer layers to colonize the root endophytically. (Vanita *et al.,* 2003) found that a plant growth promoting effect was clearly visible in all examined inoculated plants, although there was no significant difference between plants colonized by *Azospirillum* strain Sp7 or *Azospirillum* strain ORS 571. The microscopic examination of root sections showed that Azospirillum strain Sp7 has the potential to penetrate and grow inside the roots as an endophyte unlike *Azorhizobium* strain ORS 571. *Azospirillum* therefore, creates potentially better symbiosis in the form of *Para-nodules* and promotes a higher level of nitrogen fixation for better growth and development of plants with reduced requirement for chemical fertilizers.

Performance of lac Z marked thermotolorant mutant of Azospirillum in field

The thermotolerant diazotrophs were effective on pearlmillet, cotton, mustard and wheat. *Azospirillum* was more effective than *Azotobacter*. For example, ALP-3 inoculation of *A. lipoferum* in wheat in the field improved tillering and grain yields although the effects were statistically not significant. Addition of FYM at recommended dose with 30 kg N did not increase the yield over 30 kg N alone. Similar results were obtained at 60 kg N dose also. Addition of N as top dressing did not improve performance. Yield increases were of the order of 5% at 90 and 120 kg N/ha, where as it was 15% at lower doses. Efficacy of *Azotobacter* inoculation in the presence of various levels of fertilizer nitrogen and FYM (15 t/ha) was evaluated. Nitrogen saving per hectare by inoculation with *Azotobacter* was about 8-15 kg/ha in wheat and 6-17 kg N/ha in maize. The population of *azotobacter* in rhizosphere ranged from 18×10^4 cells/g soil in /Azotobacter+FYM treatment. There were no significant differences in tillering or 1000 grain weight upon inoculation.

Azotobacter strains for maize-wheat system

Azotobacter are free living microorganisms, and they grow in the rhizosphere and fix atmospheric nitrogen non-symbiotically and make it available to particularly cereals. In addition, these bacteria produce growth-promoting substances enhancing the plant growth and finally yield. A highly efficient strain of *Azotobacter* is grown in the laboratory either as shake culture or using fermenter. It has been reported that rice and wheat was increased from 5-31% and 16-30% respectively, following application of *Azotobacter* in the field. From the field experiments with various *Azotobacter* inoculants on onion, brinjal, tomato, cabbage, sugarcane, oat, barley, maize and potato indicate 7-12% increase in crop yields. *Azotobacter* species increase plant yield primarily by fixing molecular nitrogen in soil, but it is also reported to synthesize auxins, vitamins, growth substances and antifungal antibiotics which have beneficial effect of this inoculant on seed germination. It has been reported that inoculation of rice with *Azorhizobium caulinodans* along with the growth hormones 2,4-D or synthetic auxin.

Cynobacteria

Cynobacteria enter into a much wider range of association than rhizobia. Because of their ability to fix N_2 independently of the environment provided by the plant, they can be seen as more likely candidates than rhizobia to form productive associations with cereals. Cyanobacteria were also found in the intercellulaer spaces of *para nodules* tissue in agreement with findings for *Azorhizobium* (Yu and Kennedy, 1995) and *Azospirillium* (Zemen *et al.*, 1992). Gantar and Elhai (1999) have demonstrated that wheat *para nodules* can be readily colonized externally and to some extent internally be the cyanobacterium Nostoc spp. Strain 2S9B and that such coexistence provide favourable conditions for N_2 fixation. The contributions of cynobacterial BNF, based on acetylene reduction assay (ARA), are estimated to be 10-80 kg N/ha crop averaging about 30 kg N/ha/crop.

Alcaligenes

Diazotrophic members of this genus have been consistently isolated from the rhizosphere of wetland rice (Fujii *et al.,* 1987, Oyaizu Masuchi and Komagata, 1988, Vermeiren *et al.,* 1999). Recent studies have demonstrated the endophytic colonization of rice roots by the species *Alcaligenes faecalis* (Zhou and You, 1988, You and Zhou, 1989, You and Song, 1994) and the dominant strain appears to be *A. faecalis* A 15 (You and Zhou, 1989, You *et al.,* 1991, 1995). Experiments under sterile conditions with *A. faecalis* showed that these bacteria upon inoculation attach themselves to the rice root surface, particularly on root hairs near the axis of lateral root with main root (You and Song, 1994). Scanning electron microscopic studies revealed that the inoculated bacterial cells were able to invade rice roots through epidermis and colonize the intercellular spaces mainly in cortex and secondry xylem. More reliable evidence of the endophytic occurrence of the strain A 15 has come from the study of Vermeiren *et al.* (1998) which has shown that this strain will colonize the root surface, particularly at the root tips and lateral root junctions and will also enter the epidermal cells. Within the epidermal cells, the bacteria have been shown to express *nif H* (Vermeiren *et al.,* 1998). Inoculation of genetically engineered strain of *A. faecalis* which constitutively express *nif* A both in pot and field conditions resulted in an increase in grain yield of rice in the range of 5-8 per cent. It was shown that the inoculation with these genetically modified strains increased effective N_2fixing rate by about 13-20 per cent compared with wild type strain.

Azolla is a small water fern that assimilates nitrogen in symbiotic association with *Anabaena azollae.* The algal symbiont resides in the leaf cavity of *Azolla* and provides fixed nitrogen and other growth promoting substances to the host *Azolla* in exchange for nutrients and carbon sources. The heterocyst frequency of symbiont *A. azollae* under nitrogen limiting condition was 30 per cent. The algal symbiont showed polymorphism in different *Azolla* cultures. The growth of *Azolla* is influenced very much by season, phosphorous, growth regulators, herbicides and salinity. *Azolla* mineralized rapidly and its nitrogen is made available to the rice in a very short period, It is estimated that nitrogen contribution by *Azolla* is 40-60 kg ha^{-1} besides improving the soil health. *Azolla* inoculation at 500 kg ha on 7 DAT along with 75 KgN ha^{-1} increased the grain and straw yield of rice significantly. Inoculation of *Azolla* hybrid AH-C2 coupled with USG at 75 kg N ha^{-1} increased the grain and straw yield of CO43 rice besides increasing the plant N content and N uptake in sodic soil.

Burkholderia

A diazotrophic species of Burkholderia was isolated from rice fields in Vietnam and subsequently named *B. vietnamiensis* (Gills *et al.,* 1995). As *B. vietnamiensis* was isolated from rhizosphere macerates (root+adhering soil) however, it can not yet be discribed as an endophyte (Baldani *et al.,* 1997). On the other hand, diazotrophic strains of a species of Burkholderia (*e.g.* strains M 130 and M 209) have been isolated from the interior of rice roots, stems and leaves in Brazil. Preliminary results with optical microscopy suggest that they can enter rice roots via cracks in the root epidermis, particularly at the points of emergence of secondary roots, and subsequently massively colonize their interior (Baldani *et al.,* 1997). Rice seedlings when inoculated with some strains of Burkholderia *spp.* isolated from rice plants contributed a relatively high levels of N to rice via BNF This bacterium appears to have potential as a rice inoculum as it has resulted in significant yield increase after field in Vietnam. Van (2000) reported that the inoculation of rice with *B. vietnamiensis* TVV 75 significantly increased several yield components resulting in a final 13-22 per cent increase in grain yield. Van *et al.,* (2000) have also reported an increase in average grain weight. This increase is indicative of an effect of inoculation on very late yield components also. It shows that inoculation positively affected grain filling where nitrogen fertilizer had no effect on this late yield component. This indicates the long lasting effect of inoculated bacteria. This has already been observed in rice field inoculation studies by Omar *et al.,* (1989, 1992).

Acetobacter

Acetobacter diazotrophicus is one of the newly discovered endophytic diazotrophs isolated from leaves, stems and roots of sugarcane (Cavalcante and Dobereiner, 1988). This bacterium was isolated only from sugarcane not from the roots of other weeds growing in the same field nor in non-rhizosphere soils suggesting its specific interaction with sugarcane (Boddey *et al.,* 1991, Li and Mc Rae, 1991). This bacterium had been reported to be present in relatively high numbers inside sugarcane plants reaching upto 10^6-10^8 cells/g fresh weight of tissue (Baldani *et al.,* 1994 and Reis jr. *et al.* 1995). On the root surface, the bacteria were concentrated at the root tips and around lateral root junction and infected the plants at these sites. The confirmation that *A. diazotrophicus* colonize the intercellular spaces of sugarcane stem parenchyma comes from the work of Dong *et al.* (1994) who showed the presence of the bacteria in numbers of 10^4 cells/mL of fluid inside the apoplast. Inoculation of rice at an early seedling stage (just emergent) showed that rice seeds were not heavily colonized; however, the portion of roots just after two days of inoculation (Sevilla and Kennedy, 2000). Some but not all the roots hairs were also found colonized , and as in sugarcane *A. diazotrophicus* was also found concentrated in junctions where lateral roots were emerging (Sevilla and Kennedy, 2000). This again suggests that cracks entry is one of the ways by which *A diazotrophicus* can get into the plant. It was also common to find *A. diazotrophicus* lined up along the grooves of the epidermal cells of the root. Similar observations were reported by Rolfe *et al.,* (1997) who examined toluidine blue stained roots of rice grown with 2,4-D and inoculated with *A. diazotrophicus.* This may represent another way for bacterial entry in addition to crack entry (Sprent and Faria, 1989). Another unique characteristic of this bacterium is that its ability to excrete part of the fixed nitrogen into the medium as demonstrated by Cojho *et al.,* (1993). Sevilla and Kennedy *et al.,* (2000) suggested that *A. diazotrophicus* could enhance rice growth and that growth promotion might be related to the transfer of biologically fixed N although other factors such as auxin production could be involved. When the rice plants were grown in the presence of N, there was no significant increase in the height of inoculated plants compared with uninoculated plants, suggesting that *A. diazotrophicus* did not influence plant growth when fixed N was sufficient.

Azoarcus

This obligate endophytic diazotroph has been isolated from kaller grass *(Laptochloa fusca)* grown in saline sodic soils of Pakistan. Based on the results of a polyphasic study in which morphological, nutritional and biochemical features and several molecular biological techniques were employed, the genus *Azoarcus* was proposed with two species *A. communis* and *A. indegenous* (Reinhold-Hurek *et al.,* 1993). It appears to enter kaller grass and rice via root tips and lateral root junctions infecting the cortex region and penetrating the steel of roots and spread to upper parts of the plant via transpiration stream (Hurek *et al.,* 1994: Reinhold-Hurek, 1998a, b). In both the plants, *Azoarcus* colonize mainly the lysigenous aerenchyma of the root cortex, within which it can express the Fe protein of nitrogenase-encoding genes while within and in association with rice roots. Hurek *et al.,* (1997) have detected *nif* HDNA sequence of genus in Japan (Ueda *et al.,* 1995). The presence of *Azoarcus* within wild rice in Philippines and in Nepal has recently been confirmed (Hurek *et al.,* 2000). *Azoarcus* is also able to colonize the interior of sorghum plants (Stein *et al.,* 1997) by means of its cellulytic enzyme (Reinhold-Hurek *et al.,* 1993). The high O_2 pressure in the arial parts of the sorghum did not affect the nitrogenase activity of the highly oxygen tolerant organism, *Azoarcus*.

FUTURISTIC APPROACH

Nitrogen Fixation in pulses and cereals is being dealt at length in different corners of the world. It is very encouraging thing to note that bacterial strains used for Nitrogen fixation can fix nitrogen which is abundant in the atmosphere without any loss to the environment (position of water, air) and hence, they are eco-friendly and less expensive but still it needs to strengthen the research to screen the efficient strains that can nodulate and fix nitrogen in varied agro-climatic zone. They should have the capacity against the native bacteria.

The biological nitrogen fixation in cereals is a new interesting field and leaves ample scope for further strengthening the research on the nodule development and infection process that can be transformed from the present day level as a Lab-model to the field. There is also need to use efficient strains that can fix nitrogen as it makes shift from associated symbiosis to the symbiosis between nitrogen fixing bacteria and cereal under varying agro-climatic zone. The use of physiological tool for inducing nodules can only be beneficial if there can be transformation to its genetic advancement for using any biotechnological tool. If we get success in achieving the nodulation in cereals and its symbiosis in nitrogen fixing bacteria it will revolutionize in achieving this novel target. A group of multidiscipliany scientist may prove beneficial. It needs more comprehensive research focusing on the make shift on lab to land and improve the symbiotic relationship of genetically advance cereals with efficient strain under field condition.

Genomics is revolutionizing research and commerce in the life science, and offers to support the development of a new paradigm in the genetic improvement of legumes that will substantially augement the impact of BNF technologies. Following the whole genome sequencing of *Arabidopsis,* it became clear that a number of model hubs would be required throughout the plant kingdom to stimulate rapid impacts across important crop groups. Based on the remarkable synteny between grass genomes, the sequencing of the rice genome is expected to stimulate a cascade of substantial scientific implications for a wide variety of cereal crops. Amongst the legume species, the most economically import crop, soybean, has failed to take a similar lead. Instead the USA, EU and Japan have committed tens of millions of dollars for the large-scale genome sequencing of two new model systems: *Medicago truncatula* and *Lotus japonicus.* These two species have quickly become model system for structural and functional genomics in the legumes. These species were initially chosen for their importance to research on symbiotic N_2 fixation. However, these two species have quickly become models for the analysis of a wide variety of agronomically important traits. This research will generate a global understanding of the expression, regulation, dynamics and evolution of an array of agronomic characters including those related to symbiotic associations. By virtue of genetic synteny, this wealth of knowledge in the model legume crops will faciliate leap frogging progress in the lesser studied legumes that are só important to agricultural systems, particular in marginal cropping regions. However, a critically important issue here is that developing countries have the opportunity to carry out their own biotechnology research to resolve their own problems in their own way.

REFERENCES

Amutha, K. and Kannaiyan, S.1999. Effect of 2, 4-D, NAA and inoculation with *Azorhizobium caulinodans* on induction of lateral rootlets and paranodules in rice, *Indian J. Micróbiol.,* **39**: 125-128.

Arayangkoon, T., Schomberg, H.A. and Heaver, R.W., 1990. Nodulation and N_2 fixation of guar at high root temperatura. *Plant Soil,* **126**: 209-213.

Baldani, V.L.D., Baldani, J.I. and Dobereiner, J. 1983. Effects of *Azospirillum* inoculation on root infection and nitrogen incorporation in wheat, *Can.J. Microbiol.,* **29**: 869-881.

Baldani, J.I., Caruso, L, Baldani, V.L.D., Goi, S.R. and Dobereiner, J. 1997. Recent advances in BNF with non-legume plants. *Soil Biol Biochem,* **29**: 911-922.

Baldani, V.L.D., Baldani, J.I. and Dobereiner, J. 1987. Inoculation of field grown wheat *(T. aestivum)* with *Azospirillum* spp. in *Brazil. Biol. Fertil. Soils,* **4**; 37-40.

Baldani, V.L.D. and Dobereiner, J. 1980. Host plant specificify in the infection of cereais with *Azospiriilum* spp. *Soil Biol, Biochem.,* **12**: 433-439.

Baldani, V.L.D., Gori, S.R., Baldani, J.I. and Dobereiner, J. 1995. Localization of *Herbaspiríllum* species and *Burkholderia* spp. in rice root system. In: International Symposium on Microbial Ecology. **7** : pp. 133. Brazilian Society for microbiology, Santos, Sao Paulo.

Baldani, V.L.D., James, E. K., Baldani, J.I. and Dobereiner, J. 1993. Colonization of rice by the nitrogen fixing bacteria *Herbaspirillum* spp. and *Azospirillum brasilese.* In : New horizons in Nitrogen Fixation (Eds. R. Palacios, J. Mora and W.E., Newton), Kluwer Academic Publishers, The Netherlands, pp. 705.

Boddey, R.M. and Dobereiner, J. 1995. Nitrogen fixation associated with grasses and cereals. Recent progress and perspectives for future. *Fertilizer Res.,* **4**: 1-10.

Boddey, R.M., Baldani, V.L.D., Baldani, J.I. & Dobereiner, J.1986. Effect of inoculation of *Azospirillum* species on the nitrogen assimilation of field grown wheat. *Plant Soil,* **95**: 109-121.

Boddey, R.M., Urquiaga, S., Reis, V. and Dobereiner, J. 1991. Biological nitrogen fixation associated with sugarcane. *Plant Soil,* **137:** 111-117.

Bergersen, F.J.(1982) Root Nodules of Legumes: Structure and Functions. Research Studies Press, Chichester.

Bethenod, O., Prioul, J.L and Deroche, M.E. 1984. Short-term inhibition of acetylene reduction activity by low CO_2 concentrations around the nodulated root of various legumes. *Physiol,* Veg., **22**: 565-570.

Catroux, G. and Arnand, F. 1991. Compatibility of a soybean peat inoculant with some seed applied fungicides and microgranular insecticides. *Toxicol. Environ. Chem.,* **30:** 229-239.

Chalk. P.M. 1991. The contribution of associative and symbiotic nitrogen fixation to the nitrogen nutrition of non-legumes. *Plant Soil,* **132**: 29-39.

Christiansen-Weniger, C. 1996 Endophytic establishment of *Azorhizobium caulinodans* through auxin-induced root tumors of rice. *Biol. Fetil. Soils,* **21**:293-302.

Christiansen-Weniger, C. 1997. Ammonium excreting *Azospirillum brasilense* C3: GUS A inhabiting induced root tumors along stem and roots of rice. *Soil Biol. Biochem.,* **23**: 943-950.

Christiansen-Weniger, C. and Van Veen, J.A. 1991. Ammonium excreting *Azospirillum brasilense* mutant enhancing nitrogen supply of a wheat host. *Applied Environ. Microbiol.,* **57**: 3006-3012.

Christiansen-Weniger, C. and Vanderleyden, J. 1994. Ammonium excreting *Azospirillum spp.* Become intercellularly established in maize para nodules. *Biol. Fertil. Soils*, **17**: 1-8.

Cocking, E.C, Kothari, S.L. Batchelor, CA, Jain, S., Webster, G., Jones, J., Jatham, J. and Davey, M.R. 1995. Interaction of rhizobium with non-legume crops for symbiotic nitrogen fixation in *Azospirillum* and related microorganisms. NATO ASI series. Vol. G-37, eds. I. Fendrik, M. Del Galio, J. Vanderleyden and Mode Zamaroczy, pp. 197-205. Springer Verlag, Berlin, Germany.

Cocking, E.C., Webster, G., Bachelor, C.A. and Davey, M.R, 1994. Nodulation of non-legume crops-A new look. Agro-Food Industry, Hi-Tech. January/February 21-24.

Cojoh, E.H., Reis, V.M., Schenberg, A.C.G. and Dobereiner, J, 1993. Interactions of *Acetobactor diazotrophicus* with an amylolytic yeast in nitrogen free batch culture. *FEMS Microbiol Lett.,* **106:** 341-346.

Dart, P.J. 1986. Nitrogen fixation associated with non-legumes in agriculture. *Plant Soil,* **90:** 303-304.

Dong. Z., Canny, M.J., Mc Cully, Mm.E., Roboredo, M.R., Cabadilla, C.F., Ortega.E. and Rodes, RR. 1994. A nitrogen fixing endophyte of sugarcane stems. A new role for the apoplast. *Plant Physiol.,* **105:** 1139-1147.

Eady, A., Kahn, D. and Hawkins, M. 1981. Metabolic control of *klebsiella pncumoniar* mRNA degradation: significance to nif gene regulation by NH_4+. In *Current Perspectives in Nitrogen Fixation* (eds. A.H. Gibson and W.E. Newton), C.S.I.R.O. Canberra, pp. 443.

Ek-Jander, J. and Fahraeus, G. 1971. Adaptation of *Rhizobium* to subarctic environment in Scandinavia. *Plant Sci.,* (Spl. Vol.): 129-137.

Elanchezhian, R. and Panwar, J.D.S. 1999. Effect of 2, 4-D *and Azospirillum brasilense* on nodulation and nitrogen fixation in wheat. *Indian J. Agric. Sci.,* **69**(1): 58-61.

Elkan.G.H. 1992. Current Issue in Symbiotic Nitrogen Fixation. *Plant and Soil* **186:** 85-87.

Ferreira.M.C.B., Fernandez, M.S. and Dobereneier, J. 1987. Role of *Azospirillum brasilense* nitrate reductase in nitrate assimilation by wheat plants. *Biol. Fertil. Soil,* **4**: 47-53.

Fujii, T., Huang, Y.D., Higashitani, A., Nishimura, Y., Lyama, S. Hirota, Y., Yoneyama, T. and Dixon, RA. 1987. Effect of inoculation with *Klebsiella oxytoc* and *Enterobacter cloacae* on dinitrogen fixation by rice-bacteria associations. *Plant Soil.,* **103:** 221-226.

Finn, G.A. and Brum, W.A. 1982. Effect of atmospheric CO_2 enrichment on growth, non structural carbohydrate content and root nodule activity in soybean. *Plant Physiol.,* **69:** 327-331.

Fyson, A. and Sprent, J.L. 1982. The development of primary root nodules on *Vicia faba* grown at two root temperatures. *Ann. Bot.,* **50:** 681-692.

Gantar, M. and Elhai, J. 1999. Colonization of wheat para-nodules by the N_2 fixing cyanobacterium Nostoc sp. Strain 2S9B. *New Phytol.,* **141**: 373-379.

Gaskins, M.H., Umali-Gracia, M., Tien.T.M. and Hubbell, D. H. 1977. Nitrogen fixation and growth substances produced by *Spirïllum lipoferum* and their effects on plant growth. *Plant Physiol.,* **59**:128.

Gills, M., Van, V.T., Bardin, R., Goor, M., Hebber, P., Williams, A., Segers, P., Kerster, K., Heulin, T. and Fernandez, M.P. 1995. Polyphasictaxonomy in the genus *Burkholderia* leading to an amended description of the genus and position of *Burkholderia vietnamiensis* isolates from rice in Vietnam. *Internat. J.Systemic Bacteriol.,* **45**: 274-289.

Golebiowska, J., Kaszsubiak, H. and Pajouska, M. 1967. Adaptation of *Rhizobium* to thiram. *Acta Microbiol. Pol.,* **16**: 153-158.

Graham, P.H., Viteri, S.E. and Mackie, F. 1982. Variation in acid soil tolerance among strains of *Rhizobium phaseoli. Field Crops Res.,* **5**: 121-128.

Hardy, R.W.F. and Havelka, U.D. 1976. Photosynthate as major factor limiting nitrogen fixation by field grown legumes with emphasis on soybean. *In:* Symbiotic N_2 fixation (Ed. P.S. Nutman). pp. 421-436. Cambridge Univ. Press, Cambridge, U.K.

Hom, S.S., Hennecke, H. and Shanmugam, K.T. 1980. Regulation of nitrogenase biosynthesis in *Klebsiella pnecumoniae* : Effect of nitrate. *J.Gen. Microbiol,* **117**: 169-179.

Hartwig, U.A. and Nosberger, J. 1994. What triggers the regulation of nitrogenase activity in forage legume nodules after defoliation. *Plant Soil,* **169**: 109-114.

Hartwig, U.A. Heim, 1., Lusher, A. and Nosberger, J. 1994. The nitrogen sink is involved in the regulation of nitrogenase activity in white clover after defoliation. *Physiol. Plant,* **92**: 375-382.

Hernandez-Armenta, R., Wien, H.C. and Eagelsham, A.R.J. 1989. Maximum temperature for nitrogen fixation in common bean. *Crop Sci.,* **29**: 1260- 1265.

Hungria, M. and Franco, A.A. 1993. Effects of high temperature on nodulation and N_2 fixation by *Phaseolus vulgaris. Plant Sci.,* **149**: 95-102.

Hunt, S. and Layzell, D.B. 1993. Gas exchange of legume nodules and the regulation of nitrogenase activity . *Ann. Rev. Plant Physiol. Plant Mol. Biol.,* **44**: 483-511.

Hurek, T., Egener, T. and Reinhold-Hurek, B. 1997. Divergence in nitrigenase of *Azoarcus* spp. Proteobacteria of the B subclass. J. *Bacteriol.,* **179** : 4172-4178.

Hurek.T. and Reinhold-Hurek, B. 1994. Identification of Azoarcus species grass-associated diazotrops by analysis off partial 16S rRNA sequence. *In:* Nitrogen Fixation with Non-legumes (N.A. Hegazi, M., Fayez, and M. Monib, eds.), pp. 59-68, American University in Cairo Press, Cair. Egypt.

Hurek, T., Reinhold-Hurek, B.. VanMontagu, M.and Kelleberg, 1994. Root colonization and systemic spreading of *Azoacus* sp. Strain BH 72 in grasses. *J. Bacteriol.,* **176** : 1913-1923.

Hurek, T., Tan, Z., Mathan, N., Egener, T., Englehard, M., Prasad, G., Ladha, J.K. and Reinhold-Hurek, B. 2000. Novel nitrogen fixing bacteria associated with the root interior of rice. In: The Quest for Nitrogen Fixation in Rice (Eds. J. K. Ladha and P.M. Reddy). International Rice Research Institute, Manila, Philippines, pp. 47-61.

Ingestad, T. 1980. Growth, nutrition and nitrogen fixation in grey alderat varied rates of nitrogen addition. *Physiol, Piant.,* **50**: 353-364.

Ingestad, T. 1982. Relative addition rate and external concentration: driving variables used in plant nutrition research. *Plant Cell Environ.,* **5** : 443-453.

Kishinevsky, B.D., Sen, D. and Weaver, R.W. 1992. Effect of high root temperature on *Bradyrhizobium-peanut* symbiosis. *Plant Sci.,* **143:** 275-282.

Ladha, J.K., Barraquio, W.L. and Watanabe, I. 1982. Immunological techniques to identify *Azospiríllum* associated with wetland rice. *Canadian J.Microbiol.,* **28** : 478-485.

La Favre, A.K. and Eaglesham, A.R.J. 1987. Effects of high soil temperatures and starter nitrogen on the growth and nodulation of soybean. Crop *Sci.,* **27***:* 742-745.

Layzell, D.B., Rochmann, P. and Canvin, D.T. 1984. Low temperatures and nitrogenase activity in soybean. Can. *J. Bot.,* **62**: 965-971.

Li, R.P. and Mc Rae, J.C. 1991. Specific association of diazotrophic Acetobacter with sugarcane. *Soil Biol.Biochem.,* **23**: 999-10002.

Lie, T.A., 1974. Environmental effects on nodulation and symbiotic N_2 fixation. In: The Biology of Nitrogen Fixation (A. Quispel, ed.), pp. 555-582, North Holland Publishing Co. Inc., Amsterdam.

Lowerdorf, H.S. and Alexander, M. 1983. The identification of *Rhizobium phaseoli* strains that are tolerant or sensitive to soil acidity. *App. Environ. Microbiol.*, **45:** 737-742.

Lynch, D.H. and Smith, D.L. 1993. Soybean nodulation and N_2 fixation as affected by exposure to a low root-zone temperature. *Physiol. Plant.,* **88:** 212-220.

Munns, D.N. and Keyser, H.H. 1981. Tolerance of rhizobia to acidity, aluminium and phosphate. *Soil Sci. Soc. Am. J.,* **34**: 519-523.

Naidu, V.S.G.R. and Panwar, J.D.S. 2003. Nitrogenase activity and endophytic establishment of *Azorhizobium caulinodanse* in auxin induced root tumors of rice .Abstract: 2nd International Congress of Plant Physiology, N.Delhi, pp. 620.

Nambiar, P.T.C, and Dart, P.T. 1983. Factors influencing nitrogenase activity of root nodules of groundnut. *Peanut Sci.,* **10**: 26-29.

Okon, Y. and Labandera-Gonzalez, C. 1994. Agronomic application of *Azospirillum.* An evaluation of 20 years worldwide field inoculation. *Soil Biol. Biochem.,* **26:** 1591-1601.

Omar, N., Berge, O., Shalaan, S.N., Hubert, J.L. Heulin, T. and Balandreau, J. 1992. Inoculation of rice with *Azospirillum brasilense* in Egypt. Results of five different trails between 1985-1990. *Symbiosis*, **13:** 261-289.

Omar, N., Heulin, T., Weinhard, P., Alaa-el-Din, M.N. and Balandreau, J. 1989. Field inoculation of rice with in vitro selected plant growth promoting rhizobacteria. *Agronomia,* **9**: 603-808.

Oti-Boateng, C. and Salisbury, J.H, 1993. The effect of exogenous amino acid as acetylene reduction activity of *Vicia faba* cv. Forid. *Ann. Bot.,* **71**: 71-74.

Oyazu-Masuchi, Y. and Komagata, K. 1988. lsolation of free living nitrogen fixing bacteria from the rhizosphere of rice. *J. Gen. Appl. Microbiol.,* **34:** 127-164.

Pankhurst, C.E. and Layzell, D.B. 1984. The effect of bacterial strain and temperature changes on the nitrogenase activity of *Lotus pedunculatus* root nodules. *Physiol. Plant,* **62:** 404-409.

Panwar, J.D.S. 1991. Effect of VAM and *Azospirillum brasilense* on photosynthesis, N metabolism and grain yield in wheat. *Indian J. Plant Physiol.,* **34**(4): 357-361.

Panwar, J.D.S. 1992. Effect of VAM and *Azospirillum* inoculation to water status and grain yield in wheat under water stress conditions. *Indian J. Plant Physiol.,* **35**(2): 157-161.

Panwar, J. D.S. 1993. Effect of VAM and *Azospirillum* inoculation to water status and grain yield in wheat under water stress conditions. *Indian J. Plant Physiol*., **36**(1):41-43.

Panwar, J.D.S., Chandra R. and Sirohi G.S., 1988. Effect of temperature on growth, nodulation and nitrogen fixation in summer mung (*Vigna radiata* (L) Wilczek). *Ann. Plant Physiol.* **2**(2): 121-129.

Panwar, J.D.S. and Naidu V.S.G.R. 2003. Physiological studies in relation to nodulation, nitrogen fixation and yield in rice using plant growth regulators and bacteria Abstract: 2nd International Congress of Plant Physiology, N.Delhi, pp. 363.

Panwar, J.D.S., Pandey, M. and Abrol, Y.P, 1990. Effect of *Azospirítlum* on growth and yield of wheat under low fertility conditions. . *Indian J. Plant Physiol.,* **33**: 157-189.

Panwar, J.D.S., and Sirohi, G.S. 1989. *Azospirillum*-An important biofertilizer for improving crop productivity. *Farmers and Parliament.,* **24**: 23-24.

Parsons, R., Stanforth, A., Roven, J.A. and Sprent, J.I. 1993. Nodule growth & activity may be regulated by a feedback mechanism involving phloem nitrogen. *Plant Cell Environ.,* **16**: 125-136.

Patriquin, D.G. and Dobereiner, J. 1978. Light microscopy observations of tetrazolium-reducing bacteria in the endorhizosphere of maize and other grasses in Brazil. *Can.J. Microbiol.,* **24:** 734-742.

Pereira, J.A.R., Cavalcante, V.A., Baldani, J.I. and Dobereiner, J. 1988. Field inoculation of sorghum and rice with *Azorhizobium* spp. and *Herbaspirillum seropedicae. Plant Soil,* **110**: 269-274.

Piha, M.I. and Munns, D.N. 1987. Sensitivity of common bean symbiosis to high soil temperature. *Plant Sci.*, **98**: 183-1.

Raman, R. 1992. Effect of soil acidity factors on aspects of symbiotic N_2 fixation of Lens *culinarís* in acid soils. *J. Gen. Applied Microbiol.,* **38**(5): 391-406.

Reis Jr. F.B., Clivares, F.L., Oliveria, A.L.M., Reis, V.L.M., Baldani, J.I. and Dobereiner, J. 1995. Infection and colonization of *Acetobacfer diazotrophicus* in sugarcane plant lets. *In:* International Symposium on Sustainable Agriculture for the Tropics: The Role of Biological Nitrogen Fixation. Angra dos Reis, Rio de janero, pp. 225.

Rennie R.J. and Kemp, G.A. 1981. Dinitrogen fixation in *Phaseolus vulgaris* at low temperatures. Interaction of temperature: growth stage and time of inoculation. *Can. J. Bot.,* **60:** 1423-1427.

Ridge, R.W., Ride, K.M. and Rolfe, B.G. 1993. Nodule like structures induced on rice seedlings by addition of the synthetic auxin, 2,4-dichlorophenoxy acetic acid. *Australian J. Plant Physiol.,* **20:** 705-717.

Rinhold-Hurek, B. and Hurek, T. 1998. Life in grasses: diazotropic endophytes. *Trenos, Microbiol.,* **6**: 139-144.

Rinhold-Hurek, B. and Hurek, T. .Gillis.M., Hoste, B., Van Canneyt, M., Kersters, K. and De Ley, J. 1993. *Azoarcus* gen. Nov. nitrogen fixing proteobacteria associated with roots of kallar grass *(Leptochloa fusca L. Kunth.)* and description of two species *Acetobacter diazotrophicus* sp. Nov. and *Azoarcus communis* sp. Nov. *Internat. J. Systematic Bacteriol.,* **43:** 574-584.

Roger, P. A. and Watanabe, I. 1986. Technologies for utilizing biological nitrogen fixation in wetland rice : Potentialities, current usage and limiting factors. *Fert. Res.,* **9**: 39-77.

Rolfe, B.G. Djordjevic, M.A., Weinman, J.J., Mathesius, V., Pittock, C., Gartner, E., Reid, K.M,, Dong, Z., McCully, M. and Mclver, J. 1997. Root morphogensis in legumes and cereals and the effect of bacterial inoculation on root development *Plant Soil,* **194** : 131-13.

Rossum, D.V., Muyotcha, A., DeHoop, B.M., Van Verseveld, H.W., Stouthamer, A.H. and Boogered, F.C. 1994. Soil acidity in relation to groundnut-*Bradyrhizobium* symbiosis. *Plant Soil,* **163**(2): 165-175.

Ryle, G.J.A., Powell, C.E. and Davidson, A. 1992. Growth of white clover, dependent on N_2 fixation, in elevated CO_2 and temperature. *Ann.* Bot., **70**: 221-228.

Sabry, R.S., Saleh, S.A., Batchelor, C.A., Jones, J., Jotham, J., Webster, G., Kothari, S.L., Davey, M.N. and Cocking, E.C.1997. Endophytic establishment of *Azorhizobium caulinodans* in wheat. Proc. Roy.Soc.London *Biol. Sci.,* **264**: 341-346.

Saikia, S.P., Srivastava, G.C., Jain,V. 2003. Effect of *Azospirillum* on maize productivity. Abstract: 2nd International Congress of Plant Physiology. pp. 385.

Sarwar, M. and Kremer, R.J. 1995. Determination of bacterially derived auxins using a micro-plate method. *Letters in Applied Microbiology* **20**: 282-285.

Schulman, H.M., Lewis, M.C., Tipping, C.M. and Bordeleau, L.M. 1988. Nitrogen fixation by three species of leguminaceae in Candian High Arctic Tundra. *Plant Cell Environ.,* **11:** 721-728.

Scott, J. and Howieson, J. 2000. Pulse research and industry development in Western Australia: Chickpea Nodulation Survey. *In:* Crop updates, pp. 32.

Serraj, R., Vasquez-Diaz , H. and Drevon, J.J. 1998a. Effects of salt stress on nitrogen fixation, oxygen diffusion and ion distribution in soybean, common bean and alfalfa. *J. Plant Nutr.,* **21**: 475-488.

Serraj, R., Vasquez-Diaz , H., Hernandez, G. and Drevon, J.J. 2001. Genotypic difference in response of nitrogenase activity (C_2H_2 Reduction) to salinity and oxygen in commmon bean. *Agronomia,* **21**: 645-651.

Sevilla, M. and Kennedy, C. 2000. Colonization of rice and other cereals by Acetobacter diazotrophicus, an endophyte of sugarcane. *In :* The Quest for Nitrogen Fixation in Rice (Eds. J.K. Ladha and P.M. Reddy). International Rice Research Institute, Manila, Philippines, 151-165.

Singh, G.R., Sharma, S.R. and Anil Kumar, T.B. 1986. Effects of biloxazol, carbandazijm and triadimefon on nodulation and nitrogen fixation in cowpea. *Zentralbi Mikrobil,* **141**: 35-37.

Smith, R.L., Boutan.J.H., Schank, S.C.,Queseiberry, K.H., Tyler, M.E., Matim, J.R., Gaskins, M.H. and Litell, R.C. 1976. Nitrogen fixation in grasses inoculated with Spirillum lipoferum. *Science,* **193**:1003-1005.

Soam, S.K. and Agrawal, P.K. 1989. Effect of dimethoate on nodulation and nitrogenase activity of nodules in pea and blackgram. *Indian J. Agril, Sci.,* **59**(12): 824-825.

Sprent, J.L. and de Faria, S.M. 1989. Mechanisms of infection of plants by nitrogen fixing organisms. *Plant Soil,* **110**: 157-165.

Stein.T., Hayen-Schneg, N. and Fendrik, I. 1997. Contribution of BNF by *Azoarcus* spp. BH 72 in *Sorghum vulgare. Soil Biol. Biochem.,* **29**: 969- 971.

Tarrand, J.J., Krieg, N.R. and Doberreiner, J. 1978. A Taxanomic study Of the *Spirillum Lipoferum* group with description of a new genus, *Azospirillum* gen. Nov., and two species *Azospirillum lipoferum* (Beijer inck) sp. Nov., and *Azospirillum brasilense* spp. Nov. *Canadian J. Microbiol.,***24**: 967-980.

Tien, T.M., Gaskin, M.H. and Hubbell, D. H. 1979. Plant growth substances produced by *Azospirillum brasilanse* and their effect on the growth of pearlmillet. *Appl. Enviro. Microbol.,* **37**:1016-1022.

Tilak, K.V.B.R. and Subba Rao, N.S. 1987. Association of *Azospirillum brasilense* with pearlmillet. Biol. Fertil. Soils. 4: 97-102. pearlmillet. *Appl. Environ. Microbiol.,* **37**: 1016-1022.

Ueda, T., Suga, Y., Yahiro, M. and Matsuguchi, T. 1995. Remarkable N_2- fixing bacterial diversity detected in rice roots by molecular evolutionary analysis of n/f H gene sequence. *J. Bacteriol.,* **177**: 1414-1417.

Van, V. T., Berge, O., Ngo Ke, S., Balandreau, J. and Henlin, T. 2000. Repeated beneficiai effects of rice inoculation with a strain of *Burholderia vietnamiensis* on early and late yield components in low fertility sulphate soils of Vietnam. *Plant Soil,* **218**(1-2): 273-284.

Van Bulow, J.G.D. and Dobereiner, J. 1975. Potential for nitrogen fixation in maize genotypes in Brazil. *Proc. Natl. Acad. Sci.,* USA, **72**:2389-2393.

VanHoorn, J.W., Katerji, N., Hamdy, A. and Mastrorüli, M. 2001. Effect of salinity on yield and nitrogen contribution from the soil. *Agric. Water Management,* **51:** 87-98.

Vermeiren, H., Willems, A., Schoofs, G., deMot, R., Keijirs , V., Hai, W. and Vanderleyden, J. 1999. The rice inoculant strain *Alcaligenes faecalis A* 15 is a nitrogen fixing *Pseudomonas stutzeri. Systematic and Appl. Microbiol.,* **22**: 215-224.

Vermeiren, H., Vanderleyden, J. and Hai, W. 1998. Colonization and nifH expression in rice roots by Alcaligenes faecalis A 15. In: Nitrogen Fixation with Non-legumes (Eds. K.A. Malik, M.A. Mirza and J.K. Ladha), Kluwer Academic Publishers, Dordrecht (Netherlands), pp. 287-305.

You, C.B. and Song W. 1994. Establishment of an effective nitrogen fixing association between rice and diazotroph Alcaligenes spp. In: Biological Nitrogen Fixation: Novel Associations with Non-legume Crops (Eds. N. Yanfu, I.R. Kennedy and C. Tingwei), Quigdao Ocean University Press, Quingdao, China, pp. 7-27.

You, C.B. and Zhou, F.Y. 1989. Non-nodular endorhizosperic nitrogen fixation in wetland rice. *Canadian J. Microbiol.*, **35**:403-408.

You, C.B., Lin, M., Fang, J and Song, W. 1995. Attachment of *Alcaligenes in* rice roots. *Soil Biol. Biochem.,* **27**: 463-466.

You, C.B., Song, W., Wang, H.X., Li, J.P., Lin, M. and Hai, W.L. 1991. Association of *Alcaigenes faecalis* with wetland rice. *Plant Soil,* **137**:81-85.

Yu, D.and Kenndy, I. R. 1995. Nitrogenase activity (C_2H_2 reduction) of *Azorhizobium* 2, 4-D induced root structures of wheat. *Soil Biol. Biochem*., **27**: 459-462.

Walsh, K.B. and Layzell, D.B., 1986. Carbon and nitrogen assimilation and partitioning in soybeans exposed to low root temperature. *Plant Physiol.,* **80:** 249-255.

Waughman, G.J. 1977. The effect of temperature on nitrogenase activity. *J. Exp. Bot,* **28**: 949-960.

Zemen, A.M.M.., Tchan.Y.T., Elmerich, C. and Kennedy, I.R. 1992. Nitrogenase activity in wheat seedlings bearing para nodules induced by 2,4-D and inculated with *Azospirillum . Research in Microbiology Institute, Pasteur, Elsevier,* **143**: 847-855.

Zhou, F.Y. and You, C.B. 1988. Interaction between diazotrophic bacteria *Alcaligenes faecalis* and host plant rice. *Scientia Agricultura Sinica,* **21**(3): 7-13.

Zhang, F. and Smith, D.L. 1994. Effects of low root zone temperatures on the early stages of symbiosis establishment between soybean and *Bradyrhizobium japonicum. J. Exp. Bot.,* **45:** 1467-1473.

Zhang F. and Smith D.L. 1995. Preincubation of *Bradyrhizobium japonicum* with genistein accelerates nodule development of soybean at suoptimal root-zone temperatures. *Plant Physiol.,* **108**: 961-968.

Zhang. F. and Smith, D.L. 1997. Application of genistein to inocula and soil to overcome low spring soil temperature inhibition of soybean nodulation and nitrogen fixation. *Plant Soil,* **192**: 141-151.

Section V: Sustainable Crop Productivity

Developments in Physiology, Biochemistry and Molecular Biology of Plants, 2005
Eds.: Bandana Bose and A. Hemantaranjan
Vol., 1, pp. 159-180, New India Publishing Agency, New Delhi
E-mail: spjain_niph@rediffmail.com web: www.bookfactoryindia.com

CHAPTER - 8

PGPR : A NEW GROUP OF BIOFERTILIZERS FOR SUSTAINABLE CROP PRODUCTIVITY

J.D.S. PANWAR AND K. SWARNALAKSHMI

INTRODUCTION

To keep pace with the ever-increasing demand for crop yield by the exploding population, use of chemical fertilizers and pesticides in agricultural production has become indispensable. Production cost of agriculture has skyrocketed owing to the abandoning of green, organic farmyard manure and use of chemicals. However, the problems associated with chemical pesticides are abundant. The excessive application of chemical fertilizers causes soil sealing, fertility diminishing and residual problems. The leftover chemical fertilizers and pesticides have seriously affected the quality of agricultural products, people's health and caused environmental pollution. The over use of fertilizers has also damaged the soil's original micro-ecological balance and deteriorated the diseases spread by soil. While some pesticides must be abandoned because of their unacceptable non-target effect, there will always be a need in agriculture for safe and selective chemicals to limit the effects of pests. More significantly, it is becoming increasingly more difficult and expensive to find new kinds of synthetic chemical pesticides. Implications of chemical pesticides in ecological, environmental and human health problems have increased public awareness and concern regarding the continued use of agrichemicals that are damaging to human or environmental health thereby obviating the search for effective alternative approaches which have minimal deleterious effects, more environmentally friendly, and will contribute to the goal of sustainability in agriculture. In this line Plant growth-promoting rhizobacteria (PGPR) present immense potential and promise as effective substitutes for chemicals. PGPR enhance plant growth and survival, control soil-borne fungi, and induces systemic resistance to phytopathogens and is used as inoculants for biofertilization, phytostimulation and biocontrol.

The term rhizosphere was introduced by Hiltner in 1904, and is defined as a volume of soil surrounding plant roots in which bacterial growth is stimulated (Sorensen, 1997). The rhizosphere has attracted much interest since it is a habitat where several biologically important processes and interactions take place. The rhizosphere is populated by a diverse range of microorganisms, and the bacteria colonizing this habitat are called rhizobacteria (Schroth and Hancock, 1982). The colonizing microorganism form symbiotic, associative, neutralistic or parasitic relations with the plant, which in turn is dependent on the type of microorganism, available nutrients, plant defense responses and environment. The symbiotic association between legumes and *Rhizobium* is a well known phenomenon which contributes to biological nitrogen fixation (BNF) and soil fertility. Some of the free living rhizobacteria utilize nutrients secreted by the plant roots and influence the plant in a direct or indirect way resulting in growth stimulation. This set of beneficial bacteria termed as plant growth promoting rhizobacteria (PGPR), was coined by Kloepper *et al.* (1978), that comprises of a broad-spectrum of bacterial genera. Additionally, PGPR also protect the plant from infection by phytopathogenic organisms within the vicinity of the root system.

Root exudates are believed to determine which microorganisms colonize roots in the rhizosphere (Kunc and Macura, 1988). It is now known that plant roots also generate electric signals; it has been shown that zoospores of oomycetic pathogens take advantage of these signals to guide their movements towards the root surface (Gow *et al.,* 1999). It has been estimated that about 30% of plant photosynthate production is released via root exudation (Smith *et al.,* 1993) which attracts microorganisms (Jaeger *et al.,* 1999; Lemanceau *et al.,* 1995).

Plant growth colonizing bacteria can function as harmful, deleterious rhizobacteria (DRB) or beneficial, plant growth promoting rhizobacteria. Rhizobacteria that inhibit plant growth have been described as deleterious rhizobacteria (Suslow and Schroth, 1982). However, there are no published experiments supporting the existence of DRB. Plant growth promotion by rhizobacteria can occur directly and indirectly (Glick, 1995; Persello-Cartieaux *et al.,* 2003). There are several ways by which plant growth promoting bacteria can affect plant growth directly. These include fixation of atmospheric nitrogen, solubilization of phosphorus, production of siderophores that solubilize and sequester iron, or production of plant growth regulators (hormones) that enhance plant growth at various stages of development (Fig. 1) (Enebak and Carey, 2000). Indirect growth promotion occurs when PGPR promote plant growth by improving growth restricting conditions (Glick *et al.,* 1999). This can happen directly by producing antagonistic substances, or indirectly by inducing resistance to pathogens (Glick, 1995). A bacterium can affect plant growth by one or more of these mechanisms, and also use different abilities for growth promotion at various times during the life cycle of the plant (Glick *et al.,* 1999).

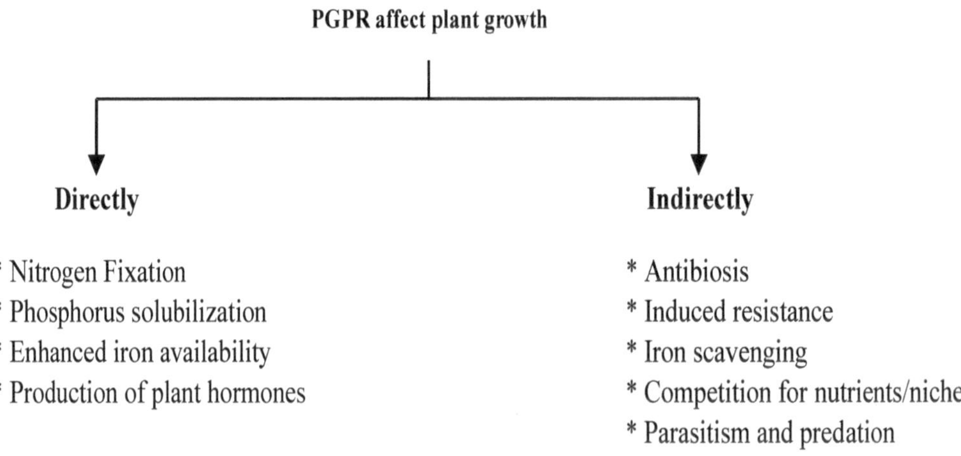

Fig. 1. Plant growth promoting bacteria affect plant growth directly and indirectly

MECHANISM OF PGPR IN DIRECT PLANT GROWTH PROMOTIONS

The ways by which PGPR can influence plant growth directly may differ from species to species as well as strain to strain. Symbiotic plant colonizers such as rhizobia mostly contribute to plant growth by nitrogen fixation. Free living rhizobacteria usually do not rely on single mechanisms of promoting plant growth (Glick *et al.,* 1999). In addition to nitrogen fixation, several PGPR are also able to provide the plant with sufficient iron in iron-limited soils (Wang *et al.,* 1993), or other important minerals, *e.g.,* phosphate (Singh and Singh, 1993).

Organic substances capable of regulating plant growth produced either endogenously or applied exogenously are called plant growth regulators. They regulate growth by affecting physiological and morphological processes at very low concentrations (Arshad and Frankenberger, 1998). Several microorganisms are capable of producing auxins, cytokinins, gibberellins, ethylene (ET), or abscisic acid (ABA). Auxins are produced by several rhizobacterial genera, *e.g., Azospirillum, Agrobacterium, Pseudomonas* and *Erwinia* (Costacurta and Vanderleyden, 1995).

Production of IAA by *Azospirillum* appears to be the most likely explanation for the observed increase in seedling's root length, root hairs, root branching and root surface areas. However, it is highly unlikely that IAA alone explains yield increases that have been reported on a large number of crops (Bashan and de-Bashan 2004). Other plant growth regulators and N-fixation are other attributes of *Azospirillum* that likely contribute to yield increases and enhanced nutrient uptake of inoculated plants. Bashan and Levnony (1990) suggested additive hypothesis to explain growth and yield promotion with *Azospirillum* (Panwar and Sirohi, 1989; Panwar *et al.,* 2001). According to this hypothesis, growth promotion is the result of multiple mechanisms working together. Hence, when a single mechanism is tested, the full growth promotion response that occurs with inoculation of *Azospirillum* is not seen. Rather, the observed plant growth promotion results from the sum effect of the multiple mechanisms *i.e.* the levels of N-fixation are not considered adequate to account for the observed growth promotion of plants. In addition, *Azospirillum* strains produce IAA and other plant growth regulators in culture and association with the plant which may contribute growth promotion.

Glick and Coworkers (1998) have defined the role of PGPR-produced IAA and the response of the host plant. With this system, bacterial production of IAA may stimulate plant cell proliferation or elongation and results in plant production of ACC (1-aminocyclo propane-1-carboxylate). ACC is a immediate precursor of ethylene, and ethylene inhibits root elongation of seedlings. IAA produced by *Pseudomonas putida* GR 12-2, together with endogenously produced plant IAA, activates ACC synthase, leading to production of ACC. ACC is converted to ethylene by ACC oxidase. However, the plant produced ACC is taken up by the PGPR strain and is hydrolyzed by ACC deaminase that converts ACC to ammonia and alpha-ketobutyrate. This conversion lowers the pool of ACC, thereby, reducing the level of ethylene and ethylene's inhibition of root elongation, resulting in increased root length (Fig.2). This model also suggests that ACC serve as a unique source of nitrogen for the bacterium. Mutant strains lacking ACC activity were enabling to promote root elongation of *canola* seedlings (Glick *et al.,* 1998).

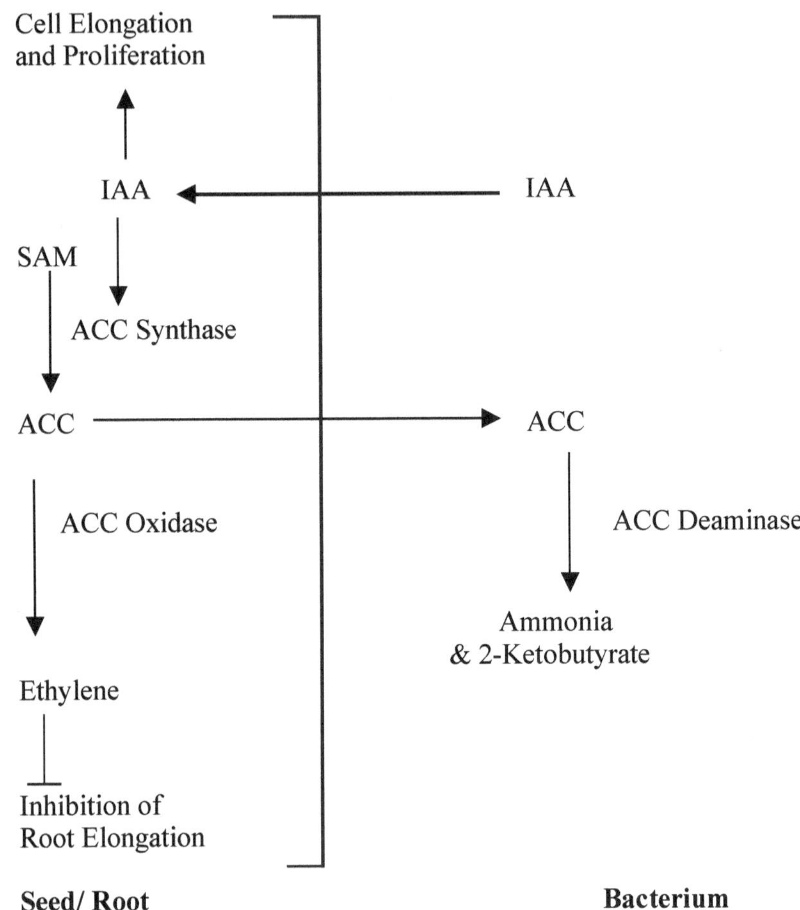

Fig 2. Model explaining how PGPR stimulation of plant root elongation (Hall *et al.*, 1996 and Glick *et al.*, 1998).

Cytokinins and gibberellins are produced in the rhizosphere by several bacteria *e.g. Azospirillum*, *Agrobacterium* and *Pseudomonas* genera (Gaudin *et al.*, 1994). Cytokinins promote root formation, but a minor overproduction of cytokinins instead leads to inhibition of root development, and severely deficient cytokinin mutant plants do not survive (Binns, 1994). Cytokinins are believed to be the signals involved in mediating of environmental stresses from roots to shoots (Jackson, 1993). Nevertheless, more studies need to be conducted before cytokinin signalling can be fully understood.

In addition to that, co-immobilization of *Chlorella vulgaris* and *Azospirillum brasilense* cd in alginate beads under synthetic wastewater culture conditions significantly increased the removal of ammonium and soluble phosphorus compare over imobilization of the microalgae alone (de-Bashan *et al.*, 2002). Co-immobilization of the two microorganisms (also with *Chlorella sorokiniana*) reaching removal of up to 100% ammonium, and 92% phosphorus within 6 days (De-Bashan *et al.*, 2003).

MECHANISM OF PGPR IN INDIRECT PLANT GROWTH PROMOTION

PGPR are known to control a wide range of phytopathogens like fungi, bacteria, viruses, nematodes *etc.,* by biocontrol mechanism which may be by producing antibiotics, siderophores, variety of enzymes, by competition or by induction of systemic resistance against pathogens throughout the entire plant system.

Antibiotics

According to Pierson and Pierson (1996); Wood and Pierson (1996) an auto-inducer signal required for induction of antibiotic synthesis. Also, other environmental sensors such as the regulating proteins Gac A and Apd A can influence the production of secondary metabolites involved in pseudomonad biocontrol (Corbel and Loper, 1995; Haas *et. al.,* 2002). In addition, house-keeping factor sigma^{70} and the stress related sigma^{s} have critical role in production of antibiotic metabolites in disease suppression (Schnider *et al.*, 1995).

Various antibiotics can be produced by rhizobacteria like *Bacillus* and *Pseudomonas,* like bacilysin, iturin like lipopeptides, diacetylphloroglucinol and phyrrolnitrin, HCN, phenazine-1-carboxylate (Thomshow *et al.,* 1990). The most widely studied group of rhizospheric bacteria with respect to the production of antibiotics is that of the fluorescent pseudomonads. The first antibiotics described as being implicated in biocontrol were phenazine derivatives produced by fluorescent pseudomonads (Weller and Cook, 1983). Their role has been elucidated by transposon insertion mutations which results in a defect in production of phenazine-1-carboxylate, thus reducing disease suppressive activity (Pierson and Pierson, 1996).

Siderophores

Iron is abundant in the Earth's crust but most of it is in the highly insoluble form of ferric hydroxide, and thus unavailable to organisms in soil solution. Some bacteria have developed iron uptake systems (Neilands and Nakamura, 1991) which involve a siderophore-an iron binding ligand-and an uptake protein, needed to transport iron into the cell. It has been suggested that the ability to produce specific siderophores and/or to utilize a broad spectrum of siderophores, may contribute to the root colonizing ability of *Pseudomonas* strains. The production of siderophores that chelate, and thereby scavenge, the ferric iron in the rhizosphere, may result in growth inhibition of other microorganism whose affinity for iron is lower (Kloepper *et al.,* 1988).

However, siderophore mechanism will only be relevant under conditions of low iron availability. As soil pH decreases below 6, iron availability increases and siderophores become less effective (Neilands and Nakamura, 1991). Optimal suppression of pathogens occurred at levels between $10^{-19} - 10^{-24}$ M. The critical level of iron at which a siderophore producing strain of *Pseudomonas putida* suppressed the growth of *Fusarium oxysporum* was found to be $<10^{-16}$ M (Neilands and Nakamura, 1991). Since the synthesis of each siderophore generally requires the activity of several gene products (Mercado-Blanco *et al.,* 2001), it is difficult to genetically engineer bacteria to produce modified siderophores.

Parasitism

An additional mechanism by which biocontrol agents can reduce plant disease is parasitism on pathogens, mostly fungi. Digestion of the parasite cell wall is accomplished by several excreted enzymes including proteases, chitinases and glucanases. Individually, all these enzymes display antifungal activity, but they often act synergistically with antibiotics (Lorito *et al.,* 1993; Lorito *et al.,* 1994). De Meyer *et al.* (1999) reported that rhizosphere colonization by *Pseudomonas aeruginosa* 7NSK 2 activated phenylalanine ammonia lyase (PAL) in bean roots and increased the salicylic acid levels in leaves. The *Pseudomonas fluorescens* strain was found to produce antibiotics (Phenazine and 2, 4 diacetyl phloroglucinol), HCN and lytic enzymes *viz.,* chitinase and glucanase. These bacterial metabolites were associated with direct inhibitory action against several *Rhizoctonia, Fusarium, Pythium, Phytophthora* and *Colletotrichum* pathogens (Samiyappan, 2003).

Competition for Nutrients and Niches

In addition to the above described and commonly reported antibiosis mechanisms there are other ways by which rhizobacteria can inhibit pathogens. One example concerns competition for nutrients and suitable colonization niches on the root surface. Such mechanisms are often overlooked, in part because they are difficult to study in biological systems. Competition for nutrients supplied by root exudates is probably a significant factor in most interactions between PGPR and pathogens. Populations of bacteria established on a plant root could act as a sink for the nutrients in the rhizosphere, hence reducing the nutritional element availability for pathogen stimulation or subsequent colonization of the root. This mechanism is most probably often used by fluorescent pseudomonads due to their nutritional versatility, and because of their high growth rates in the rhizosphere (Sorensen, 1997).

Biocontrol by Inducing Systemic Resistance

Systemic resistance refers to an increased level of resistance at sites within that plant system to those at which induction had occurred. How bacteria trigger systemic resistance is still largely unknown. Two general plant defense mechanisms are induced systemic resistance (ISR) and systemic acquired resistance (SAR). ISR is rhizobacterially mediated systemic resistance that does not involve any damage to plant. By contrast, SAR is induced by foliar pathogens and results in activation of resistance mechanisms in uninfected parts (Fig. 3). Thus, in systemic acquired resistance a first infection predisposes the plant to resist further attacks.

Signalling molecules like salicylic acid (SA), jasmonate (JA) and ethylene (ET) when accumulating coordinate the defense responses and, when applied exogenously, are even sufficient to induce resistance. It has been shown that these signalling molecules activate specific sets of defense-related genes: SA induces genes encoding pathogenesis-related proteins (PRs) (Uknes *et al.,* 1992). These proteins have anti-microbial activity (Kombrink and Somssich, 1995). ET is involved in the expression of the genes encoding *Hel* (a heveine-like protein) (*Potter et al.,* 1993), *ChiB* (basic chitinase) (Samac and Shah, 1994), and *pdf1.2* (a plant defensin) (Penninckx *et al.,* 1996). Also JA has been shown to activate the

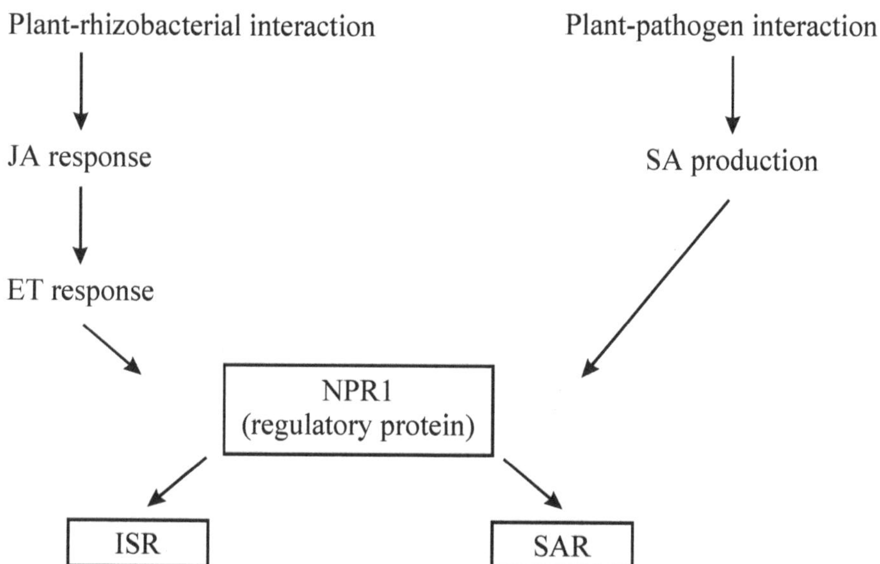

Fig. 3. Signalling pathways in induced resistance (Pieterse and Van Loon, 1999).

genes encoding these three proteins, all of which possess antifungal activity (Penninckx *et al.,* 1996). SA induces SAR and controls activity of regulatory (activator) protein NPR1. Besides that, JA and ET act synergistically in inducing genes for several PR proteins (Norman-Setterblad *et al.,* 2000).

Various *Pseudomonas* spp., strains induce broad-spectrum disease resistance in plants in a bacterial strain/plant species-specific manner. In *Arabidopsis thaliana*, specific strains of non-pathogenic fluorescent *Pseudomonas* bacteria develop an induced systemic resistance (ISR) that unlike pathogen-induced systemic acquired resistance (SAR), is independent of salicylic acid (SA) but requires sensitivity to jasmonic acid (JA) and ethylene (ET). Different bacterial determinants appear to be involved in triggering ISR *e.g.* lipopolysaccharides, siderophores, and flagella. Combining ISR and SAR increased resistance against pathogens that are resisted by both JA/ET and SA-dependent defenses.

ISR operates in the *Pseudomonas fluorescens* treated plants by activating the genes encoding pathogenesis related (PR) proteins and genes involved in the phenyl propanoid metabolic pathway. Major induced PR-proteins include chitinase, β-1,3 glucanases, thaumatin like proteins (TLP) and peroxidases. The phenyl propanoid pathway products *viz.,* phenolics and lignin which were induced in *Pseudomonas fluorescens* treated plants showed enhanced resistance to several pests and plant pathogen. Major enzymes such as phenylalanine ammonia lyase (PAL), peroxidase (PO) and polyphenol oxidase (PPO) were also induced (Samiyappan, 2003).

Rhizobacterially mediated systematically induced resistance is the reset of process that does not involve any obvious damage to the plant. For induction of ISR in radish, 10^5

CFU of the inducing *Pseudomonas* sp strain per gram of root tissue was reported as a minimum threshold (Raazijmakers *et al.,* 1995). It has been shown for *Pseudomonas* strains that an outer membrane lipopolysaccharide (LPS) with strain specific O-antigenic carbohydrate side chains is the elicitor involved in induced resistance (Leeman *et al.,* 1995). The purified siderophore of *Pseudomonas fluorescens,* pseudobactin 374, could induce resistance in radish (Leeman *et al.,* 1996). It has also been shown that treatment of radish plant roots even with nanogram quantities of salicylic acid was sufficient to induce systemic resistance against *Fusarium oxysporum* (Leeman *et al.,* 1996).

Recently, Ryu *et al.* (2003) reported that PGPR strains *Bacillus subtilis* and *Bacillus amyloliquefaciens* released a blend of volatile organic compounds (VOCs) that promote growth in *Arabidopsis* seedlings and induce resistance against *Erwinia carotovora* subsp. *Carotovora.* The volatile components 2,3 butanediol and 3-hydroxy-2-butanone (also referred to as acetoin) were released exclusively from PGPR strain trigger the greatest level of growth promotion and induced diseased resistance.

Several phytopathogenic bacteria produce harpins, a class of acidic, glycine-rich, heat stable proteinaceous elicitors of hypersensitive response (HR) in various plant pathogen interactions (Wei *et al.,* 1992). Harpins have also been shown to trigger SAR (Strobel *et al.,* 1996). Their role in pathogenesis and defense signalling has been difficult to access since it may be masked by massive production of pectinolytic enzymes in pathogens such as *Erwinia carotovora.* Recently the role of harpins as elicitors of plant defense responses has gained strong support. Apparently, the elicitor harpin HrPN produced by *Erwinia* acts in concert, and cooperatively, with oligogalacturonides (OGAs) in plant defense signalling in *Arabidopsis* (Kariola *et al.,* 2003).

ROLE OF PGPR IN CROP PLANTS

PGPR strains *Bacillus, Azospirillum, Azotobactors, Pseudomonas* spp. *etc*. are found to improve seed germination, roof/shoot length, nodulation, biomass, yield and nutrient *mobilization* in various crops (Table.1) The beneficial bacteria also protect the crops from the attack of soil borne pathogen among which *Fusarium* spp., *Rhizoctonia solani, Rhizoctonia bataticola* and *Sclerotium rolfsii* are highly destructive (Table 2). Pyrrolnitrin, Phenazine-1-carboxylic acid and 2,4-diacetlyphloroglucinol produced by *Pseudomonas* spp were the major antibiotics involved.

PGPR: Usefulness under biotic stress

Combined inoculation of other PGPR with rhizobia was observed to exert positive effects on the growth of legumes. Co-inoculation of PGPR alongwith *Rhizobium* increased the root nodulation. The beneficial effects of co-inoculation are strain dependent. The basic mechanisms involved in this synergistic activity were not completely known and remains a challenge. One possibility is that PGPR, by altering the host's secondary metabolism and/or creating antibiosis in the rhizosphere, out compete the pathogens and eliminate competition of *Rhizobium* with deleterious microorganisms for colonization of the plant parts. Alteration of the plant flavonoid metabolism was proposed as another mechanism of synergistic activity of PGPR and rhizobia. The applied rhizobacteria stimulate formation of additional infection sites that may later be occupied by rhizobia, thus increasing the noduation. Inoculation of *Azospirillum* sp produces large quantities of auxins and stimulates the formation of epidermal cells that become root hair cells or additional infection sites for rhizobial colonization. Unlike chemical fertilizers, PGPR exert a beneficial effect on rhizobial nodulation of legumes and should be exploited for the economical benefit of subsistence farming systems. A complete understanding of the mechanisms of the synergistic activity between PGPR and *Rhizobium* facilitates the better utilization and improvement of bioinoculants in legume based cropping systems.

Table: 1. Plant growth promoting rhizobacteria (PGPR) in enhancing the growth of crop plants.

Crop	Bacterial isolate	Effects on growth and yield
Wheat	*Bacillus subtilis*	Seed germination and grain yield
Pigeonpea	*Bacillus substilis AF1*	Increase in shoot and root length and biomass
Mungbean, pea	*Azotobacter*	Increase in yield
Chickpea	*Pseudomonas sp*	Enhanced seed germination, root and shoot dry weight, yield
	Azospirillum brasilense	Nodulation
	Azotobacter	Number of shoots/plant, yield, seed protein
Soybean	*Bacillus cereus UW 85*	Increase in root growth, nodulation and yield
Fababean	*Azospirillum brasilense*	Increase in growth of root and shoot and nodulation
Groundnut	*Bacillus subtilis,* *Bacillus amyloliquefaciens,* *Pseudomomas aeruginosa,* *Pseudomonas fluorescens*	Increase in root and shoot length, biomass and yield.
Sunflower	*Azospirillum* *Azotobacter*	Increase in yield
	Pseudomonas	Higher seed yield
Black pepper ginger, turmeric, spice crop	*Pseudomonas fluorescens* *Phosphobacteria IISR 4000*	Nutrient mobilization

PGPR's and their effectiveness has been established on foot rot and slow decline of black pepper and rhizome rot of ginger and dump rot of cardamom. Strains of *Pseudomonas fluorescens* were found to increase the growth and vigor of these crops apart from suppressing the soil borne diseases (Sarma *et al.,* 2000).

In coconut and cocoa mixed crop increased activity of *Beijerinckia* sp., *Aspergillus niger* and *Pseudomonas* sp., which also include growth promoting substance like IAA and GA (Nair and Subba Rao 1977) which reflected on the yield increase in the crop mixture. Similar increase with nitrogen fixers and phosphorus solubilizers were reported in coconut-tree spices (Cinnamon, clove, nutmeg) mixed crop system (Iyer, 1983). The role of *Pseudomonas fluorescens* mediated nutrient flux in the soil microcosm in plant growth promotion was studied with the higher uptake of nutrients by the bacterized plants. Significant uptake of nitrogen (N) and Potassium (K) was noticed in the treated black pepper (Dibypaul *et al.*, 2003).

Table: 2. Plant growth promoting rhizobacteria (PGPR) in control of root diseases of crop plants.

Crop	PGPR strain used	Disease
Rice	*Pseudomonas fluorescens* *Pseudomonas putida* *Bacillus polymyxa* *Bacillus pumulus* *Bacillus coagulans* *Enterobacter agglomerans* *Bacillus cereus*	Suppression of plant pathogens, blast, sheath blight, sheath rot, stem rot, bacterial blight, tungro disease, downey mildew
Sorghum	*Pseudomonas fluorescens*	Head mould
Pigeonpea	*Bacillus brevis, Bacillus subtilis AF 1, Pseudomonas fluorescens*	Wilt
Chickpea	*Bacillus subtilis, Pseudomonas aeruginosa, Pseudomonas fluorescens*	Wilt
Pea	*Azotobacter chroococcum, Azospirillum lipoferum, Pseudomonas fluorescens*	Root rot
Soybean	*Bacillus megaterium*	Damping-off
Cowpea	*Bravibacterium linens, Pseudomonas fluorescens*	Stem and root rot
Bean	*Pseudomonas capacia*	Root rot
Groundnut	*Bacillus subtilis AF 1*	Crown rot
Lentil	*Pseudomonas fluorescens*	Wilt
Urdbean	*Pseudomonas aeruginosa, Pseudomonas fluorescens*	Root rot
Mustard	*Trichoderma viride*	White rust
	Bacillus amylolique faciens	Bud rot
Coconut	*Pseudomonas fluorescens*	Leaf rot

PGPR: Usefulness under abiotic stress

Microbial colonization of plant roots is affected by biotic and abiotic factors such as root exudates, competition, inorganic nutrients, soil pH and temperature *etc*. Among the agriculturally important abiotic stresses, drought, heat, soil salinity, and alkalinity rank as the most detrimental and cause loss of crop productivity all over the world.

Kulkarni and Nautiyal (2000) conducted a study to examine the growth response of a *Rhizobium* sp. NBR 1330 isolated from root nodules of *Prosopis juliflora* growing in alkaline soil. The individual stress survival limit of the strain in a medium containing 32% NaCl was 8h and at 55°C up to 3h also conducted an ecological survey (Kulkarni *et al.,* 2000) to characterize 5000 *Rhizobium* sp., *Sesbania* strains of diverse geographical origin, isolated from the root nodules of *Sesbania aculeata* growing in neutral (pH 7) and alkaline (pH 8.5 and above) soils. The rhizobia from the alkaline soil showed significantly higher salt tolerance than those isolated from neutral soil. High temperature was tolerated efficiently by *Rhizobium* sp. NBRI 2505 Sesbania, in the presence of salt at higher pH. Studies were conducted by Rehman and Nautiyal (2002) to elucidate the nature of drought tolerance in *Rhizobium* sp. NBRI 2505 sesbania and its transposan Tn5 induced mutant *Rhizobium* sp. NBRI 2505 sesbania T 112 containing single Tn5 insertion, selected after screening about 1000 clones. Our results demonstrate positive effect of calcium on the survival of *Rhizobium* sp. sesbania under acidic stress conditions. Mutant strain T 112 produced ineffective symbiosis with the plant host in the presence of 2.5 and 5% PEG indicating that drought tolerance is required for effective symbiosis.

Several soil bacteria produce osmolytes to protect themselves against frequent fluctuations in osmotic conditions *e.g. Paenibacillus polymyxa* (previously *Bacillus polymyxa*) and *Bacillus subtilis* produces glycine betaine (Lucht and Bremer, 1994). This compound lowers the water potential outside the cell wall and inhibits intercellular spaces. The increased concentration of these compounds could be sensed by the plant as a local dehydration. Consequently, dehydration genes such as ERD 15 and RAB 18 would be activated in plants in order to tolerate drought.

Tripathi *et al.* (1998) have studied in detail salinity stress responses in the plant growth promoting rhizobacteria *Azospirillum* sp. They have shown that in order to adapt to the fluctuations in soil salinity/osmolarity, the bacteria accumulate compatible solutes such as glutamate, proline, glycine betain, trehalose *etc.* Proline seems to play a major role in osmoadoptation. With increase in osmotic stress the dominant osmolyte in *Azospirillum brasilense* shifts from glutamate to proline. Accumulation of proline in *Azospirillum brasilense* takes place both uptake and synthesis.

Rangarajan *et al.* (2001) analysed populations of *Pseudomonas* for their biochemical characters and genetic diversity using molecular tools including RAPD and PCR-RFLP. It was observed that increasing salinity caused a predominant selection of salt tolerant species, in particular *Pseudomonas pseudoalcaligenes* and *Pseudomonas alcaligenes,* irrespective of the host rhizosphere. Rangarajan *et al.* (2003) have recently isolated indigenous *Pseudomonas* strain from rhizosphere of rice cultivated in the coastal agri-ecosystem and screened for *in vitro* antibiosis against *Xanthomonas ozyzae* p.v. *oryzae* and *Rhizoctonia solani.*

Phosphorus (P) is one of major essential macronutrients for biological growth and development. P in soil is immobilized or becomes less soluble either by absorption, chemical precipitation or both. Seed or soil inoculation with phosphate-solubilizing bacteria (PSB) is known to improve solubilization of fixed soil phosphorus and applied phosphates, resulting in higher crop yields. However, the establishment and performance of these microbes are affected severely under stress such as high salt, pH and temperature prevalent in degraded ecosystems such as alkaline soils with tendency to fix phosphorus. Therefore, it would be desirable to have PSB capable of solubilizing P under stress conditions. Gaind and Gaur (1991) have described thermotolerant phosphate solubilizers microorganism and their interaction with mungbean. Nautiyal *et al.,* (2000) characterized NBRI 2601 isolated from the rhizosphere of chickpea and alkaline soils which has the capability to solubilize phosphorus in the presence of 10% salt, pH 12 or 45°C.

PSB that could tolerate high salinity have also been isolated (Kumar, 2003). P-starvation inducible quinoprotein glucose dehydrogenase (Gcd) has been shown to be responsible for the high gluconic acid secretion resulting efficient P solubilization. Gluconic acid producing PSB show P-solubilizing ability on a mixture of low concentrations of several sugars as a consequence of the broad substrate specificity of the Gcd enzyme. Since, Gcd in certain bacteria is under catabolite repression by TCA cycle intermediates, the P-solubilizing phenotype of these organism is adversely affected by the presence of organic acids such as succinate, malate *etc*. Also, expression of PQQ synthase gene has been demonstrated to be sufficient to confer P-solubilization ability in transgenic bacteria. Alternatively, mineral phosphate solubilizing genes have been cloned from *Synechocystis* PCC 6803 in *Escheritia coli* from genomic DNA library (Kumar, 2003). Attempts are also being made to develop P-solubilization ability by the over expression of phosphoenol pyruvate carboxylase gene. Conditional repression of TCA cycle genes by antisense RNA could also help in developing PGPRs with P-solubilization ability.

GENETICS OF PGPR STRAINS

Rhizospheric Competence

Efficient PGPR should survive in the rhizosphere, make use of nutrients exuded by the plant root, proliferate, be able to colonize the entire root system and compete with indigenous microbes (Bolemberg and Lugtenberg, 2001). Some of the genes reportedly influencing the rhizospheric colonization are : (a) The *rhi* genes specifically expressed in rhizosphere (Rainey and Preston, 2000); (b) Flagellar syntheses, motility and chemotaxis genes *chvA* and *chv B* (Matthysse and McMahan, 1998); (c) The *nuo4* gene encoding part of NADH dehydrogenase; (d) A site-specific recombinase (*sss*) (Dekkers *et al.,* 1999); (e) The *att* genes (Matthysse and McMahan, 1998); (f) bacterial cellulose synthesis (*cel*) genes; (g) genes for outer membrane protein OprF that functions as adhesion; (h) The genes of *pot* operon, encoding proteins involved in putrescine uptake and the *PyrR* protein; (i) Genes required for synthesis of the lipopolysaccharide O-antigens; (j) the pilus genes, (k) The genes *htrB* and *secB*

(Dekkers *et al.,* 1999); and (l) The genes involved in the synthesis of amino acids and vitamin B1, nutrient acquisition, utilization of major exudates of carbon sources. Synthesis of bacterial agglutinins, homologues of a calcium-binding protein hemolysin, role of Levan in the aggregation of root-adhering soil, stress related proteins, a potential multi-drug efflux pump (Bolemberg and Lugtenberg, 2001) are some other factors influencing the rhizospheric colonization (Bolemberg *et al.,* 2002).

Bio-fertilization

The gene products of both the *nif* and the *fix* genes involved in nitrogen fixation have been well documented in some PGPR. Soil microorganisms possessing phytase activity contribute to plant phosphorus nutrition (Idriss *et al.,* 2002). Another beneficial effect due to bacterial phytase activity in the rhizosphere is elimination of chelate-forming phytate, which binds nutritionally important minerals (Reddy *et al.,* 1989). Siderophores help by chelating iron from the rhizosphere to make it bio-available.

Bio-stimulation

The prominent PGRs and their analogues produced by PGPR are auxins, cytokinins, and gibberellins. Besides having nitrogen fixing ability, *Azospirillum* spp. secrete PGRs. Some PGPR release a blend of volatile components that promote growth (Ryu *et al.,* 2003). Though genes for PGR synthesis have been identified, in several cases, the complete bio-synthetic pathways and their regulatory cascades leave many details wanted yet. Few of the genes involved in biosynthesis of PGR are: (a) Auxins : indole-3-acetic acid (IAA : CYP79B2, CYP 79B3, *nit1* to *nit4*); (b) Cytokinins : *trans-zeatin (ipt), cis-zeatin (ZOGI, ZOXI*); and (c) Glibberellins : GA_3 (CYP88A Cytochrome P 450 Gene, *ga3-1, ga3-2*).

Bio-control

Important among the group of PGPR are the broad-spectrum antibiotic producing fluoresent *Pseudomonas* spp. This group of rhizobacteria is an ideal candidate for biocontrol applications (Mark *et al.,* 2003). *Pseudomonas putida* is a metabolically versatile saprophytic soil bacterium that has been certified as a bio-safe host for the cloning of foreign genes. Ten genes (*pltA to pltG, pltL, pltM,* and *pltR)* required for the biosynthesis of pyoluteorin were identified within a 24-kb genomic region of *Pseudomonas fluorescens* Pf-5 (Nowak-Thompson *et al.,* 1999). Genes responsible for the production of HCN have been identified in the hydrogen cyanide synthase gene operon *hcnABC*. Genes (*phlA, to phlF*) required for 2,4-diacetylphloroglucinol synthesis are encoded by a 6.5-kb fragment of genomic DNA (Bangera and Thomashow, 1999). Introduction of alien genes such as *cry1Ac7, chiA,* and ACC-deaminase gene into rhizobacterial strains enhanced biocontrol and/or plant growth promotion (Bolemburg and Lugtenberg, 2001).

Rhizobium leguminosarum, Azospirillum spp. are some of the lesser characterized candidate-PGPR for genetic enhancement. One strategy to enhance rhizosphere competence is the development of genetically modified (GM) microbial inoculants (Walsh *et al.,* 2001). The focus is on developing strains in which the timing of the metabolite production is altered. If the relevant biosynthetic genes are uncoupled from their regulatory controls, it could facilitate early production of antifungal metabolites. However, public concerns as to the biosafety of GM bacterial strains in the environment must be addressed satisfactorily before the ultimate utilization of the GM-PGPR.

Bio-remediation

Bacteria engage in important metabolic activities in the environment, including element cycling and the degradation of biogenic and xenobiotic pollutants. They can also act as bio-indicators and bio-catalysts. The *bph* genes together with *Pex* promoter boosted the contaminant poly-chlorinated biphenyl (PCB) degradation capabilities of *Pseudomonas* (Brazil *et al.*, 1995). Addition of such capabilities can enhance the usefulness of PGPR and broaden their spectrum of utility.

Gene Regulation

Control of gene expression at the transcriptional level is a primary mechanism for modulating the production of secondary metabolites. Key factors in the regulation of the biosynthesis of most anti-fungal microbials are global regulation that simultaneously affects multiple biosynthetic pathways along with pathway-specific regulators that control linked biosynthetic genes, and the quorum sensing (Bolemburg and Lugtenberg, 2001). Global regulation is directed by the *gacS/gacA* genes. The quorum sensing (QS) involves the production of *N-acyl* homoserine lactone (AHL) signal molecules by an AHL-synthase. At a threshold concentration of AHL, reached only when a certain density of bacterial cells is present, the AHL will sufficiently bind and activate a transcriptional regulator. The cinR and cinI appear to be at the top of a regulatory cascade or network influencing several AHL-regulated QS loci. *CinR* regulates *cinI* expression in a cell-density dependent manner, and *cinI* expression is positively autoregulated by its own product. The *colS* and *colR* genes regulate the colonization in several cases (Dekkers *et al.*, 1999). The phytopathogen *Pythium* produces diffusible factors that can down-regulate genes important for the survival of the bio-control agent *Pseudomonas* (Walsh *et al.*, 2001). The fungal toxin fusaric acid (FA) produced by the phytopathogen *Fusarium* can repress expression of the gene *phlA*.

Plant Growth Promoting 'meta-bug'

Notwithstanding the state of art of the gene technology, it would be a far cry to wish for internalization of all the relevant traits of plant growth promotion into the host plant genome. Alternatively, the PGP 'meta-bug' approach would involve developing a 'core-competent' bug envisaging rhizosphere colonization and survival related functions, and stacking it with the genes for bio-stimulation, bio-fertilization, bio-control, *etc*. Hypothetically, the core-competent bug would have as many compatible genes relating to rhizospheric competence as possible, loaded into its genophore. To list is a few genes, *rhi, col, che, chv, nuo4, att, pot, htr, sec, etc.* This core competent bug then can be decorated with genes for production of phyto stimulatory agens [*aux (nit, CYP); ipt/ZOG, ZOX; ga3]*; bio-control agents [phz, *phl, plt, hcn, etc.*]; bio-fertilizing activities (*nif, fix, phy, etc.*); and bio-remediation effectors (*bph, pex, etc.*). With appropriate fine tuning and balancing of regulatory systems involved, such as *gacS/gacA, cinR/cinI, colR/cols*, the meta bug should be ideally a 'single window' delivery system, capable of not only establishing itself in wide range of rhizospheres, but also competent in efficient plant growth promotion. A meta-bug each targeted for different agro-ecology could be developed in the most appropriate strain/s of *Pseudomonas, Rhizobium, Azospirillum, etc.*

Near-Isogenic Strain Consortia (NISC) for Plant Growth Promotion

The development of near-isogenic strains (NIS) of the core-competent bug that vary only for the gene stack related to one specific relevant functionality, seems more feasible than the earlier approaches discussed above. The core-competent strain with its optimized rhizospheric competence can serve as a nuclear strain which would be diversified into several NIS, each equal in their rhizospheric competence, but specialized in one single target function of bio-fertilizing or bio-stimulating or bio-controlling or bio-remediation. When these NIS are made into a inoculum consortium, say the NIS consortium (NISC), each of the constituent of the consortium would not out-compete the others for the rhizospheric establishment but complement functionally for PGP. Different NISC with varied constituent-NISs are possible according to the requirements of the target ecology. Thus, the NISC approach also gives a leadway for tailoring of NISC in an ecology specific, need-based style.

PGPR RESEARCH IN INDIA

In India, the phosphate solubilizing and nitrogen fixing bacteria have been developed as effective bioinoculants and are being used for seed inoculation of important crops like wheat, rice, barley, millet, soybean, sunflower, potato, sorghum, maize and fruits and vegetables. It seems that the potential of rhizobacteria particularly symbiotic nitrogen fixing rhizobia and free living azotobacters to increase crop yield was known in India. The seed inoculants for solubilization of natural phosphate in soils has been recommended in former Soviet Union and India as commercial bioinoculants like 'Phosphobacterin' and 'Microphos', 'Nitragin', 'Azotobacterin' respectively and their appliction in fields have been shown to increase crop yields. The bioinoculatns of *Rhizobium*, *Frankia* and *Bradyrhizobium* for legumes, *Azotobacter* and *Azospirillum* for cereals as well as BGA for paddy application has been developed by Indian Agricultural Research Institute (IARI), New Delhi. These biofertilizers have been constantly used by Indian farmers and thus helped for country's sustainable agriculture. Significant improvement observed due to bacterization with PGPR (which includes genera *Acetobacter, Acinetobacter, Arthrobacter, Azotobacter, Azospirillum, Bacillus, Berkholderia, Enterobacteria, Klebsiella, Proteus, Pseudomonas, Rhizobium, Serratia and Xanthomonas*) during various field trials indicates the possibility evolving eco-friendly input for farmers with limited sources. In the 10th plan (2002-2007) major thrust by Government of India has been given on exploiting biodiversity of nitrogen fixers and PGPR biofertilizer technology improvement.

Indian Institute of Spices Research (IISR), Calicut has standardized fluorescent PGP *Pseudomonas* for large scale production of spices, especially, black pepper and transferred the technology to entrepreneurs for distribution. In order to harness potential benefits of biofertilizers in commercial agriculture, the consistency of their performance must be improved. This requires research in many diverse areas as these biological systems involve complex interactions among the host, other rhizosphere micro flora and fauna and the environment.

Currently all agricultural universities and IARI are involved in research on PGPR and biocontrol by rhizobacteria. In addition to these, many research groups in Indian universities are doing extensive research on occurrence, distribution, mechanism and role of PGPR strains for field applications. Critical research work done at ICRISAT (International Crops Research Institute for Semi Arid Tropics), Patancheru, Hydrabad has been a constant source of our knowledge of biological nitrogen fixation and other aspects of PGP and bacterial associations with crop plants. So in the country like India research on diversity of PGPR and their prospect as future fertilizers to increase soil productivity has bright future and important role in the development of country's sustainable agriculture to meet demand of food for growing population.

CONCLUSIONS AND FUTURE PROSPECTS

More attention is needed to orient our future research on some of the areas *viz*., PGPR survival potential, identification of crop specific strains, mixture formulation, biochemical/ molecular markers for identification of effective PGPR and genes activated in octadecanoid pathway and Phenyl propanoid pathway due to bacterial treatment in order to enhance the performance of biocontrol bacterial inoculants for managing plant pest and diseases in sustainable manner. Most PGPR strain selected from the rhizosphere do not fulfill their initial promise. Failures are often due to poor rhizosphere competence, and to difficulties associated with the instability of bacterial strains in long term culture. The intimate relationship between endophytic bacteria and their hosts could make them good candidates for successful biocontrol agents. This would in part circumvent the problem of selecting bacteria that display high rhizosphere competence which is often considered obligatory for successful seed or root bacterization treatment.

It should be of paramount importance to study the microbial succession in various crop combinations, nutrient availability at various depths, soil ecology to understand population dynamics of microorganisms. Microbial consortia concept need to be coupled with an ideal and quality product development which the farming community can handle with ease and thereby realize the benefits of the PGPR technology for sustainability in the perennial cropping systems.

Inconsistent performance of biocontrol agents in the field thus far has plagued researchers and their efforts to exploit them for commercial application. Therefore, there is a compelling need to identify efficient PGPRs to be used either alone or consortia could be priority, so as to ensure consistent performance in the farmer's field. Since the soils in the mixed cropping system harbour both fungal and nematodal pathogens and these pathogens are not spatially segregated and form a component of the same ecosystem, identification of PGPR with multiple mode of action to suppress pathogens is logical.

It needs to be remembered nevertheless, that most of the world's rice farmers who live in Asia are resource-poor. Therefore, cost effective formulations of biocontrol agents that perform consistently in their fields when made available, either by them or as part of an integrated disease management (IDM) package will benefit these small income group rice farmers with increased grain harvests.

In India, development of biological control technologies involving PGPR for pest and disease management is a newly emerging field. The major criteria for consideration of successful biocontrol technology involving PGPR are identification of effective agents, formulations, mass production technology and delivery systems. In addition, keeping in view of the current and future restrictive legislation against chemical pesticides, the process of commercial formulation with good delivery system should be implemented with compatibility to industrial and commercial development methods and field applications.

Further research is needed to study environmental factors affecting in situ synthesis of PGRs in the rhizosphere, relations between concentrations of PGRs in the rhizosphere and plant responses, the influence of exogenous PGRs in a nutrient-limited matrix, stability and kinetics of PGR production and microbial degradation in soil, physicochemical adsorption and chemical transformations of PGRs that promote their synthesis in the rhizosphere soil, agronomic practices that promote the synthesis and stability of PGRs by the soil-root environment and enhance the availability and uptake of PGRs by plant roots, and interaction of rhizosphere PGRs with other biologically active substances and their effects on microbial activities.

An understanding of these aspects can aid in the utilization of microbial PGRs for the betterment and benefit of sustainable agriculture.

Basic research on the molecular mechanism of biological control, ISR and growth promotion, need to be intensified in order to identify the genes controlling these. Since strain improvement for bio-efficacy is important to reduce the cost of the products. The discovery that specific volatile organic compounds produced by PGPR can promote plant growth opens a new research direction for mode of action work. A complete understanding of the mechanisms of the synergistic activity between PGPR and *Rhizobium* is needed for the better utilization and improvement of bioinoculants in legume based cropping systems.

REFERENCES

Arshad, M and Frankenberger, W.T.J. 1998. Plant growth regulating substances in the rhizosphere : Microbial production and functions *Adv. Agron.,* **2**: 145-151.

Bangera, G.M. and L.S. Thomashow. 1999. Identification and characterization of a gene cluster for synthesis of the polyketide antibiotic 2,4-Diacetylphloroglucinol from *Pseudomonas fluorescens* Q2-87. *J. Bacteriol.,* **181**: 3155-3163.

Bashan, Y and de-Bashan, L.E. 2004. *Azospirillum*-plant relationships : II, interaction with plants, agricultural and environmental advances (1997-2002). *Can. J. Microbiol.,* in press.

Binns, A.N. 1994. Biochemical, genetic, and molecular approaches. *Annu. Rev.Plant.Phys.Plant. Mol. Biol.* **5**: 173-196.

Bolemberg, G.V., M.M. Camacho-Carvajal, T.F.C. Chin-A-Woeing, L.C. Dekkers, A.M. Cees, J.J. van den Hondel, I. Kuiper, A.L. Lagopodi, G.E.M. Lamers, I. Mulders, A.F.J. Ram, S. deWeert, A.H.M. Wijfejes, and B.J.J. Lugtenberg 2002. Rhizosphere colonization by biocontrol *Pseudomonas* spp. In : Proc. 5th Intl. Plant Growth Promoting Rhizobacterial Workshop, 29 Oct. – 3 Nov. 2002, Cordoba, Argentia.

Bolemburg, G.V. and B.J.J. Lugtenberg. 2001. Molecular basis of plant growth promotion and biocontrol by rhizobacteria. *Curr. Opin. Plant Biol.,* **4**: 343-350.

Brazil, G.M., L. Kenefick, M. Callanan, A. Haro, V. doLorenzo, D.N. dowling, and F.O'Gara 1995. Construction of rhizosphere pseudomonad with potential to degrade polychlorinated biphenyls and detection of *bph* gene expression in the rhizosphere. *Appl.Environ. Microbiol.,* **6**: 1946-1952.

Chanway, C.P. and Holl, F.B. 1994. Growth of outplanted lodgepole pine seedlings one year after inoculation with plant growth promoting rhizobacteria *Forest Sci.,* **4**: 238-246.

Corbell, N and Loper, J.E. 1995. A global regulator of secondary metabolite production in *Pseudomonas fluorescens* pf-5. *J. Bacteroil,* **7**: 6230-6236.

Costacurta, A and Vanderleyden, J. 1995. Synthesis of Phytohormones by plant associated bacteria. *Crit. Rev. Microbiol.,* **2**: 2001-2013.

De Meyer, G., Capieau, C., Audenaert, K., Buchola, A., Metraux, J.P. and Hofte, M. 1999. Nanogram amounts of salicylic acid produced by the rhizobacterium *Pseudomonas aeruginosa* TNSK 2 activate the systemic acquired resistance pathway in bean. *Mol. Plant Micro.b Interact*., **12**: 450-458.

de-Bashan, L.E., Hernandez, J.P. and Bashan, Y. 2003. Microalgae growth promoting bacteria as "helpers" for microalgae : a noval approach for removing ammonium and phosphorus ions from wastewater. In : *Phosphate solubilizing bacteria.* (Ed) Velazquez, E. Kluwer Academic publishers, Dordrecht, The Netherlands (in press).

de-Bashan, L.E., Moreno, M., Hernandez, J.P. and Bashan, Y.2002. Removal of ammonium and phosphorus ions from synthetic wastewater by the microalgae *Chlorella vulgaris* coimmobilized in alginate beads with the microalgae growth promoting bacterium *Azospirillum brasilense Water. Res.,* **36** : 2941-2948.

Dekkers, L.C., C.C. Phoelich, and B.J.J. Lugtenberg. 1999. Bacterial traits and genes involved in rhizosphere colonization. In : *Microbial Biosystems.* C.R. Bell, M. Brylinsky, P. Johnson-Green (eds.), Atlanta Canada Society for Microbial Ecology, Halifax, Canada.

Enebak, S.A. and Carey, W.A. 2000. Evidence for induced systemic protection to *Fusarium* rust in Loblolly pine by plant growth promoting rhizosphere *Plant Disease,* **84**: 306-308.

Gaind, S. and Gaur, A.C. 19991. Thermotolerant phosphate solubilizing microorganism and their interaction with mungbean. *Plant and Soil*, **133**: 141-149.

Gaudin, V., Vrain, D and Louanin, L 1994. Bacterial genes modifying hormonal balance in plant. *Plant Physiol. Biochem.* **3**: 11-29.

Glick, B.R. 1995. The enhancement of plant growth by free-living bacteria. *Can. J. Microbiol.,* **1**: 109-117.

Glick, B.R., Jacobson, C.B., Schwarze, M.M.K. and Pasternak, J.J. 1998 1-Aminocydopropane-1-carboxylic acid deaminase mutants of the plant growth promoting rhizobacterium *Pseudomonas putida* GR 12-2 do not stimulate canola root elongating. *Can. J. Microbial*, **4**: 911-915.

Glick, B.R., Patten, C.L., Holguin, G and Penrose, D.M. 1999. *Biochemical and Genetic Mechanisms used by Plant Growth-promoting Bacteria.* London : Imperial College press.

Gow, N.A.R., Campbell, T.A., Morris, B.M., Osborne, M.C., Reid, B., Shepherd, S.J. and Van West, P. 1999. Signals and Interaction between Phytopathogenic Zoospores and Plant Roots. In : *Microbial signalling and communication* (Ed) R.R. England, G. Hobbs, N.J. Bainton and D. Roberts. Cambridge University press.

Haas, D., Keel, C. and Reimmann, C. 2002. Signal transduction in plant beneficial rhizobacteria with biocontrol properties. *Antonie van Leeuwenheek* **8**: 385-395.

Hall, J.A., Peirson, D., Ghosh, S and Glick, B.R. 1996. Root elongation in vartous agronomic crops by the plant growth promoting rhizobacterium *Pseudomonas putida* GR 12-2 *Isrl. J. Plant. Sci.,* **4**: 37-42.

Idriss, E.E., O. Makarewicz, A. Farouk, K. Rosner, R. Greiner, H. Bochow, T. Richter, and R. Borriss. 2002. Extracellular phytase activity of *Bacillus amyloliquefaciens* FZB45 contributes to its plant growth promoting effect. *Microbiol.*, **148**: 2097-2109.

Iyer, R. 1983. Rhizosphere microflora in crop mixed coconut soil. *Philippine J. Coconut Studies* :42-45.

Jackson, M.B. 1993. Are plant hormones involved in root to shoot communication? *Adv. Bot. Rev.,* **1**: 103-186.

Jaeger, C.H., Lindow, S.E., Miller, W., Clark, E and Lirestone, M.K. 1999. Mapping of sugar and amino acid availability in soil around roots with bacterial sensors of sucrose and tryptophan. *Appl. Environ, Microbiol.*, **5**: 2685-2690.

Kariola, T., Palomaki, T.A., Brader, G. and Palva, E.T. 2003. *Erwinia carotovora* sub sp. *carotovora* and *Erwinia* – derived elicitors HrpN and PehA trigger distinct but interacting defense responses and cell death in *Arabidiopsis. Mol. Plant Microbe Interact.,* **6**: 179-187.

Kloepper, J.W., Lifshitz, R. and Novacky, A. 1980. *Pseudomonas* inoculation to benefit plant production. *Anim. Plant Sci.* 60-64.

Kombrink, E and Somssich, I.E. 1995. Defense responses of plants to pathogens. *Adv. Bot. Res.* **2**: 1-34.

Kulkarni, S. and C.S. Nautiyal 2000. Effects of soft and pH stress on temperature-tolerant *Rhizobium* sp. NBRI 330 nodulating *Propsopis juliflora. Curr. Microbiol.,* **40**: 221-226.

Kulkarni, S., Surange, S and Nautiyal, C.S. 2000. Crossing the limits of *Rhizobium* existence is extreme conditions. *Curr. Microbiol.*, **41**: 402-409.

Kumar, G.N. 2003. Improvement of PGPR by transformation. Prospects and challenges. 6th International PGPR workshop held at IISR calicut, India. 5-10 Oct, 46.

Kunc, F and Macura, J. 1998. Mechanisms of adaptation and selection of micro-organisms in the soil. In : *Soil Microbial Associations*, pp 281-299. (Ed) V. Vancura and R. Kunc Amsterdam, Elsevier.

Leeman, M., den Ouden, F.M., Van Pelt, J.A., Drik, F.P.M., Steiil, H., Bakker, P.A.H.M. and Schippers, B. 1996. Iron availability affects induction of systemic resistance to *Fusarium* wilt of radish by *Pseudomonas fluorescens. phytopathol*., **6**: 149-155.

Lorito, M., Hayes, C.K., Di Pietro, A and Harman, G.E. 1993. Biolistic transformation of *Trichoderma harzianum* and *Gliocladium virens* using plasmid and genomic DNA. *Curr. Genet.,* **2**: 349-356.

Lorito, M., Hayes, C.K., Zoina, A., Scala, F., Del Sorbo, G., Woo, S.L. and Harman, G.E. 1994. Potential of genes and gene products from *Trichoderma* sp. and *Gliocladium* sp. for the development of biological pesticides. *Mol. Biotechnol.,* **2**: 209-217.

Lucht, J.M. and Bremer, E. 1994. Adaptation of *Escherichia coli* to high osmolarity environments. Osmoregulation of the high affinity glycine betaine transport system. *Pro. FEMS. Microbiol. Rev*., **4**: 3-20.

Mark, G.L., J.P. Morrissey, and F.O'Gara. 2003. Designing improved GM bacteria for application in environmental biotechnology. Pages 11-24 In : Ecological Impacts of GM Dissemination in Argo-Ecosystems. Lelly T., E. Balzs, and M. Tepler (eds.), Facultas, Austria.

Matthyssee, A.G. and S. McMahan. 1998. Root colonization by *Agrobacterium tumefaciens* is reduced in *cel, attB, attD* and *attR* mutants. *App. Environ. Microbiol.*, **64**: 2341-2345.

Mercado-Blanco, J., Van der Drift, K.M., Olsson, P.E., Thomas-Oates, J.E., Van Loon, L.C. and Bakker, P.A. 2001. Analysis of the pms CEAB gene cluster involved in biosynthesis of salicylic acid and the siderophore pseudomonine in the biocontrol strain *Pseudomonas fluorescens* WCS 374. *J. Bacteriol*., **8**: 1909-1920.

Nair, S.K. and Subba Rao, N.S. 1977. Microbiology of the root region of coconut and cocoa under mixed cropping. *Plant soil,* **46**: 511-519.

Nautiyal, C.S., Bhaduria, S., Kumar, P., Lal, H., Mondal, R and Verma, D. 2000. Stress induced phosphate solubilization in bacteria isolated from alkaline soils. *FEMS Microbiol. Lett.***, 182**: 291-296.

Neilands, J.B. and Nakamura, K. 1991. Detection, determination, isolation, characterization and regulation of microbial iron chelates. In : *CRC Handbook of Microbial Iron Chelates.* (Ed) G.Winkelmann. London : CRC press.

Norman-Setterblad, C., Vidal, S. and Palva, E.T. 2000. Interacting signal pathways control defense gene expession in *Arabidopsis* in response to cell wall degrading enzymes from *Erwinia carotovora Mol. Plant Microbe Interact*., **3**: 430-438.

Nowak-Thompson, B., N. Chaney, J.S. Wingh, S.J. Gould and J.E. Loper. 1999. Characterization of the Pyoluteorin biosynthetic gene cluster of *Pseudomonas fluorescens* Pf-5. *J. Bactriol.,* **181**: 2166-2174.

Panwar, J.D.S. and Sirohi, G.S. 1989. *Azospirillum.* An improtant biofertilizer for improving crop productivity. *Farmers and Parliament*, **24**: 23-24.

Panwar, J.D.S., Saikia, S.P. and Naidu, V.S.G.R. 2001. Biofertilizers for enhancing crop productivity and environmental security. *Indian Farming* (special issue) during 88th session of the Indian Science Congress, **50**(10) : 56-60.

Pennickx, I.A., Eggermont, K., Terras, F.R., Thomma, B.P., De Samblanx, G.W., Buchala, A., Metraux, J.P., Manners, J.M. and Broekaert, W.F. 1996. Pathogen-induced systemic activation of a defensive gene in *Arabidopsis* follows a salicylic acid-independent pathway. *Plant cell,* **8**: 2309-2323.

Persello-Cartieaux, F., Nussaume, L. and Robaglia, C. 2003. Tales from the underground: molecular plant-rhizobacteria interactions. *Plant Cell Environ.,* **2**: 189-199.

Pierson, L.S. and Pierson, E.A. 1996. Phenazine antibiotic production in *Pseudomonas aureofaciens* : Role in rhizosphere ecology and pathogen suppression. *FEMS. Microbiol. Lett.,* **2**: 101-108.

Pieterse, C.M., and Van Loon, L.C. 1999. Salicylic acid-independent plant defence pathways. *Trends Plant Sci.,* **4**: 52-58.

Potter, S., Uknes, S., Lawton, K., Winter, A.M., Chandler, D., DiMaio, J., Novitzky, R., Ward, E. and Ryals, J. 1993. Regulation of a heveine like gene in *Arabidopsis. Mol. Plant Microbe Inberact.,* **6**: 680-685.

Raazijmakers, J.M., Leeman, M., Van Oorschot, M.M.P., Van der slwis, I., Schippers, B and Bakker, P.A.H.M. 1995. Dose-response relationships on biological control of *Fusarium* wilt of radish by *Pseudomonas* spp. *Phytopathol.,* **8**: 1075-1081.

Rainey, P.B. and G.M.. Preston. 2000. *In vivo* expression technology strategies : Valuable tool for biotechnology. *Curr. Opin. Biotechnol.,* **11**: 440-444.

Rangarajan, S., Loganathan, P., Saleena, L.M. and Nair, S. 2001. Diversity of pseudomonads isolated from three different plant rhizospheres *J. Appl. Microbiol.*, **91**: 742-749.

Rangarajan, S., Saleena, L.M., Vasudevan, P and Nair, S. 2003. Biological suppression of rice diseases by *Pseudomonas* spp. under saline soil conditions. *Plant Soil*, **251**: 73-82.

Reddy, N.R., M.D. Pierson, S.K. Sathe, and D.K. Salunke. 1989. *Phytases in Cereals and Legumes*. CRC Press, Boka Raton.

Rehman, A. and Nautiyal, C.S. 2002. Effect of drought on the growth and survival of the stress-tolerant bacterium *Rhizobium* sp. NBRI 2505 sesbania and its drought-sensitive transposan Tn5 mutant. *Curr. Microbiol.,* **45**: 368-377.

Ryu, C-M., Farag, M.A., Hu, C-H, Reddy, M.S., Wei, H.X., Pare, P.W. and Kloepper, J.W., 2003. Bacterial volatiles promote growth in *Arabidopsis. Proc. Natl. Acad. Sci.,* **100**: 4927-4932.

Samac, D.A. and Shah, D.M. 1994. Effect of chitinase antisense RNA expression on disease susceptibility of *Arabidopsis* plants. *Plant Mol. Biol.,* **2**: 587-596.

Samiyappan, R. 2003. Molecular mechanisms involved in the PGPR mediated suppression of insect pests and plant pathogens attacking major agricultural and horticultural crops in India. *6th International PGPR workshop*, 5-10 Oct., India.

Sarma, Y.R., Rajan, P.P., Paul, D., Beena, N and Anandaraj, M. 2000. Role of Rhizobacteria on disease suppression in spice crops and future prospects. In : *Seminar on Biological control with plant growth promoting Rhizobacteria for sustainable agriculture*. University of Hyderabad, Hyderabad.

Schnider, U., Keel, C., Blumer, C., Troxler, J., Defago, G and Haas, D. 1995. Amplification of the housekeeping sigma factor in *Pseudomonas fluorescens* CHAO enhances antibiotic production and improves biocontrol abilities. *J. Bacteriol.*, **7**: 5387-5392.

Schroth, M.N. and Hancock, J.G. 1982. Disease-suppressive soil and root colonizing bacteria. *Science*, **2**: 1376-1381.

Singh, H.P. and Singh, T.A. 1993. The interaction of rock phosphate, *Bradyrhizobium*, *vesicular-arbuscular mycorrhizae* and phosphate-solubilizing microbes on soybean grown in a sub-Himalayan mollisol. *Mycorrhiza*, **4**: 37-43.

Smith, J.L., Papendick, R.L., Bezdkek, D.F. and Lynch, J.M. 1993. Soil organic matter dynamics and crop residue management. In : *Soil Microbial Ecology*, 65-94. (Ed) F.B. Melting, New York.

Thomashow, L.S., Weller, D.M., Bonsall, R.F. and Pierson, L.S. 1990. Production of the antibiotic phenazine-1-carboxylic acid by fluorescent *Pseudomonas* species in the rhizosphere of wheat. *Appl. Env. Microbiol.*, **56**: 908-912.

Tripathi, A.K., Mishra, B.M. and Tripathi, P. 1998. Salinity stress responses in the plant growth promoting rhizobacteria *Azospirillum* spp. *J. Bio-Sci.*, **23**: 463-471.

Walsh, U.F., J.P. Morrissey, and F.O'Gara. 2001. Pseudomonas for biocontrol of phytopathogens: from functional genomics to commercial exploitation. *Curr. Opin. Plant Biol.*, **12**: 289-295.

Wei, Z.M., Laby, R.J., Zumoff, C., Bauer, D.W., He, S.Y., Collmer, A and Beer, S.V. 1992. Harpin, elicitor of the hypersensitive response produced by the plant pathogen. *Erwinia amylovora. Science,* **7**: 85-88.

Developments in Physiology, Biochemistry and Molecular Biology of Plants, 2005
Eds.: Bandana Bose and A. Hemantaranjan
Vol., 1, pp. 181-201, New India Publishing Agency, New Delhi
E-mail: spjain_niph@rediffmail.com web: www.bookfactoryindia.com

CHAPTER - 9

TOWARDS AN UNDERSTANDING OF THE FACTORS AFFECTING PRODUCTIVITY OF FRUIT CROPS

V. K. SINGH

INTRODUCTION

India has a big challenge in fruit production in the current backdrop of decreasing per capita land holding and static fruit productivity. In recent years, China has surpassed and secured top position in the fruit production in the world accounting for 4,44,651 mt production followed by India and Brazil. Now Chinese share in the fruit production is around 13 per cent while India contributes 10.2 per cent, Brazil 8 per cent and USA 6 per cent of total fruit production.

India is bestowed with a wide variety of agro-climatic conditions and enjoys an enviable position in the horticultural map of the world. Almost all types of fruit crops (tropical, sub-tropical and temperate) are grown in one or other part in the country. The major tropical and sub-tropical fruits grown in India include mango, banana, papaya, orange, mosambi, guava, grape, pineapple, coconut, sapota, ber, pomegranate, litchi, aonla, bael, jackfruit *etc.* Cashew nut cultivation has a big potential and its production, productivity and export has increased significantly in recent decades. Indian grape has recorded highest productivity per unit area in the world. The productivity of fruit per unit area in India has increased nearly from 10 t/ha to 12 t/ha in almost one decade. However, the average productivity in most of the fruit crops is far below than satisfactory except grape and banana in few states. It is paradox to record that India's contribution in the overseas market is almost (0.11 per cent) negligible (Chadha, 2000). However, other countries like Mexico, Philippines and Venezuela, which produce far less, export 4 per cent of their total production.

As per recommended dietary allowances (RDA), minimum per capita consumption of fruit is 90 g. Accordingly, a minimum production of 60 million tonnes of fruit is required to meet the need of present population of the country. Further, WTO regime has necessitated the increased production of quality fruit for export as well as to compete in internal market with the imported fruits (Rathore, 2001). Therefore, to achieve the target of required fruit production in future, the aim of fruit research strategies should be to understand the physiological basis of fruit productivity in order to garner higher productivity of quality fruits per unit area.

FACTORS FOR LOW PRODUCTIVITY OF FRUIT CROPS

Inadequate Conservation of Genetic Resource

In order to launch a successful varietal improvement programme, it is necessary to have a rich germplasm bank in any crops. Interestingly, large numbers of fruits owe their origin in the Indian sub-continent. Many factors like increasing urbanization, decline of old plant material, unscrupulous exploitation of wild resources are causing havoc to survival of indigenous and rare species of fruit crops like *Mangifera* spp, *Citrus* spp, *Artocarpus* spp, *Emblica officinalis*, *Aegle marmelos*, *Syzyzium cuminii, Prunus* spp, *Pyrus* spp. and many other important fruit species. Therefore, systematic efforts to conserve these materials in *in situ, ex situ* or conservation in a conventional gene bank are essentially required. The documentation of germplasm is another key issue that must be addressed. Their systematic study, location of specific genes and their incorporation need to be exploited. Molecular characterization of important germplasm will go a long way in systematic exploitation of the material.

Inferior Planting Material

One of the important aspects of successful development of commercial orcharding is the use of high quality planting material, which in turn manifests through production of quality fruits. Lack of genuine planting material remains one of the most important factors for low productivity of many fruit crops in India. Most of the fruit nurseries are not properly monitored on quality issue. Still majority of nurserymen are using the age-old techniques of propagation. It is surprising that multiplication in mango and guava are still being done by approach grafting on non-descript rootstock resulting in inferior plants with variable performance. Despite development of several new propagation techniques in mango, nurserymen are still adhere to old practice of inarching, which greatly affects rate of multiplication. Similarly, crops like guava, litchi and citrus are being multiplied through air layering which is a sluggish and cumbersome. Provision of separate mother block propagated from elite clones and their scientific management is still a dream. Therefore, multiplication should be done only from the mother plants of established superiority. It would be desirable to establish elite orchards of important fruit crops in the fruit growing state for the supply of authentic plant materials. Clonal selection with established cultivars would be an important aspect of this programme. (Singh, 1996). Micropropagation has shown promise in multiplication of banana. The low cost technology of banana tissue culture should be passed on to nurserymen and the banana tissue culture material along with package of practice should be made available to farmers. This technology can be tried with other fruit crops as well.

Lack of Intensive Orchard Creation

Most of the fruit orchards are at present planted at low density and such orchards provide low returns with long gestation period. Lack of dwarfing rootstock and non-availability of precocious cultivars are the main reasons for the low-density plantation. Therefore, transforming fruit industry through high/medium density planting is the dream of day for number of fruits. High density planting increases productivity and fruit quality, shortens juvenility, gives high early returns and provides better use of natural resources like light, water, nutrient and easy harvest. To have a full physiological control of the tree in the high density, it would be essential to have dwarf tree. A shallow canopy (1.5 – 2.0 m depth) is needed in high density to achieve maximum efficiency for trapping solar energy through foliage and channel metabolites for quality fruit production. In India, high density planting have been successfully demonstrated for increased yield in apple, banana, pineapple, papaya and mango, but most of the orchards are still under the traditional low density system, resulting in low average productivity (Giesen, 1990; Ram *et al.*, 1997; Awasthi and Mehta, 2000). High density technology developed in mango utilizes vigorous seedlings stock of varying genotype rather than dwarfing stock like in many temperate fruits. Growth control is primary requirement of high density orchard, which can be achieved by dwarfing rootstock, pruning, precocious scion cultivars, use of chemical *etc.* Recent advancement in tree physiology has shown that growth retardant has tremendous potential to control growth of tree with or without dwarfing rootstock and scion cultivars, which is prerequisite of high density orchard. In mango, dwarfing rootstock or scion cultivar is not available, but chemical like paclobutrazol was found (Singh, 2000 personal communication) effective to control the tree growth by reducing the xylem to phloem ratio with higher yield. Besides this, tree is also managed dwarf through training and pruning and growth is restricted within planting distance provided between the trees. Closer the density, higher the productivity has been the general guiding principle. High density orchard having closer planting (3.0 X 3.0 m) in mango for regular crop is practiced through training and annual pruning after crop harvest and induction of flowering through paclobutrazol in alternate bearing cultivars like Langra and Dashehari. The training helps to develop proper frame of trees in early stages of growth while pruning helps to curtail growth and maintain tree vigour on sustainable basis for regular fruiting year after year. In Israel, productivity of mango has been doubled by adopting high density planting technology. Mango tree training technique for high density for hot tropics has been developed by Campbell and Wasielewski , (2000). Plant density in papaya plays a vital role in productivity per unit area. Increasing plant density can enhance the yield per unit area. It is generally grown at planting density of 1400 to 1700 plants / ha. A plant population of 2500 / ha is recommended for high density planting (Awasthi and Mehta, 2000). With the development of dwarf cultivars, *viz.,* Pusha Nanha and Ranchi Dwarf, it is possible to plant papaya still closer. Ranchi Dwarf planted @ 2922 plant per hectare yielded 98.05 t / ha fruits and Pusha Nanha planted at 1.25 X 1.25 m (6400 plant / ha) yielded 60 – 65 t / ha as compared to traditional yield of 15 – 20 t / ha (Ram, 1983).

Alternate and Erratic Bearing

Alternate and erratic flowering and fruiting along with the low fruit set and excessive fruit drop is one of major problems in fruit crops in general and mango in particular. Each fruit tree in commercial groves does not bear equal crops year after year. Climatic variations in particular year, genetic nature of cultivar as well as physiological changes occurring in trees with the progress of time are the main reason for erratic bearing and accounting for low productivity. Alternation in cropping habit of mango is used as a synonym for poor yields. In mango, growth flushes are very erratic and occur up to three to four times per year on individual shoots, depending upon cultivar and growth condition and they tend to flower only after 9 – 10 months after attaining physiological maturity (Pandey, 1989). In north India, March-April flush is the major one that accounts for over 80 per cent of annual growth and its shoots has the maximum potential for becoming new fruiting shoot (Hanumashetty, 1978). The other flushes are minor ones and are not appreciably related to flowering. Therefore, biennially / irregularity in flowering would ensue because of inability of one shoot to bear vegetative growth and flower in the same year. Several chemicals and plant growth regulators like Chlormequat Chloride (Maiti *et al.*, 1972) Ethephon, KNO_3, Salicylic acid and triazoles particularly paclobutrazol have been found effective to regulate the flowering and bearing in mango, apple, citrus *etc.* (Bondad *et al.*, 1978; Lopez, 1984; Kulkarni, 1988; Singh, *et. al.*, 2001; Singh *et al.*, 1998; Davenport *et al.*, 1993; Singh *et al.*, 2001). Amongst them paclobutrazol is being widely used to increase flowering, enhance yield and control the alternate bearing habit in commercial mango orchards of India (Burondkar and Gunjate, 1993; Singh and Saini, 2001), China (Tongumpai *et al.*, 1989), Australia (Rowley, 1990); and South Africa (Hilliere and Rudge, 1991; Voon *et al.*, 1991). Triazole having anti-gibberellin activity induce flowering even in the 'off' year of bearing by regulating the synthesis of gibberellins (Singh and Saini, 2001). There have been numerous studies on the inhibitory effect of GA_3 on flowering of fruit crops (Singh, 1961; Andrews and Le Fook, 1985; Chen, 1985; Mullins, 1985, Oosthusyse, 1995). Abundant axillary flowering and cauliflory panicles were also observed specially on the trees which received higher dose of paclobutrazol which simply reflect the suppression of apical dominance and activation of axillary meristem during the floral cycle. Paclobutrazol is also being commercially used to advance harvesting of the mango cultivars by about a month (Singh and Saini, 2001). However, its application through judicious nutrient management also needs to be integrated for continued and sustainable production.

Papaya (*Carica papaya* L.) is one of the most important crops grown in tropics and sub-tropics. The major bottleneck in papaya cultivation is its inherent heterozygosity, dioceous nature and susceptibility to number of viral diseases. Sex reversal under varied environmental condition is also one of the reasons for low productivity of papaya (Bose *et al.*, 2001). Papaya is a polygamous plant and has many sex forms. There are three basic sex forms, hermaphrodite or bisexual, pistillate or "female", and staminate or "male" (Storey, 1941). Amongst these, only female is stable whereas flowers of hermaphrodite and male vary in sex expression under different environment conditions. Planting duration has been observed to determine the sex expression as high percentage of female (66.0%) produced during November planting closely followed by September (65.0 %) planting in cultivar Pusa Delicious. Sex in (*Carica papaya* L.) papaya is determined by three homologous genes complexes on sex chromosomes (Horovitz, 1954; Storey, 1953). The genes are so tightly linked that no crossing-over occurs among them; thus the complexes are transmitted to offspring with pleiotropic effect on phenotypic expression. Sex in papaya cannot be identified unless they flower but the ratio can be predicted provided it is pollinated under controlled condition. It was estimated that the average yield of papaya could vary from 37.23 t / ha with a maximum of 6.0% male plants present to 19.96 t/ha with 50% male plant. The male plants serve only as pollenizer and hence, it would be adequate to leave one male plant for every 20 female plants. Thus, removal of male plants allowing the robust growth of female plants is a potential strategy to increase the production of papaya. The profitable productive life of papaya is two and half years under northern Indian conditions provided the crop is well managed, therefore, they should be replaced by the new plantation for getting profitable yield.

Several chemical and plant growth regulators have been used to improve fruit production via producing more normal and female flower. Hermaphrodite trees sprayed with 2,3-dichloroisobutyrate (DCIB, 2 to 6 g/L) and 2,3-dichloropropionate (Dalapon, 2 to 6 g/L) produced more normal and female like (Carpelloid) flowers than untreated trees (Lange, 1961). TIBA, NAA or IAA were also found to induce the flower significantly (Dedolph, 1962). The papain yield, which is considered as one of the important products of papaya, could be increased four-fold compared with control by the application of 200 ppm ethrel (Chacko *et al.*, 1972).

Poor Orchard Efficiency

In most of perennial fruit crops, predomination of old, dense and senile orchards with low productivity (30 – 35%) has become a matter of serious concern. Decline in productivity in old and dense orchard is largely due to poor photosynthetic efficiency coupled with several other compounding factors. Compared to temperate fruit orchards, canopies of tropical and sub-tropical fruit orchards like mango have a higher proportion of 'shade' to 'sun' leaves. The maximum photosynthetic rates for sun-leaves of trees occurred at 60% of full sun light (PPF approximately 1200 μ mol quanta m^{-2} s^{-1}) (Whiley, 1993; Schaffer *et al.*, 1994). For overcoming this problem, rejuvenation technology in mango was developed at Central Institute for Subtropical Horticulture, Lucknow (India) that provides new productive life to existing old and unproductive orchards (Lal and Padaria, 2001). The technique aims at pruning of undesired branches for inducing development of umbrella like open canopy of healthy shoots. Open canopy ensures better light penetration and interception improves photosynthetic efficiency, flowering and fruiting potential of shoots. Pruned trees attain canopy of healthy shoots in two years and from third year onward they start bearing fruits. The new leaves produced after pruning develops capacity to supplement the demand of fruit load. Lower gibberellin levels were recorded from leaves of pruned trees of mango where as pruning augments levels of cytokinins (Shanmugavelu and Saidha, 1993). Lower levels of gibberellins and higher levels of cytokinins favour flowering. Guava has also proved to be very responsive towards pruning for better harvest capacity. In mulberry and apple when shoots were decapitated, photosynthesis in mature basal leaves increased by 36% whereas photosynthesis was reduced in corresponding unpruned trees (Mika, 1986). Thus this technology was found helpful in giving new productive life of the orchards and have potential for the improvement of yield. Reducing the size of petiole and commercial exploitation of micropropagation, development of virus tolerant cultivars will go long way in utilization of full potential of papaya. In apple, pruning of dense tree, thinning of spurs has been demonstrated to improve the tree vigour and fruit quality. Similar to mango rejuvenation technique need to be standardized in aonla and cashew nut plantation. Strategically timed, selective and non-selective pruning need to be commercialized in improving quality production.

PHYSIOLOGICAL APPROACHES FOR IMPROVING FRUIT PRODUCTIVITY

Crop physiology in general and molecular physiology in particular is gaining grounds in solving some of the important problems related to productivity of fruit crops. Use of bio-regulators and chemicals has revolutionized production of few fruits. Various traits of plants, which directly or indirectly affect productivity, can easily be manipulated with the use of newly synthesized or naturally occurring molecule. Molecular biology enables us to insert single genes or two or three genes into the crop to give it new and advantageous characteristics without affecting other traits of the crops. The technology has a wide application especially in tropical and sub-tropical countries. Some of the traits of fruit crops that are being modified through genetic engineering include viral, diseases, stress resistance, increasing shelf life, regular flowering and fruiting, resistance to insect, could acquire great importance. Thus, genetic engineering could in time to come provide the basis for an adequate food supply in the form of high yield and resistant plants. Nevertheless, the success of such a strategy depends upon the four important criteria *i.e.* gene isolation, gene transfer systems, gene expression and regeneration. Molecular physiology has varied application in the field of agriculture. The molecular physiology most probably carries solutions to the difficult problem with which farmers are faced. The following are some of the most important objectives having important bearing on increasing fruit productivity.

Regulation of Flowering

The flowering process is of vital importance to fruit crop productivity as yield is directly dependent upon its success or failure. The erratic and irregular flowering in most of fruit crops causes low orchard efficiency. From the point of view of size and complexity of organization trees represent the highest development of the plant kingdom and to interpret their form and behaviour in physiological terms is the ultimate challenge for the plant physiologist. Many of the problems to be solved are common to herbaceous species, but there are certain facets of plant physiology, which are of particular relevance to the culture of woody perennials. Among these are the phenomenon of juvenility, apical dominance, dormancy, mechanism controlling balance between vegetative and reproductive development, transport and storage of water and metabolites, relation between tree and its aerial as well as subterranean environments and above all the complex hormonal interaction responsible for the co-ordination of growth in the whole tree, are important to understand the process for improvement in production of fruit crops. These phenomenons are quite common in mango, which is perennial and evergreen tree. Various environmental factors such as photoperiod, temperature, water stress and combination of these during floral induction period are important for the induction of flowering. Biennial bearing and erratic flowering in some sub-tropical fruit crops are posing serious problem and has been a major bottleneck in the expansion of the fruit industry especially mango. Although, flowering of fruit crops like mango, citrus can be regulated by certain flower inducing chemicals, including paclobutrazol and potassium nitrate. This approach has been ineffective at some places having different environment conditions. Flowering has thought to be regulated by carbohydrate and nitrogen ratio with high levels being conducive to flowering. Ringing of branches, a practice known to increase C-N ratio of shoots also gives supporting evidence in flower induction in mango. However, it was further emphasized that flowering in Baramasi variety of mango takes place at a wide range of C-N ratio without any relation to C-N status of the shoots. Thus C-N ratio did not always govern the fruit bud formation. In mango, it was observed that more than 30 leaves are required for normal development of fruit, if fruit depends on current photoassimilates. Singh and Saini (2001)) have indicated that the regular bearing cultivars such as Amrapali has potential to develop maximum photosynthetic efficiency and ribulose 1,5-biphosphate carboxylase activity in 20-25 days old leaves whereas irregular bearer Chausa takes about 60-75 days to develop maximum photosynthetic efficiency. Moreover, the long-term use of flower-inducing chemicals on fruit crop is yet to be investigated. Recently, activity of LEAFY gene having flower-meristem from *Arabidopsis thaliana* (Weigel and Nileson, 1995) was found to cause flower initiation in transgenic poplar (*Populus* sp.). Thus, it could be possible to overcome the problem of erratic flowering in fruit crops by introducing this gene in plant species, which are erratic bearer. The regulation of flowering to optimize yield is one of the important approach which has been discussed below.

Mango (Mangifera indica)

In fruit crops, particularly in mango, the initial fruit set is high, but the ultimate retention till harvest is very low (1.12 – 3.66%) and less than 1% fruits reach maturity (Mukherjee, 1949). High level of ethylene and abscisic acid during the initial phase of fruit growth cause the heavy immature fruit drop (Malik, 1999). It was also reported that polyembryonic mango cultivars consistently yield more than the monoembryonic mangoes (Campbell, 1961). These differences in levels of production may be related to higher production of ethylene in monoembryonic mangoes with a high rate of immature fruit drop than the polyembryonic one. It was recently reported that the climacteric feature of fruits is reflected in the sensitivity to ethylene of plantlets *in vitro* (Jana *et al.*, 2002). Thus climacteric fruits (very sensitive to ethylene) can be differentiated from non-climacteric fruits, which were only slightly sensitive to the ethylene *in vitro* conditions. Controlling ethylene production at the initial fruit set could block the high rate of immature fruit loss. The inhibition of ethylene synthesis in tomato has been demonstrated to be effective for controlling fruit ripening (Oeller *et al.*, 1991). This strategy involves genetic transformation with gene that control ethylene production in the antisense configuration *e.g.* ACC synthase and alternative oxidase. At this time the constitutive expression of these genes in transformed plants is under the control of the 35 S promotors from cauliflower mosaic virus. Thus, it would be highly desirable if a flowering specific promotor could be isolated so that ethylene synthesis could be selectively blocked only at the time of early fruit set. Efforts made to introduce ABA are not encouraging because of ABA is regulated by family of genes. Thus, this hi-tech strategy will certainly bring the gene revolution for the fruit crop for getting quality higher fruit yield.

Grape (Vitis vinifera)

Various growth promoters and inhibitors were found to affect flowering of grape. Time of flowering could be delayed by GA (Weaver and Pool, 1971) and strongly enhanced by

H_2CN_2(Shulman, 1983), pentachlorophenol, dinitroorthoeresol (DNOC), thiourea and other dormancy breaking agents. GA application was found to increase the formation of anlagen, but to reduce their differentiation to inflorescences. GA also induced latent buds to develop the current season growth. Growth inhibitors such as MH, ethephon had no effect on floral differentiation. On the other hand, growth retardants such as CCC and Amo-1618, promoted flowering (Coombe, 1967; Weaver, 1975). Cytokinins seem to be a major regulator of grapevine flowering. Application of cytokinins affected inflorescence (Srinivasan and Mullins, 1981) and flower differentiation (Mullins, 1967). Cytokinins were also found to be responsible for the development of young inflorescence and prevention of their abscsion. Srinivasan and Mullins extensively studied the involvement of cytokinin in floral induction, differentiation and development. They showed that infloresence abscission on cutting could be prevented by cytokinin, even when rooting occurred after bud opening (Mullins and Rajsekharan, 1981).

Guava (Psidium guajava)

Flowering in guava may be regulated either by suspending the vegetative growth flush through cultural treatments (root exposure and pruning), withholding irrigation, and fertilization with hand or through chemicals. The main objective in suspending the vegetative growth is to provide rest to plant, which results in accumulation of food reserve in large quantity for enhancing flowering in the next season. Some chemicals such as NAA, 2,4-D, NAD, ethephon and MH have been tried to achieve this objectives. NAA has been found to be most effective chemical for removal of flower bud (Chundawat *et al*., 1975; Pandey *et al*., 1980).

Pineapple (Ananas comosus)

Pineapple is easily induced to flower by application of growth regulators. Pineapple grown on a field scale is induced to flower and fruit production can be timed precisely and harvesting is easier. As with natural flowering, pineapple must attain a minimum plant size before induction can occur. Many chemicals such as acetylene, calcium carbide, ethephon, ethylene NAA or its sodium salt are being used to regulate flowering in pineapple.Forcing success is influenced by temperature and by the same cultural factors that affects natural flowering.

Regulation of Photosynthesis

Poor photosynthetic efficiency in fruit crop is one of important factors responsible for low yield and inferior orchard efficiency (Flore and Lakso, 1989). For normal fruit development a balance photoassimilates is required. Definite leaf number (area): fruit ratio is necessary for photoassimilates to the fruit development The optimum leaf number: fruit ratio has already been determined for apple, grapes and peach (Harley *et al*., 1942; Purohit *et al*., 1979). There has been great interest in the possibility of genetic engineering of genes in order to change and improve the process of photosynthesis. Techniques for specifically manipulating the chloroplast genome have not so far been developed, but site directed mutagenesis is proving to be a valuable approach for altering the first enzyme for photosynthesis system, ribulose biphosphate carboxylase oxygenase (Rubisco). This enzyme fixes CO_2, also catalyses an alternative reaction involving oxygenation of sugar biphosphate. Both carboxylase and oxygenase reactions occur at the same site and compete with each other (Govindjee, 1982). This oxygenase reaction is energetically wasteful, especially at limiting light intensities in the orchard of fruit crops, which in turn decrease the photosynthetic efficiency of the tree resulted in low yield. The amount of light available is a function of climate and cannot be manipulated, and potential net photosynthetic efficiency of a crop is inherent and cannot be altered without genetic manipulation. Modification of Rubisco to eliminate or decrease oxygenase activity should, in theory enhance photosynthetic efficiency. Rubisco is therefore, prime candidate for genetic engineering. Increasing affinity of enzyme for CO_2 may also enhance carboxylase activity. Turnover rate of the enzyme can also be improved. Genes of both subunits of Rubisco have been isolated and sequenced from many plants and can now be subjected to *in vitro* mutagenesis to obtain desired results. Recently, Rubisco gene from pea npt II has been identified that can be introduced in the fruit crop for increasing orchard efficiency through increased photosynthetic rate. Alternatively, C_3 plants may be converted to C_4 type photosynthesis, which is more efficient (Castresana *et al.*, 1988). For optimizing biological yield of the orchard, it would be desirable to manipulate Rubisco enzyme genetically so as to increase carboxylase activity and decrease oxygenase activity, thereby increasing photosynthetic activity of orchard.

Regulation of Tree Size and Canopy Architecture

Lack of dwarfing rootstock in promotion of high-density orchard is one of most important reason for low production of fruit crops. The ability to influence the development and productivity of tree fruits rests in genetic or cultural techniques. Breeding to improve fruit production has so far had limited success. Among several agro-techniques, *viz*; high density planting, control of tree size and canopy management are few important technologies to achieve high productivity per unit area both in short duration and perennial crops. In India, high density planting is being recommended in fruits like pineapple, banana, papaya, citrus and mango. Nevertheless, high-density planting systems were recommended long ago but the growers are yet reluctant to adopt the technique. Delay in acceptance of the high density planting system can be attributed to the lack of a reliable and universally acceptable method of tree vigor control and higher initial capital investment. In mango, although dwarf and compact tree have been identified (*e.g.* 'Amrapali' in India and other selections in Thailand and Pakistan) but the fruit quality of these selections are generally considered to be inferior to that of existing cultivars. The polyembryonic 'Sabre' cultivar has reportedly been effective as a dwarfing rootstock in South Africa although in Israel it was unsuccessful. Therefore, a hi-tech strategy for high density planting in horticultural crops in India would call for the introduction of dwarfing gene for the management of tree size and canopy shape. The use of somatic embryogenesis to propagate selection of polyembryonic mango that have potential as dwarfing rootstock could create a research opportunity that has not been yet exploited. Genetic transformation of embryonic cultures of mango scion cultivars with the dwarfing gene ro / c (Oono *et al.*, 1987) could be an effective strategy for size control. Ideally, the dwarfing gene in 'Amrapali' and other dwarf genotypes could be isolated, cloned and transferred to other mango cultivars.

Biotic Resistant Fruit Cultivars

Different fungal, bacterial and viral diseases reduce crop yields. Although, plant breeders are trying to breed crop plants with in-built disease resistance, this has not always been possible because of species barriers in the transfer of genes that impart disease resistance. Moreover, plants have evolved several defense mechanisms that have enabled them to survive microbial attack. Among the most important barriers to microbial infection are repellents such as alkaloids, saponins and waxes. Plant also responds to microbial infection by the activation of genes that produce antimicrobial compounds like phytoalexins (Anderson, 1978). In addition, some proteins in plants have antibiotic or antifungal activity and especially lectins and chitinases have been shown to have antifungal role. Chitinase is not present in plant but is an important constituent of fungal cell-wall (Wessels and Sietsma, 1981). The function of the chitinase is to hydrolyse the fungal cell wall. Therefore, if chitinase activity could be induced or genes for it inserted in to susceptible plants it would lead to resistance to fungi. *In vitro* selection of embryogenic mango cultures for resistance to the phytotoxin produced by *Colletotrichum* spp. has also demonstrated that the host resistant response is probably linked with the production of chitinase (Jayasankar, 1995). Thus, transformation of mango with constitutively expressed chitinase and other pathogen-resistant genes could be very useful strategy for disease control in mango. In papaya, ring spot virus (PRSV) and leaf curl virus (PLCV) is the main limiting factor for its cultivation. Recently, in Hawaii a genetically transformed papaya cultivar 'Rainbow' resistant to ring spot virus has been produced (Westwood, 1999) and has paved the way for production of virus free papaya. Work on transgenic papaya conferring resistance to leaf curl has just been initiated in India. In citrus the *citrus-tristeza* virus (CTV) is the most devastating citrus virus that limits the quality production of citrus fruits. The CTV-viral coat protein (CP) gene was identified and studies have shown that this gene can be introduced in to the transgenic plants and citrus cultivar free from this virus can be developed (Bose *et al.*, 2001). The transgenic grapevines are also developed through this technique that are resistant to grapevine fanleaf nepovirus (GFNL V). Bt gene, which is getting the higher place for the resistance to many serious pests in cotton, maize and potatoes, is underway for the field test in perennial plants.

Abiotic Stress Resistant Fruit Cultivars

Many environmental factors produce stress conditions for plants such as heat, drought, excess moisture, high salt content of the soil *etc.* As a defense against these stress conditions, various reactions are initiated by the plants, the importance of which is not yet fully understood. In response to heat shock, for example, characteristic proteins are synthesized in many crop plants, precise function of these is not yet known. They are, however, certainly involved in a protective mechanism and are partly transported to the chloroplast and incorporated in to them. Stress-related fluctuations in the specific compound also occur.

Mango, guava and papaya are most important tropical and sub-tropical fruit crops and their full potential is seldom reached because of limitations on physiological; and morphological process imposed by environmental stress. The most important factors limiting productivity are drought, heat and salinity. Heat and drought stress are intimately interrelated. As a result of drought stress, the stoma close down, evaporative cooling is reduced and plant temperature frequently increases to levels where a vital process such as photosynthesis is affected. Soil salinization may arise from intrinsic soil components, use of low quality water for irrigation or excessive use fertilizers. It was estimated that salinization imparts between $4x10^{8}$ to $9x10^{8}$ ha of land, an area that is three time greater than the land currently being used for agriculture (Pasternak, 1982). In mango, Whiley and Shaffer (1997) observed that fruit size of those trees were 34 % smaller, which were subjected to drought stress (1.2 mpa). Fruit quality is also related to water availability; anything that reduces photosynthesis may delay or reduce the colour development of fruit. It was also noticed that fruit harvested from the tree under stress condition contains low amount of starch and will soften earlier (Singh, 2004 Unpublished data).

A general response of plants to a number of abiotic stresses is *in vivo* accumulation of toxic oxygen like hydrogen peroxide, and superoxide radicals that may cause irreversible damage to cellular membrane. The reactive oxygen generated during normal photo respiration in plants is detoxified by an enzyme called superoxide dismutase (SOD). Different isoforms of SOD exist in the cytosol, chloroplast and mitochondria of plant cells and the expression of some forms is dramatically induced during stress condition. Constitutive expression of SOD activity was thought to improve the tolerance of plants to drought, frost hypoxia, exposure to ozone, sulphur dioxide and calcium deficiency. The cytosolic and chloroplastic Cu, Zn – SOD gene of tomato has been introduced into potato resulting in expression of this enzyme in potato (Perl *et al.*, 1993). Research over past two decade has provided a better understanding of molecular biology of stress response in plants. This has led to identification of several genes and gene products that are induced upon exposure to plants to various abiotic stresses *viz.*, drought, salinity and low/high temperatures (Abdin *et al.*, 2001; Grower *et al.*, 2003). Consequently, genetic engineering has been applied to transfer candidate genes from diverse source to susceptible crop plants for developing transgenic resistant to abiotic stresses (Tayal *et al.*, 2004). Recently, transgenic tomato over expressing the At NHX1 gene accumulated high sodium in leaves but not in the fruits and exhibited increased tolerance to salt stress (Zhang and Blumwald, 2001). Recently, transgenic tomato especially expressing the arabidopsis CBF1 gene showed enhanced resistance to drought, chilling and oxidative stress (Hsich *et al.*, 2002). Transgenic potato and tomato with osmotin and Cod A gene are also potential prospects for elevated tolerance to abiotic stress (Babu and Bansal, 1998; unpublished data). Useful work has been done on genes related to salt stress in annual plants (Claes *et al.*, 1990) that can be exploited for the fruit crop. In some fruit crops like citrus, soil salinity significantly limits its production. First time salt sensitive cultivar of citrus recently transformed with the halotolerance gene HAL 2 using *Agrobacterium tumefaciens* was found successful for the resistance of salt stress.

Physiological Disorders

There are number of eco-physiological disorders in fruit crop which limits production and quality of fruit. Among them biennial bearing, malformation, spongy tissue, recurrent flowering, internal necrosis in mango are the . Fruit cracking in citrus, litchi, grape, banana, granulation in citrus are very common disorders, which causes significant losses in fruit production.

Biennial Bearing and Recurrent Flowering

It is a serious problem in mango as most of the commercial cultivars of mango flowering in alternate year. Flowering and fruiting in mango is a complex phenomenon and thus it is not possible to pinpoint a single factor responsible for biennial bearing (Fig. 1). The work on mango hybridization has shown that regular bearing character can be transmitted to the F_1 hybrids. Therefore, there are more chances of tackling this problem through hybridization. A number of cultural practices have been tried to reduce the intensity of biennial flowering and fruiting, however, success has been achieved more to overcome this problem by the application of paclobutrazol (Singh and Saini, 2001). Recently recurrent flowering is noticed in some commercially variety of mango which is characterized by the emergence of new lateral panicles from the base point of early emerged panicles leading to sever fruit drop from the main panicles (Bondaad *et al.*, 1978)). Recurrent flowering not only deprives the farmer from early season premium prices but also reduces the total return anticipated from the orchard. Foliar application of GA @ 200 ppm at 50 per cent flowering was found to minimize the recurrent flowering (Fig. 2).

Fig. 1. Biennial bearing in mango

Fig. 2. Recurrent Flowering

Internal Necrosis

Internal necrosis is also a problem at preharvest of some of fruits like mango (Fig. 3). Such fruits cannot be marketed and affects the economy of growers. First, water soaked greyish spots develop on the lower side of the fruit. Later, the spots enlarge and develop into dark brown necrotic area. The internal tissue start disintegrating and the pericarp and mesocarp is disintegrated exposing the flesh. Fruit also split exposing the internal tissue, which gives and appearance of rotting of tissue. Yellow coloured droplets come out. Stone, also show browning. Affected fruits drop easily. the incidence of internal necrosis can be minimized by the application of borax at the rate of 500 g/tree and foliar spray of borax (1%). The first spray should be done positively at pea stage followed by two more sprays at 15 days interval.

Fig. 3. Internal necrosis in mango

Malformation

Mango malformation, either vegetative or floral, is very common in northern India where temperature becomes low during flowering where as its incidence southern part was sporadic (Fig. 4 and 5). However, its incidence is showing increasing trend in some parts of South India. Therefore, recently it attracted national concern in India since it is a prominent bottleneck in mango cultivation owing to the extensive economic losses. Various biotic and abiotic factors are reported to be associated with the causation of mango malformation. Several works have been done on causes and control of this malady but the results are still inconclusive. Recently, *mangiferin* a natural source for resistance to malformation was identified (Singh and Prasad, 2004). Signal molecule and oxidative enzymes for scavenging active oxygen species during the development of malformation were identified in resistant and susceptible cultivars of mango. However, recently integrated approach for the management of floral malformation was developed on susceptible cultivar of mango Amrapali by pruning, spraying with chelated copper, zinc phosphomidon, carbendazim and NAA. Significant reduction in the incidence of malformation was obtained by this integrated approach. This integrated strategy appears to be promising and will be useful to control the mango malformation (Singh *et al.*, 2001).

Fig. 4. Vegetative Malformation

Fig. 5. Floral Malformation

Spongy Tissue

Alphonso mango, which is the main export cultivar, suffers from a serious malady known as spongy tissue or internal breakdown in the ripe fruits (Fig. 6). This disorder renders the fruit unfit for consumption and hence it has become a bottleneck in export and expansion of its cultivation in the state of Maharashtra, Gujarat where it is grown commercially. There are many biochemical changes associated with spongy tissue, however no conclusive results have been obtained to control this malady. Convicting heat arising from soil and intense solar radiation are reported to be the main cause for this disorder and mulching with paddy straw and dry leaves were found effective for its control (Katrodia and Seth, 1989).

Fig. 6. Spongy tissue in mango

Softening of Fruits

The texture and softening of harvested fruit are of considerable physiological importance and of horticultural concern. From a horticultural stand point, texture serves as an important determinant of quality in harvested fruit and thus influences the methods by which commodities are harvested and handled in market channels. Post harvest softening is a major factor limiting the shipping, storage and shelf life of plumbs, mango, apple and other fruits. It is characterised by loosening of pulp tissues and gelly formation with browning of tissue around the stone by disintegration and breakdown of tissue during the advance stage of maturity of the fruit. The browning of the tissue start in the vascular tissue dose to the stone and gradually spreading outwards and fruits from outside look normal, but inside around the stone flesh becomes gelly like dark yellow and off taste (Fig. 7). This malady renders the fruit unfit for consumption, reducing in nutritional level and consumer acceptance may also arise with these disorders and hence it has become a bottleneck in export and expansion of mango industry.

Fruit Cracking

Fruit cracking is another eco-physiological disorder, which causes even as high losses as 75 per cent in litchi, citrus and grapes. Significant losses also occur in banana and mango (Fig. 8). The cracked fruits deteriorate rapidly and often suffer secondary infestation by disease and pest and fruits ultimately become non-marketable and cause great loss to the

Fig. 7. Softening of tissue in mango cv. Dashehari

Fig. 8. Fruit cracking in mango

grower. Preharvest hormonal spray like 2,4,5-T, NAA, nutritional spray like, K, Ca, Zn, B, Cu, Mo and Mn and maintenance of adequate soil moisture during the dry period were found helpful to reduce the cracking in the fruit crops (Singh and Singh, 1993).

Fruit Drop

In some of the fruit crops like mango, citrus in spite of profuse number of panicles and very high initial fruit set the ultimate retention and harvestable and marketable produce is phenomenally low primarily due to heavy fruit drop (Fig. 9). It is assumes an important facet having a direct bearing on the economic of the crop and is also a bottleneck in the improvement of mango through hybridization. In mango fruit drop is very high at the early stages of development but continues at a low rate till the fruits attain more than half size. The initial set varied from 23.5 fruits per panicle to 74.5 fruits, however, despite this heavy initial number, fruit retention at harvest varied between 0.4 and 0.8 fruits per panicle. Besides many factors for the fruit drop, the main factor at early drop is due to the low level of auxins and gibberellins and high level of ABA and ethylene. Investigations of the pattern of movement and distribution of ^{14}C photosynthates at various stages of flowering and fruit development have shown that the immediate and direct cause for the heavy drop of fruit at pea and marble stages is due to the competition of developing fruits for the photoassimilates among themselves and with the newly emerging vegetative growth. The extent of fruit drop can be reduced significantly by mulching and irrigation during fruit development, use of plant growth regulators and anti-transpirants, timely and effective control measures against major pests and disease, proper nutrition *etc.*

Granulation

Granulation is the major problem in citrus, which causes great economic loss to the fruit growers. This disorder was also called dry end, Kaosarn and crystallisation in several countries. In the affected fruit the juice vesicles become hard and enlarged and contain a

Fig. 9. Fruit drop in mango

viscous jelly like substance instead of free running juice with an increase in moisture, inorganic matter and polysaccharides and decrease in acidity and sugars. Although, no successful method to control granulation has yet, been found, certain measures for reducing its incidence and intensity have been suggested. Sprays of lime, zinc sulphate and Bordeaux mixture singly or in combination reduced granulation (Awasthi and Nautiyal, 1973). Some growth regulators like 2,4-D, GA and NAA @ 5 per cent, 20 ppm and 300 ppm respectively were found to reduce the incidence of granulation in citrus crops (Kaur *et al.*,1988).

Occurrence of Post-bloom Vegetative Flush

Heavy flower and fruit drop is a serious problem in mango. This is attributed to several causes, such as genetic, hormonal, insect pests, diseases, degeneration of embryo, lack of pollination and competition between fruit lets. However, in some mango cultivars like Alphonso, Langra *etc.* produced heavy vegetative flush during flowering and fruit set. Yields from such trees were invariably very poor in spite of profuse flowering. Flushing tendency was observed more pronounced in young orchards. The possible way in which the vegetative growth affected fruit retention could be the competition between the source and sinks. Although, seeds in growing fruits are generally considered as powerful sinks for mobilization of the photosynthates, in this case, the post bloom vegetative flush may become a powerful sink than the seed and fruits. Such type of phenomenon is also common in apple and pear. Removal of the post bloom vegetative flush was found the best remedy for remarkable increase in fruit retention and yield. Removal of the post bloom vegetative flush was found the best remedy for remarkable increase in fruit retention and yield.

Due to the above problems in the fruit crops the productivity of majority of fruit crop is below the potential level to meet the requirement of present population of the country. However, WTO necessitated the increase production of quality fruits for export as well as to compete with the imported fruit. Therefore, the need of the day is to exploit hi-tech technologies through the application of holistic approach for the improvement in quantity and quality of product.

CONCLUSION AND FUTURE PROSPECTS

The last decade has been very important from the point of view of emergence of horticulture as important field of diversification. Fruit industry today also forms an integral part of food and nutritional security. However, the productivity of majority of fruit crops is below the potential level to meet the requirement of present population of the country. Long standing problem of biennial bearing and erratic flowering in mango, citrus and other tropical and subtropical fruits could be overcome through introducing the flower meristem LEAFY gene or by the gene of regular cultivar through genetic engineering. Introduction of dwarfing gene for the management of tree size and canopy shape is the need of the day. Genetic transformation of embryonic cultures of mango scion cultivars with the dwarfing gene ro/c has paved the way for the production of dwarf tree technology for the control of fruit ripening using antisense configuration either with ethylene control or polygalactouronase (a cell wall hydrolase) has been found effective to extend the storage life of harvested fruit.

Therefore, the need of the hour is to understand the physiological and genetic basis of low productivity and to exploit new interventions, which can lead to improvements in quantity and quality of produce. The use of hi-tech technologies such as molecular physiology and biotechnology is solving some of the problem confronting the fruit based industries needs no emphasis. Some of the report on flowering process which is of vital importance to fruit crop productivity are contradictory due to different approaches utilized for the same experiment, as a result, it is difficult to draw definite conclusion with respect to role of specific factors on flowering, fruit set and retention. Thus to avoid different physiological barriers extensive research is needed in these areas in controlled environments. In order to increase the photosynthetic efficiency of the fruits crop it is the need of the day to manipulate Rubisco enzyme genetically for maximising its affinity to CO_2 which would lead to higher production of fruits. Understanding of molecular physiology and biotechnology is paving the way for genetic control of tree, shape, flowering, post harvest problems, disease and pest resistance etc. During the past few decades several developments in hi-tech agro-techniques of horticultural crops like high density planting, use of plant growth regulator and chemicals like paclobutrazol, KNO_3, ethephon, salicylic acid, rejuvenation of old, senile and unproductive orchard, integrated nutrient management have taken place which improved the fruit production manifold. However, much of the commercial application of these technologies is yet to become reality. There is need to utilize these technologies to optimize the fruit production in the country.

REFERENCES

Abdin, M.Z.,Rehman, R.U., Israr, M., Srivastava, P.S. and Bansal, K.C. 2002. Abiotic stress related genes and their role in conferring resistance in plants. *Ind.J.Biol.*, **1**: 225-244

Anderson, A.J. 1978. Isolation from three species of *Colletotricbum* of glucon – containing polysaccharides that elicit browning and phytoalexin production in bean., *Phytopathology,* **68**: 189-194.

Andres, I. and Le Fook, U. 1985. Effect of growth regulators on flowering pattern, flower suppression and fruit set in mango (*Mangifera indica* L.). *Joint Proceeding* 21st *Annual Meeting of the Carribbean Food Crops Society and 32nd Annual meeting of the American society for Horticulture Science*. Tropical Region Port of Spain, Trinidad, pp. 61-65.

Awasthi, R P and Mehta, K 2000. Strategies for developing high density planting in Horticulture crops. In : *Souvenir of Nat. Seminar Hi-tech Hort.,* Bangalore (India), 26-28 June, pp. 29-33.

Awasthi, R.P. and Nautiyal, J.P. 1973. Effect of different frequencies of irrigation on granulation in sweet orange. *J. Res*., PAU, **10** : 329-330.

Bondad, N.D., Blanco, A.E. and Mercado, E.L. 1978. Foliar sprays of potassium nitrate for flower induction in mango (*Mangifera indica* cultivar Pahutan) shoots. *Philippines J. Crop Sci,* **3** : 251-256.

Bose, T.K., Mitra, S.K. and Sanyal, D. 2001. *Fruits : Tropical and Subtropical,* **1**: 496-555.

Burnondkar, M.M. and Gunjate, R.T. 1993. Control of vegetative growth and induction of regular and early cropping in 'Alphonso' mango with paclobutrazol. *Acta Hort.*, **341**: 206-215.

Campbell, C.W. 1961. Comparison of yield of polyembryonic and monoembryonic mangoes. *Proc. Florida State Hort. Soc.*, **74**: 363-365.

Campbell, R.J. and J. Wasielewaski 2000. Mango tree training techniques for the hot tropics. *Acta. Hort*., **509** : 641-651.

Castresana, C., Stanalonia, R., Malik, V.S. and Cashmore, A.R. 1988. C_3 plants converted to C_4 type of photosynthesis. *Plant Mol. Bio.,* **10**: 117.

Chacko, E.K., Randhawa, G.S., Minon, M.A. and Negi, S.P. 1972. Effect of ethrel on papain production in papaya. *Curr. Sci.*, **41**: 455.

Chadha, K.L. 2000. An overview of Hi-tech horticulture: opportunities and constraints. *Nat. Symposium* on *Hi-tech Horticulture* held at IIHR, Bangalore (India) during June 26-28, 2000 1-30.

Chaplin, G.R. 1989. Advances in post harvest physiology of mango. *Acta Hort*., **231**: 639-648.

Chen, W.S. 1985. Flower induction in mango (*Mangifera indica* L.) with plant growth substance. *Proc. Nat. Sci. Council, Part B, Life Sciences*. Taipei, Republic of China, **9:** 9-12.

Choudhury, R.S. 1957. Identification of sex in papaya seedling stage. *Indian. J. Hort.*, **14**: 179-185.

Chundawat, B.S., Gupta, O.P. and Godra, N.R. 1975. Crop regulation of Banarsi Surkh, a guava cultivar. *Haryana J. Hort. Sci*., **4**: 23-25

Claes, B., Dekeyser, R., Villarroel, R., Den-Bulcke, M.B., Bauw, G., Bontagu, M. and Caplan, A. 1990. Characterization of a rice gene showing organ specific expression in response to salt stress and drought. *Plant Cell,* **2**: 19-27.

Coombe, B.G. 1967. Effect of growth retardants on *Vitis vinifera*. *Vitis*, **6**: 278-287

Davenport, T.L. 1993. Floral manipulation in mangos. In : *Proc. Conference* on *Mango in Hawaii* (Eds. Chia, L.E. and Evans, D.O.), Cooperative Extension Service, University of Hawaii, Honolulu 54-60.

Dedolph, R.R. 1962. Effect of benzothiazole-2-oxyacetate on flowering and fruiting of papaya. *Bot. Gaz.*, **124**: 75-78.

Flore, J.A. and Lakso, A.N. 1989. Environmental and physiological regulation of photosynthesis in fruit crops. *Hort. Review*, **11**: 111-157.

Giesen, B. 1990. First known data on sweet cherries as hedges. *Fruittectt – Den – Haag*, **80**: 5-11.

Govindjee 1982. Photosynthesis : Development, Carbon Metabolism and Plant Productivity Vol. II. New York : Academic Press.

Hanumashetty, S.I. 1978. Studies on vegetative growth and its relationship of flowering in mango cultivars with special reference to cvs. Pairi and Totapuri. *Ph. D. thesis* U. A. S., Dharwad.

Harley, C.P., Magnest, J.R., Masure, M.P., Fletcher, L.A. and degman, E.S. 1942. Investigations on the cause and control of binnial bearing of apple trees. Tech. Bull. US Department of Agriculture, 792.

Hillier, G.R. and Rudge, T.G. 1991. Promotion of regular fruit cropping in mango with Cultar. *Acta Hort.,* **291**: 51-59.

Horovitz, S.L. 1954. Determinacion del sexo en *Carica papaya* L. Estructura hipotetica de los cromosomas sexuales, *Agron. Trop*., **3**: 229-249.

Hsich, T.H.,Lee, J.T.,Chang, Y.Y. and Chan, M.T. 2002. Tomato plants ectopically expressing Arabidopsis CBF1 showed enhanced resistance to water deficit stress. *Plant Physiol*.,**130**: 618-622.

Jana, R., Cattro, A. and D. Travaglio 2002. The climacteric feature of fruits is reflected in sensitivity to ethylene of plantlets *in vitro*. *Sci. Hort.*, **94**: 273-284.

Jayasankar, S. 1995. *In vitro* selection of mango for anthracnose resistance and characterization of the regenerates. *Ph. D. thesis.*, University of Florida, Gainesville.

Kadman, A., Gazit, S. and Ziv, G. 1976. Selection of mango rootstocks for adverse water,soil condition in arid areas. *Acta Hort*., **57**: 81-88

Kulkarni, V.K. 1988. Chemical control tree vigour and the promotion of flowering and fruiting in mango (*Mangifera indica* L.) using paclobutrazol. *J. Hort. Sci.*, **65**: 557-566.

Kaur, H., Kapur, S.P. and Chanana, Y.R. 1988. Effect of growth regulators on granulation in sweet orange cv. Mosambi (*Citrus sinensis* Osbeck). *National Seminar on Advances in Plant Growth Regulator Research*, Jodhpur. 17-19 Nov., 189.

Lal, B. and Padaria, R.N. 2001. Rejuvenation of mango orchard, CISH Publication, **9**: 1-20.

Lopez, M.R. 1984. El nitrato de potasio como promotor de la sitesis endogena de etileno y la induccion floral en mango (*Mangifera indica* L.) cv. Manila. *M. Sc. Thesis*, Universidad Autonoma Chapingo, Chapingo, Mexico.

Maiti, S.C., Basu, R.N. and Sen, P.K. 1972. Chemical control of growth and flowering in *Mangifera indica* L. *Acta Hort.,* **24**: 192-195.

Maika,A. 1986. Physiological response of fruit trees to pruning. *Horticulture Review*, **8**: 339-369.

Malik, C.P. 1999. Isolation and Quantification of Growth Hormones. In : *Advances in Plant Hormones Research* (Ed. C. P. Malik). *: Indian Scenario*, Agro-Botanica Press, 185-199.

Mukherjee, S.K. 1949. The mango and its relatives. *Science* and *Culture*, **15:** 5-19.

Mullins, P.D.F. 1985. Delaying of flowering in Haden mango tree. Technical Communications in Horticultural Science Series No. 200. Department of Agriculture, Republic of South Africa, citrus and sub tropical Research Institute, Netspruit, RSA 6-8.

Mullins, M.G. and Rajsekharan, K. 1981. Fruiting cuttings: Revised method for producing test plants of grapevine cultivars. *Am. J. Enol. Vitic.*, **32**: 35-40

Mullins, M.G. 1967. Morphogenetic effect of roots And some synthetic cytokinin in *Vitis Vinifera, J. Exp. Bot.*,**18**: 206-214

Oellar, P.W., Lu, M.W., Tavlor, L.P., Pike D.A. and Theologis, A. 1991. Reversible inhibition of tomato fruit senescence by antisense RNA. *Science*, **254**: 437-439.

Oonco, Y., Handa, T., Kanaya, K. and Uchimiya, H. 1987. The Ti-DNA gene of Ri plasmids responsible for dwarfness of tobacco plants. *Jap. J. Genetics*, **62**: 501-505.

Oosthuryse, S.A. 1995. Effect of aqueous application of GA_3 on flowering of mango trees, Mango 2000 – Marketing Seminar and Production Workshop Proceedings, Department of Primary Industries, Brisbane 75-85.

Pandey, R.M. 1989. Physiology of flowering in mango. *Acta. Hort.*, **231**: 361-380.

Pandey, R.M.,Lal, S. and Kaul, G.L.1980. Effect of chemical and flower thinning on regulation of crop of guava. *Ind. J. Hort.*, **37**: 234-239

Perl, A., Perl-Treger, R., Galili, S., Abib, D., Shalgi, D., Malkin, S. and Galum, E. 1993. Enhance oxidative stress defense in transgenic tomato expressing tomato Cu, Zn-superoxide dis-mutages. *Theoretic Appl. Genetics.*, **85**: 568-576.

Purohit, A.G., Shikhamany, S.D. and Kumar, P. 1979. Effect of leaves per branch on growth and quality of grape (*Vitis vinifera* L.). *Ind. J. Hort.*, **36**: 36-41

Ram, M. 1983. High density plantation in papaya. *Indian Hort.*, **28**: 17-20.

Ram, M. and Ray, P.K. 1992. Study on pure seed production in papaya. *Seed Research*, **20**: 81-84.

Ram, S., Singh, C P and Kumar, S. 1997. A success story of high density orcharding in mango. *Acta. Hort.*, **455**: 375-382.

Rathore, D.S. 2001. Networking for genetic resources management of horticulture crops. *Nat. Symposium Plant Genetic Res.* at NBPGR, New Delhi (India) during Feb. 1-4, 2001.

Rowley, A.J. 1990. The effect of Cultar applied as a soil drench on ' Zill' mango trees. *Acta Hort.*, **275**: 211-215.

Samson, J.A. 1980. Papaya, in *Tropical Fruits*, Longman, New York, 208.

Schaffer, B., Whiley, A.W. and Crane, J.H. (1994). Mango. In : *Handbook of Environmental Physiology. Vol. II. Subtropical and Tropical Crops* (Eds. Schaffer, B and Anderson, P C). CRC Press, Boca Raton 165-198.

Shanmugavelu,K.G. and Saidha, T. 1993. Pruning in mango. Advances in Horticulture 2: Fruit Crops. (eds. Chaddha, K.L. and Pareek, O.P.). Malhotra Publishing House, New Delhi, 681-686.

Shulman, Y., Nir, G. Finberstein, I. And Laves, S. 1983. The effect of cynamide on the release from dormancy of grapewine buds (*Vitis vinifera*). *Scientia Hort.*, **19**: 97-102.

Singh, G., Rajan, S. and Singh, A.K. 2003. Approaches and strategies for precision farming in guava. In: Precision Farming in Horticulture. (Eds. Singh, H.P., Singh, G., Samuel, J.C. and Pathak, R.K.) Pub. By NCPAH, DAC/PFDC, CISH, Lucknow, 92-113.

Singh, L.B. 1961. Biennial bearing in mango – effect of gibberellic and Malic hydrazide. *Hort. Advance*, **5**: 96-106.

Singh, R.N. 1996. Mango. Indian Council of Agricultural Research, New Delhi, Publication, pp : 114-118.

Singh, S. 1986. Pear. In: *Biotechnology in Agriculture and Forestry 1. Trees 1.* (Eds. Bajaj, YPS). Springer-Verlag, Berlin 198-206.

Singh, V.K. and Saini, J.P. 2001. Regulation of flowering and fruiting in mango (*Mangifera indica* L.) with paclobutrazol. In : *Plant Physiological paradigm for Fostering Agro and Biotechnology and augmenting environmental productivity.* (Eds. Dwivedi, R.S. and Singh, V.K.). Indian Society of Plant Physiology, New Delhi 2001, 61-68.

Singh, V.K., Saini, J.P. and Misra, A.K. 1998. Effect of plant growth regulators and other chemicals on floral malformation flowering, fruit set, yield and fruit quality of mango cv. Amrapali. *Biol. Memoirs*, **24**: 19-24.

Singh, V.K., Saini, J.P. and Misra, A.K. 2001. Response of salicylic acid on flowering, floral malformation, fruit, yield and associated bio-physical and bio-chemical character of mango. *Indian J. Hort.*, **58**: 196-201.

Storey, W.B. 1941. The botany and sex relation in papaya. I. Papaya production in Hawaiian Islands, *Hawaii Agric. Exp. Stn. Bull.*, **87**: 5-22.

Storey, W.B. 1953. Genetics of the papaya. *J. Hered.*, **44**: 70-78.

Tongumpai, P., Hongsbhanich, N. and Voon, C.H. 1989. 'Cultar' – for flowering regulation of mango in Thailand. *Acta Hort.*, **239**: 375-378.

Voon, C.H., Pitakpaivan, C. and Tan, S.J. 1991. Mango cropping manipulation with Cultar. *Acta Hort.*, **291**: 219-228.

Weaver, R.J. 1975. Effect of growth retardant sprays on fruitfulness and cluster development of Thompson seedless grapes. *Am J.Entol.Vitic.*, **26**: 47-49

Weaver, R.J. and Pool, R.M. 1971. Effect of ethephon and morphactin on growthynand fruiting on Thompson Seedless and Carignane grapes. *Am J. Enol. Vitic.*, **22**: 234-239

Weigel, D. and Nilsson, O. 1995. A developmental switch sufficient for flower initiation in diverse plant. *Nature*, **377**: 495-500.

Wessels, J.G.H. and Sietsma, J.H. 1981. Fungal cell walls a survey. In : *Encyclopedia of Plant Physiology vol. 13B Plant Carbohydrates* (eds.Tanner, W. and Loewsus, F.A.), Springer – Verlag. Berlin, 352-394.

Westwood, J. 1999. Development of genetically transformed papaya cultivar against resistant to ring spot virus. *RIS Biotech. Develop. Rev.*, **2**: 78-80.

Whieley, A.W. 1993. Environmental effects on phenology and physiology of mango – A reviews. *Acta. Hort.*, **341**: 168-176.

Zhang, H.X. and Blumwald, E. 2001. Transgenic salt tolerant tomato plants accumulate salt in foliage but not in fruit. *Nature Biotech.*, **19** (August): 765.

Section VI: Physiology and Biochemistry of Herbicides

Developments in Physiology, Biochemistry and Molecular Biology of Plants, 2005
Eds.: Bandana Bose and A. Hemantaranjan
Vol., 1, pp. 203-287, New India Publishing Agency, New Delhi
E-mail: spjain_niph@rediffmail.com web: www.bookfactoryindia.com

CHAPTER - 10

ROLE OF HERBICIDES IN IMPROVING CROP YIELDS

I.V. SUBBA RAO AND T.Y. MADHULETY

PREFERENCE OF HERBICIDES FOR WEED MANAGEMENT IN AGRICULTURE AND HORTICULTURE

Weeds are unwanted and undesirable plants which interfere with the utilization of land and water resources and pollute atmosphere with their allergic pollen thus affecting human welfare adversely. They are the plants out of place, growing where we want other plants or no plants at all. They compete with desirable and beneficial vegetation reducing the yield and quality of the produce. The effect of weeds in agriculture or horticulture is greatest. There is a scanty reliable study of worldwide damage to crop yields due to weeds. Whether the data are available or not, it is well known that losses caused by weeds exceed the losses from any category of agricultural pest, *viz.,* insects, nematodes, diseases, rodents *etc*. Weeds account for 45% while insects 30%, diseases 20% and other pests 5% of the total annual loss of agricultural produce[186].

Weeds manifest losses to agricultural produce by competing with crop plants for nutrients, soil moisture and sunlight. The extent of losses depends upon: type of weed species, severity of weed infestation, duration of weed infestation, competing ability of crop plants and climatic conditions which affect weed and crop growth. Crop yield reduction is directly correlated with weed competition so much so an increase in one kilogram of weed growth corresponds to a reduction in one kilogram of crop growth. Crop yield loss from weeds is highest in tropics. In Asian economy, proper weed control in rice could increase the grain yield by 20 to 75% or even triple the yield. Weeds remove plant nutrients more efficiently than crop plants, thrive better in a drought situation, grow faster and taller than crop plants, if left undisturbed, inhibit tillering and branching, curtail sunlight and adversely affect photosynthesis and plant productivity. Based on the extent of competition, weeds reduce crop yields by 10 to 25%. Annual losses due to weeds are responsible for a 10% reduction in total agricultural productivity of the United States accounting to a loss of US $ 12 billion[111]. Farmers, in addition, spend about US $ 3.6 billion for chemical weed control and about US $ 2.6 billion for cultural, ecological and biological methods of weed control estimating to a total loss of < US $ 18 billion.

In a developing country like India, total weed control in the entire farmland would add US $ 5 billion to the economy, at the current price level. Thus prevention of loss to yield is to enhance productivity. By complete elimination of weed competition from the entire crop land, the world's farm production could be increased by 10% to 25%. The impact of weeds on the US economy is about US $ 20 billion with agricultural sector alone accounting for

US $ 15 billion[32]. The world is annually losing 11.5% of total food production. So if all the weeds in food crops were controlled, the current world's food production would be higher by 11.5% or 450 million tons[177].

Besides crop yield losses, very heavy weed infestations render some economically important crops *viz.* pulses, vegetables, cotton, jute and forage crops unsuitable for cultivation. The quality of leafy and other vegetable crops suffers in the presence of weeds. For example *Allium canadense* (wild onion), *Avena fatua* (wild oats) and wild rice may impair the quality of onion, wheat and rice produce respectively. The leaves of *Mikania micrantha* may contaminate tea leaves during plucking and reduce the value of tea. Contamination of other noxious weed seeds greatly reduce the value of crop seed, grain and sometimes render them unsaleable and reduces market value or causes spoilage during storage.

Thus to combat weed infestation, various methods of weed management are devised *viz.,* manual, mechanical, biological and finally the chemical methods. Many organic herbicides in the past 50 years enabled substantial progress in controlling weeds the world over. To be useful, herbicides must distinguish between crop plant and weed. This may be problematic: although herbicides are designed to affect significant processes of plants such as photosynthesis and amino acid biosynthesis, which are common to both crops and weeds. Consequently, selectivity is based on differential uptake of herbicide between crop and weed, controlled timing and site of application or detoxification of the herbicides by the crop plant. Herbicides besides other pesticides certainly be a major and increasing part of the agricultural technology in the decades ahead, which will be needed to provide a greater supply of food, fiber and shelter with greatest cost-effectiveness.

TYPES OF HERBICIDES AND THEIR CLASSIFICATION

There are over 400 herbicides discovered and almost all weeds can be controlled by herbicides. High-yield agriculture largely depends on chemical herbicide usage, owing to its intensive characters and diversification of labour. Mainly organic herbicides, could be classified based on:

(a) Method of application;
(b) Chemical affinity and structural similarity;
(c) Mode of action.

(a) Method of Application

On the basis of method of application, herbicides are divided into:

(i) Soil surface applied: The herbicides that are surface applied or soil incorporated at pre-planting or those applied at pre-emergence of crops. Such herbicides kill weeds at very sprouting stage *in situ* and crop will be free from weed competition and grow vigorously.

(ii) Foliage-applied: Herbicides that are applied at post-emergence on the plant parts are included in this group. These herbicides are effective only when they are adequately absorbed by weed flora. Hence, effective weed control would be uncertain and depends on leaf morphology, plant canopy, environment after spraying (rain wash) *etc*. Thus crop-weed competition depends on appropriate method of application for effective absorption of a herbicide by the weed flora.

(b) Chemical Affinity and Structural Similarity

Based on the limitations in respect of soil or foliar applied chemicals, herbicides of effective absorption are developed from a proven herbicide chemical nucleus or family. Thus, herbicides from one chemical nucleus or family since have structural similarities and affinities, provided basis for chemical classification: ex: Acetamides, aliphatics, arsenicals *etc* (Table 1)[186].

Table.1. Classification of (organic) herbicides on the basis of chemical affinity, mode of application, and mode of action (herbicides not being widely used or which have been phased out were excluded).

Chemical Family	Soil-applied Herbicides		Foliage-applied	
	Systemic	Contact	Systemic	Contact
ACETAMIDES	acetochlor, alachlor, dimethenamid metolachlor, napropamide, pronamide, propachlor	-	Acetochlor, alachlor, dimethenamide metolachlor, napropamide, propanil	-
ALIPHATICS	-	-	dalapon	acrolein
ARSENICALS	-	-	DSMA, MSMA	cacodylic acid
BENZAMIDES	isoxaben	-	-	-
BENZOICS	dicamba	-	-	-
BENZOTHIA-DIAZOLES	-	-	-	bentazon
BIPYRIDILIUMS	-	-	-	diquat, paraquat
CARBAMATES	-	-	asulam, desmedipham, phenmedipham	-
CINEOLES	cinmethylin	-	-	-
CYCLOHEXA-NEDIONES	-	-	clethodim, cydloxidim, sethoxydim, tralkoxydim	-
DINIRTOANILINES	benefin, dinitramine, ethalfluralin, fluchloralin, isopropalin, nitralin, oryzalin, pendimethalin, prodiamine,trifluralin	-	-	-
DIPHENYL-ETHERS	fluorodifen, oxyfluorfen	nitrofen	acifluorfen, bifenox, fluoroglycofrn fomesafen, lactofen, oxyfluorfen	-
IMIDAZOLID-INONES	buthidazole	-	buthidazloe	-
IMIDAZOLINONES	imazapyrm unazaqyub, imazethapyr,	-	imazapyr, imazaquin, imazethapyr,	-
IMINES	-	-	CGA-248757	-
ISOXAZOKID-ININES	clomazone	-	-	-
NITRILES	-	-	-	bromoxynil,ioxynil
OXADIAZOLES	oxadiazon	-	-	-
OXADIAZOLIDINES	methazole	-	methazole	-
PHENOLS	-	-	-	dinoseb
PHENOXYALK-ANOIC ACIDS				
Phenoxyacetics	2,4-D	-	2,4-D, MCPA	-
Phenoxybutyrics	2,4-DB	-	2,4-DB, MCPB	-
Aryloxyphenoxy	-	-	dichlorprop, diclofop,	-
propionics	-	-	fenoxaprop, fluzaifop-P, haloxyfop, mecoprop,	

Contd...

Table :1 *Contd.*

Chemical Family	Soil-applied Herbicides		Foliage-applied	
	Systemic	Contact	Systemic	Contact
N-PHENYLPHT	-	-	quizalofop-P, flumiclorac	-
PHEYLPYRID-AZINES	-	-	pyridate	-
PHENYL TRIA-ZINONES	sulfentrazone	-	sufentrazone	-
PHTHALAMATES	naptalam	-	naptalam	-
PYRAZOLIUMS	-	-	difenzoquat	-
PYRIDAZINONES	norflurazon, metflurazon, pyrazon	-	-	-
PYRIDINECARBO-XYLIC ACIDS	-	-	clopyralid, picloram, triclopyr	-
PYRIDINES	dithiopyr, thiazopyr	-	-	-
PYRIDINONES	fluridine	-	fluridone	-
PYRIMIDINYLTH-IO BENZOATES	pyrithiobac	-	Pyrithiobac	-
QUNIOLINE CARBO-XYLIC ACIDS	quinclorac	-	quinclorac	-
SULFONYLUREAS	bensulfuron, chlorimuron, chlorsulfuron, halosulfuron, sulfometuron	-	bensulfuron, chlorimuron, chlorsulfurin, halosufuron, metsulfuron, nicosulfuron, primisulfurin, prosulfuron,tribenuron	-
TETRAHYFRO-PYRIMIDINONES	experimental compounds	-	experimental compounds	-
THIOCARBAMATES	butylate, diallate, EPTC, molinate, pebulate, thiobencarb, triallate, vernolate	-	-	-
TRIAZINES	ametryn, atrazine, cyanazine, hexazinone, prometryn, simazine,	-	-	-
TRIAZINONES	metribuzin	-	-	-
TRIAZOLES	-	-	amitrole, amitrole-T	-
TRIAZOLOPYR-IMIDINESUL-FONANILIDES	flumetsulam	-	-	-
URACILS	bromacil, terbacil, UGC-C4243	-	UCC-C4243	-
UREAS	diuron, fluometuron, linuron, methabenzthiazuron, metoxuron, isoproturon, monuron, siduron, tebuthuron	-	diuron, fluometuron, linuron, isoproturon, tebuthiuron	-
UNCLASSIFIED	bensulide, ethofumesate	-	ethofumesate, fosamine, glufosinate, glyphosate	-

(c) Mode of Action

This classification of herbicides follows as a result of differences in the physiological and biochemical actions of herbicides[186]. On this basis herbicides are broadly categorized as 1) systemic or translocated or 2) non-systemic or contact herbicides.

1. ***Systemic or translocated herbicides:*** These herbicides require development of specific anatomical characteristic features *i.e.* well developed vascular tissue throughout the plant body for effective functioning of herbicide *i.e.* a well formed translocation system.

2. ***Non-systemic or contact herbicides:*** These herbicides do not require well developed translocation system; instead require effective coverage by the herbicide. Efficacy thus would be less particular when weed flora confined to ground coverage under the crop canopy.

Chemical weed control in agriculture although reported several centuries ago, introduction of the Bordeax mixture in 1896 stimulated chemical weed control. This lead to the discovery of copper salts for selective control of broadleaf weeds in cereals followed by sulphuric acid, iron sulphate, copper nitrate, ammonium and potassium salts, sodium nitrate, ammonium sulphate *etc.* With the introduction of new and improved agricultural practices, the concept of control of weeds changed to systemic control considering absorption by roots and top growth of the plant and translocation in the plant body. The use of organic chemicals such as nitrophenols as selective herbicides leads to the discovery of 2, 4-D in 1940s which revolutionized the chemical weed control. Zimmerman and Hitchcock[186] in 1942 tested 2, 4-D as a growth regulator and for selective weed control. They established that herbicides could be effective in very small quantities if they are highly selective and systemic in action.

Based on chemical structure, most of the organic herbicides are classified as hydrocarbons as saturated and unsaturated or aromatic hydrocarbons or hetero-cyclic hydrocarbons. These structures react and influence water solubility, electrical charges and volatility. The active group of a herbicide can be altered to improve its biological activity by esterification, salt formation *etc.* Herbicides produced from alcohols, phenols or organic acids are more soluble in water while ester forms are relatively more soluble in oil and organic solvents and produce vapours and thus attain volatility.

Based on electrical phenomenon of a molecule or an ion, herbicides can be divided into polar and non-polar. The polar substances are known as hydrophilic, water soluble, while those non-polar as lipophilic substances are soluble in oil and other non-polar solvents.

They have greater affinity for their respective solvents. Hence, herbicide absorption and translocation by a plant is greatly influenced by these hydrophilic or lipophilic properties. Because of high surface tension of water molecules, hydrophilic substances (polar solutions) form large spherical droplets and do not react with waxy cuticle of the leaf surface resulting in poor wetting of the foliage, hence reduced herbicide activity.

Non-polar compounds (lipophilic substances) exhibit low water solubility but high oil solubility which readily wet the waxy cuticle of the foliage. Inspite of better penetration, their translocation in the water continuum is very slow, thus affecting herbicide activity.

Method of preparation of herbicide affects solubility, volatility and specific gravity and thus phytotoxicity. Usually herbicides are formulated by combining with a liquid or solid carrier so that they can be applied uniformly. Based on the time of application, herbicides are applied as pre-planting, pre-emergence and post-emergence in relation to crop or weed. Herbicides can be applied to soil or foliage. Herbicides, which are volatile or liable to photodecomposition, are incorporated into the soil either as blanket application over the entire soil surface or weed infested area as band treatment by treating only the narrow strips in the crop rows.

Even variations in the placement of a herbicide can affect its activity and so much so a particular crop which may be susceptible to one method of placement, may tolerate another method of placement.

Further, herbicides selective character varies with the rate and hence this rate differs from one weed or crop species to another.

The uncertainity of effective weed killing in case of contact herbicides and pre-requisite development of vascular system in weed plants for translocated herbicides, perhaps lead to the identification of (i) selective and (ii) non-selective herbicides:

(i) ***Selective Herbicides:*** This group of herbicides kills or suppresses only particular weeds without significantly injuring an associated crop or their desirable plant species. Some weeds are not injured by selective herbicides.

(ii) ***Non-selective herbicides:*** This group of herbicides kills or suppresses any type of vegetation including the crop. They are used to kill all plants indiscriminately.

Further, each herbicide or herbicide family differs in their primary site of action at more than one site as inhibitory or secondary. This modified classification[186] (Table 2) includes protein and nucleic acid biosynthesis which serve as primary sites of action of acetamides, thiocarbomates, pyridine carboxylic acids, phenoxys, benzoics and pyrazoliums. The sites are designated as I, II and III representing protein biosynthesis (inhibition), protein or nucleic acid biosynthesis (stimulation) and nucleic acid biosynthesis (inhibition) respectively. This classification aids in the management of herbicide resistance in weeds or crop plants[186].

HERBICIDAL ACTIVITY AND SELECTIVITY

Herbicidal activity refers to the degree of phytotoxic effects of a chemical on plant growth and development. The activity of a herbicide is known when it inhibits or slows or prevents the germination and other developmental processes of that plant. Obviously it is said active on sensitive plants and inactive on tolerant or resistant plants. Hence, the herbicide activity is determined by the degree of tolerance of the plant to a chemical.

Herbicidal selectivity is the phenomenon of killing the target plant species by a chemical, in a mixed plant population without harming or only slightly affecting the other plants. Selectivity character alone helped for effective weed control of target weed species even in the presence of other crop plants. This phenomenon has been more evident since the discovery of 2,4-D.

Herbicidal activity thus indicates the ability of a herbicide to control a weed while selectivity refers to its ability to control it without affecting the other plants in a mixed stand. Activity cannot discriminate selectivity.

Both activity and selectivity determine the vital functions of the plant from the very entry of herbicide into the plant involving mechanical, physical, physiological and metabolic processes. The activity and selectivity of a herbicide that determine the plant's productivity not only depends on their chemical properties alone, but on other factors also.

They are :

1. Plant Morphology

Horizontally arranged leaves intercept more of the spray and retain more herbicide than those of upright leaves. Pubescence and wax formation on the leaf surface reduces wettability of the foliage. The cuticular waxes, pectin and cellulose which vary with plant age and species influence the amount of herbicide penetrance. Since the plant surfaces are lipoidal, the absorption of lipophilic herbicides is more rapid than that of hydrophilic and polar herbicides. It indicates plants with large number and wide open stomata on upper surface of leaf become susceptible to herbicides barring their biochemical tolerance.

2. Plant Phenological Development

Stage of plant development affects the ability of herbicides to enter the plant. Weed seed germination or young seedlings of 0-30 days are more susceptible to soil-applied herbicides. Hence, productivity would be drastically improved if weed plants are exposed to toxic doses of herbicides at early stages. At the active stage of growth with active photosynthesis and movement of photosynthates, the absorption of herbicide and translocation would be more effective. Hence, most of the annual weeds since susceptible to 1 to 4 phenological growth stages could be effectively killed[186] which facilitates crop plants to grow with early seedling vigour.

3. Cultivation Method

Even tolerant weed species could be made susceptible to a particular herbicide by taking pre-planting and post-planting cultivation operations *viz.,* deep ploughing. By these operations, dormant weeds seeds, perennial stolons and roots are brought up to top soil layer, exposed to sunlight resulting in germination. Thus, they are made susceptible to herbicides applied to the soil or foliage. Repeated cultivation also exhausts food reserves of deep rooted weeds which become more sensitive to herbicides later[186].

In the spree of modern agriculture where high valued crops are of common target, selective herbicide usage is a must. This situation invariably indicates necessity of herbicide resistance in such crop.

In view of the large spectrum of weed flora where effective weed control is difficult, identifying primary site of action of herbicide to kill the weed or secondary or tertiary site for affecting physiological and biochemical disturbance leading to the death of the tissues, has become necessary as herbicide resistant weed species are developing fast with the usage of a wide variety of herbicides.

4. Cultivation Practices

Herbicide application coupled with herbicide activity in other words herbicide efficacy could be improved in case of some tolerant species to a particular herbicide by pre-planting or post-planting cultivation which brings dormant weed seeds to the top soil layer and on exposure to sunlight break their dormancy. Thus, making germinating seedlings more susceptible to herbicide, applied to foliage or soil which could have not been possible without the cultivation. The succulent roots, rhizomes or stolons of perennial plants could be brought to soil surface by cultivation and killed easily when herbicide is sprayed on exposed plant parts. This facilitates easy reach of herbicide to the site of its action. To repeat such cultivations, exposure to herbicide sprays and sunlight exhaust food reserves of deep rooted perennial weeds to make them more sensitive to herbicides. Herbicide toxicity would be enhanced when exposed to light as in the case of atrazine and lasso on nut sedge in maize[143, 144]. The constituent crop plant would get relieved of weed competition and express their productive potential.

5. Environmental Factors

Selectivity of an applied herbicide could be influenced by the environmental factors. Similarly, the response of plant to the chemical applied depends on the environmental stresses that follow the application of herbicides. Severe temperature or water stress cause herbicide more deleterious to tolerant weed species. In field, tolerant crop plants also may be injured under adverse environmental factors: temperature, rainfall, wind, relative humidity and light which affect herbicide activity, selectivity and toxicity as they influence absorption and translocation of herbicide to the site of action by the crop plant and weeds as well.

6. Herbicide Absorption

Effective phytoxic action of herbicide also depends on the absorption of threshold levels of herbicide. This is influenced by the morphology of root, stem and leaf which also determine the transpiration rate, chemical and electrical properties of plant surfaces. The rate of absorption and amount of chemical absorbed also varies from plant species and stage of plant growth. The differential absorption of herbicide by plant species determines the difference in the activity and selectivity of the same chemical.

7. Herbicide Translocation

The absorbed chemical needs to be translocated to the site of action, via the phloem or xylem unlike contact herbicide. However, contact herbicides also move to shorter distances. Bio-physical, bio-chemical processes and age of the plant casue differential movement of the chemical to the site of action, thus, determining the differential selectivity and activity.

8. Plant Physiological Difference for Efficacy of Herbicides

The herbicide after reaching site of action effect one or more of the metabolic activities related to plant growth and survival. This depends on physiological tolerance of the plant. Variation in the physiological and bio-chemical processes causes the basis for the herbicide selectivity. The physiological tolerance of species, variety or cultivar to the given herbicide largely depends on their genetic make up, taxonomic and morphological characters and physiological processes. For example 2,4-D is very effective on dicotyledonous plants than on many monocotyledonous plants. So also some corn and sorghum genotypes are more susceptible to atrazine and propazine respectively than others. Maize seedlings are more quickly killed by lasso than atrazine[143, 144].

9. Herbicide Metabolism

Activity, selectivity and phytotoxicity of any herbicide depend on the rate of metabolism of the herbicide in the plant. Any herbicide that enters the plant (weed or crop) undergoes metabolism. The extent of metabolism depends on the physical and chemical properties of the compounds, chemical composition of the plant and capacity of the plant to effect metabolic conversion of the applied herbicide. A tolerant plant inactivates herbicide in any one or many pathways. Toxicity and killing, results in plant species where detoxification takes place at a slower rate, than that of accumulation. If it is faster than the latter, it enables the plant to be tolerant. The mode of detoxification of herbicide differs among plant species and form the basis for herbicide activity or selectivity. Since herbicides are used specifically, according to the type of weed species to be killed or based on the properties of herbicide molecules, crop plants need to be developed for multiple herbicide resistance mechanisms. It could however be possible by choosing a crop plant type as described earlier, that does not intercept much herbicide due to other plant morphological characters, without sacrificing its productivity potential.

10. Plant Resistance and Tolerance to Herbicides

It is the ability of the plant to withstand the activity of a herbicide at a dosage substantially greater than normally used for its control. The susceptible weed species or crop plants develop resistance to the same herbicide to which it succumbs, over the course of its usage[134]. It is in the evolutionary process that weed plants changed their genetic composition so that the frequency of resistance alleles and resistant individuals increase[186]. As defined by FAO, "Resistance is the decreased response of a population of plants species to a pesticide or control agent as a result of their application. It has got convenient distinction between physiological and behavioristic resistance, whereas tolerance refers to the natural and normal variability to pesticides and other agents which exists within a species and can easily and quickly evolve. It refers to variability within a species and is used to compare between species. It also refers to the relatively minor or gradual differences in intra-specific variability[134]".

Resistance is the maximum tolerance that can be achieved. The resistant individuals are found in much lower frequencies than tolerant ones in natural untreated populations.

In view of developing resistance or tolerance in weed species to herbicides or at greater frequencies are necessitated. Concomitantly, crop plants need to be developed for herbicide tolerance/resistance. Such differential resistance/tolerance in weed species to some herbicides is reviewed under "structure – function relationship, the basis for the variation in the efficacy of herbicides".

STRUCTURE - FUNCTION RELATIONSHIP – THE BASIS FOR VARIATION IN THE EFFICACY OF HERBICIDES

Organic herbicides dominate modern agriculture. Though inorganic herbicides were used between 1896 and 1930s, a few of them are mixed with certain organic herbicides.

The efficacy of a herbicide depends on the chemical structure or the nucleus with which the herbicides are prepared. The following herbicides summarized in (Table 1) reveal the relationship of their structure and function leading to their efficacy in controlling weeds[5,186].

1. Acetamides (Acetanilides)

These are the prominent diverse group of herbicides. When applied PRE, the susceptible grass and broad leaved weeds failed to emerge where as susceptible monocots appear as twisted and malformed with leaves tightly rolled in the leaf whorl and unable to unroll normally. Leaves do not emerge properly from the coleoptile. Broad leaved weed seedlings have cupped or crinkled leaves with shortened leaf midribs.

(i) Acetochlor (2 –Chloro – N – (ethoxy methyl) – N – (2 ethyl – 6 methyl phenyl) acetamide) : It is applied as PPI, or PRE. Acetochlor effectively controls annual grasses, *Cyperus esculentus* (yellow nutsedge) and certain broad leaved weeds. Its residues do not persist long enough to injure crops in the following season.

(ii) Alachlor (2 – Chloro – N – (2, 6 diethyl phenyl) – N – (methoxy methyl) acetamide): It was the first chloroacetamide herbicide commercialized as LASSO. It is applied as PPI, PRE and Early-Post Emergence. Maize, sorghum, soybean, groundnut and cotton etc tolerate this herbicide while *Cyperus esculentus* and many other annual grasses such as *Echinocloa crusgalli, Digitaria* spp., *Setaria* spp., *Panicum* spp. and *Eleusine indica* and certain broad leaved weeds including *Galinsoga* spp., *Chenopodium album, Portulaca* spp., *Solanum nigrum* are susceptible.

Alachlor residues do not persist long enough to injure crops in the following season. Its field half - life is 21 days.

(iii) Butachlor: N – (Butoxy methyl) – 2 – chloro – N – (2, 6 – diethyl phenyl) acetamide: It is a chloroacetamide herbicide commercialized as MACHTE. It is specifically applied to irrigated rice within 2-5 days of transplantation and in directly seeded irrigated rice. It can be safely used for crops like groundnut, potato and soybean and other pulses grown under irrigated or assured rainfall conditions. It is very effective against annual grass weeds such as *E. crusgalli, D. sanguinalis, Setaria* spp., *Panicum* spp., and certain sedges. It is the most widely used rice herbicide in the world. When mixed with 2, 4-D, ethyl ester granular (4 G) formulation, it will be a very broad spectrum herbicide in irrigated rice. It has got 13 days of field half-life and does not injure crops in the following season.

(iv) Dimethenamide: 2 – chloro – N – [(1 – methyl – 2 - methoxy) ethyl] – N – (2,4 – demethyl – thein – 3 – yl) – acetamide : Dimethenamide is a chloroacetamide registered as FRONTIER for maize and soybean. Grass and broadleaf weeds which are susceptible, fail to emerge. Though resistant, maize and sorghum develop injury to excessive dosage and show malformed and twisted appearance. Leaves are tightly rolled in the whorl and may not unroll normally. Injured soybean seedlings show cupped or crinkled leaves. It has no POST activity on established monocot and dicot seedlings. It is effective against many annual grasses such as *Echinocloa crusgalli, Panicum dichotomiflorum* and *Digitaria* spp. as well as *Cyperus esculentus* and annual broadleaf weeds including *Amaranthus retroflexus* and *Solanum nigrum*. It does not injure other crops planted in the following season.

(v) Metolachlor: 2 – chloro – N – (2 – ethyl – 6 – methyl phenyl) – N – (2 – methoxy – 1 – methyl ethyl): Synthesized for safe use in maize, cotton, groundnut, sorghum, saflower, soybean and potato. Metolachlor works against *Cyperus esculentus* and many annual grass weeds *viz, Setaria* spp., *Echinocloa crusgalli, Digitaria* spp., *Panicum dichotomiflorum, Brachiaria* spp., *Panicum capillare etc*., including broadleaf weed species such as *Amaranthus retroflexus, Mollugo verticillata.* The residues do not persist long enough to affect crops planted in the following season.

(vi) Nepropamide: N – N – diethyl – 2 – (1 – naphthalenyloxy) propanamide: It is a substituted amide. It can be safely applied PPI or PRE in potato, radish, turnip, watermelon, saflower, tobacco, groundnut, cabbage, cauliflower *etc.* It can also work safely for fruit crops if applied PRE or E-POST. It effectively controls many annual broadleaf and grass weeds including seedlings of perennial grasses. The susceptible weeds fail to emerge after napropamide application. It primarily inhibits root growth, shoot growth and sprouting buds of *Cyperus rotundus.*

(vii) Pronamide: 3, 5 – dichloro (N – 1, 1 – dimethyl – 2 – propynyl) benzamide: It is also a substituted amide herbicide developed for weed control in cotton, sunflower, groundnut, fruit crops and vegetables. It has PRE activity against many annual broadleaf and grass weeds including *Echinocloa crusgalli, Digitaria sanguinalis, Chenopodium album, Portulaca oleracea.* It does not accumulate from repeated annual application, hence safer to crops in the following season.

(viii) Propachlor: 2 – chloro – N – (1 – methylethyl) – N – Phenyl acetamide: It is a chloroacetamide. It is effective in maize, sorghum, millets, soybean, legumes, fruit crops and ornamentals when applied as PRE. It controlled annual grass weeds like *Echinocloa crusgalli, Setaria* spp., *Panicum dichotomiflorum* and broadleaf weeds such as *Amaranthus retroflexus* and *Mollugo veticillata.* It has short persistence hence does not injure rotational crops in the following season.

(ix) Propanil: - 2 – Chloro – N – (1 – methyl ethyl) – N – phenyl acetamide : Propanil was found specifically effective in wheat by controlling several annual broadleaf and grass weeds including *Setaria* spp., *Echinocloa crusgalli, Digitaria* spp., *Eleusine indica* and *Amaranthus retroflexus.* Its residues do not injure crops planted the in the following season.

2. Aliphatics

***(i) Arolein: 2-Propenal*:** Arolein, also known as acrylaldehyde, is developed as aquatic herbicide against floating weeds and algae in irrigation canals.

***(ii) Dalapon: 2 – 2 – dichloropropionic acid*:** It is available as sodium salt. It is a safe herbicide for plantations like tea, coffee, sugarcane and orchard crops. It is effective against *Elytrigia repens, Imperata cylindrica.*

3. Arsenicals

Organic arsenicals are non-degradable in soils hence provide good control of perennial grasses. The affected plants show foliar chlorosis and necrosis. They are applied as foliage herbicides. They include cacodylic acids, DSMA and MSMA.

***(i) DSMA: Disodium methanearsonate*:** It is applied POST specifically in cotton, citrus and non-crop areas. This controls weed spp. *viz. Digitaria* spp., *Paspalum dilatatum, Sorghum halepense, Cyperus rotundus, Setaria* spp., *Ambrosis artemisifolia, Xanthium strumarium, Amaranthus retroflexus etc.*

***(ii) MSMA: Monosodium methanearsonate*:** It is foliage applied in cotton, turf and non-crop areas and specifically controls the weed spp. listed under DSMA but more effectively. It could be used to formulate mixtures with other herbicides. It has moderate to long residual activity in soil.

4. Benzamides

Isoxaben: N [3 – (1- ethyl – 1 – methyl propyl) – 5 – isoxazolyl] – 2, 6 – dimethoxybenzamide: The weeds susceptible to Isoxaben fail to emerge following PRE and POST application. Characteristically it causes stunting, reduced root growth, root hair distortion and root clubbing symptoms in susceptible broad leaf weeds as that of dinitroaniline herbicides. Foliar application to susceptible broadleaf weeds results in swelling and cracking of the stem and petiole. It controls annual grasses *viz., Echinocloa crsugalli, Digitaria* spp., *Setaria* spp., *Panicum dichotomiflorum etc.* and broadleaf weeds like *Chenopodium album, Amaranthus retroflexus, Mollugo verticillata etc.* It has got long residual activity. At recommended rates it degrades to non-phytotoxic levels during the following season, hence safe to the crop plants.

5. Benzoics

They show strong plant growth-regulating activity. They form salts *in vitro*, in plants and soil. They show symptoms as those of auxinic herbicides on susceptible spp. – twisting and curling of stems and petioles (epinasty) stem swelling (at nodes) and elongation and leaf cupping. These symptoms are followed by chlorosis, growth inhibition, wilting and necrosis.

6. Dicamba

***Dicamba: 3, 6 – dichloro – 2 methoxybenzoic acid*:** It is the major herbicide among benzoics. It is safely applied as PPS or POST in maize, sorghum, wheat, turf and non-crop areas to kill many annual broadleaf weeds. It may be leached out of the zone of activity in humid zones of soil and hence safe and offer weed free zone for succeeding crop plants.

7. Benzothiadiazoles

Bentazon: 3 – (1 – methyl ethyl) – 2, 1, 3 – benzothiadiazin – 4 (3H) – one, 2, 2, - dioxide: Bentazon as a sodium salt BASAGRAN, is a POST contact herbicide, used for control of 2, 4 – D tolerant weeds in maize, wheat, barley, sorghum, rice and soybean. It is effective against weed spp. of Amaranthaceae, caryophyllaceae, compositaceae, Cyperaceae, Ambrosiaceae, Polygonaceae and Solanaceae. It is a selective herbicide for rice. In susceptible plants chlorosis develops within 3-5 days after application followed by foliar desiccation and necrosis. Bronzing occurs in tolerant crops.

8. Bipyridiliums

These are synthesized from pyridine and related to quarternary ammonium salts. They are phytotoxic and highly effective in killing plants quickly on contact. They are readily absorbed by the foliage. They include Diquat and Paraquat.

***(i) Diquat: 6, 7 dihydrodipyrido [1, 2 – a 2, 1- c] pyrazinedium ion*:** Diquat causes rapid wilting and desiccation of plants within several hours of application in full sunlight. Complete foliar necrosis occurs, if applied POST. It is mainly applied to ponds, lakes and drainage, ditches to control algae and submersed aquatic weeds like *Utricularia* spp., *Ceratopyllum demersum* and *Elodea* spp. and floating weeds such as *Centella asiatica. Hydrocotyle* spp, *Salvinia* spp. and *Eichhornia crassipes*. Diquat can be tolerated by sugarcane when used for tassel control. As it is strongly bound to clay, it cannot be absorbed by crop plants, hence safe.

***(ii) Paraquat: 1, 1-dimethyl – 4, 4 – bipridinium ion*:** It is available as dichloride and dimethyl sulphate salt formulations. The more widely used dichloride causes rapid wilting and desiccation of plants within few hours of application causing complete necrosis within in 1-3 days. It is a POST, non-selective contact broad-spectrum herbicide in crop and, non-crop areas. It is safe for many plantation crops *viz.,* tea, coffee, rubber, coconut, fruit and orchard crops and for maize, sorghum, sugarcane, pineapple *etc.* As pre-harvest desiccant, it could be used for potato and cotton. As tightly absorbed by clay particles, it becomes completely inactive in soil.

9. Carbamates

These are carbonilic acid esters and include alkyl- and aryl- substituted derivatives. The phenyl carbamate herbicides:

***(i) Asulam: Methyl [(4 – aminophenyl) sulfonyl] carbamate*:** It is herbicidal as sodium salt. The susceptible weeds show chlorosis in young leaves and stunting followed by necrosis. It is tolerable by sugarcane, plantation and orchard crops. The susceptible plant species include *Sorghum halepense, Digitaria sanguinalis, Eleusine indica,* Perennial ferns and *Pteridium aquilinum*. It is very effective against actively growing immature weeds and enters rhizomes to ensure excellent rapid control.

(ii) Desmedipham: Ethyl {3 – [[(phenylamino) carbonyl] oxy] phenyl} – carbamate: It is specially developed for use in sugarbeets to control annual broadleaf weeds especially *viz., Sonchus arvensis, Solanum nigrum, Chenopodium album, Abrosia artemissifolia* and *Amaranthus retroflexus.*

(iii) Phenmedipham: 3 – [(methoxy carbonyl) amino] phenyl (3 – methyl phenyl) carbamate: It is developed for weed control in sugarbeets and vegetable crops. It specifically controls *Setaria* spp. and annual broad leaf weeds including *Solanum nigrum, Protulaca oleracea etc.*

10. Cineoles

Cinmethylin: exo – 1 – methyl – 4 – (1 – methylethyl) – 2 – [(2 – methyl phenyl) methoxy] – 7 – oxabicyclo – [2.2.1] heptane: Cinmethylin, a benzyl derivative of the monoterpene has specific activity of inhibition of mitosis in meristematic regions of susceptible plants. It is effective in transplanted rice as PRE and as PPI or PRE can be effective in soybean, cotton, groundnut and certain vegetable and tree crops. Primarily it controls annual grasses, annual broad leaf weeds and sedges.

11. Cyclohexanediones (Cyclohexenones)

They are very effective in controlling many annual and perennial grasses. When applied POST, plant growth gets ceased as young and actively growing leaves are affected first. The leaf turns brown and mushy at the node. Old leaves turn purple, orange or red before becoming necrotic. But when applied PRE, the primary root growth is inhibited and leaves fail to emerge from the coleoptile.

(i) Clethodium: (E, E) – (±) – 2 – [1 – [[(3 – chloro – 2 – propenyl) oxy] imino] propyl] – 5 – [2 – (ethylthio) propyl] – 3 – hydroxy – 2 – cyclohexen – 1 – one: Clethodium registered as SELECT, is effective for cotton and soybean for control of many annual and perennial grasses. It has no activity against broadleaf weeds and sedges. It is non-persistent in soil and safe to any of the following season crops.

(ii) Cycloxidim: 2 – [1 – (ethoxyimino) butyl] – 3- hydroxy – 5 – (2H – tetrahydrothiopyran – 3 – yl) – 2 –cyclohexen – 1 – one: It is commercially available as FOCUS and LASER. It is applied as POST for control of annual and perennial grasses but without affecting broadleaf weed species. The tender emerging grasses are controlled immediately after application. It has got very short residual life and safe for subsequent crop plants.

(iii) Sethoxydim: 2 – [1 – (ethoxy imino) butyl] – 5 – [2 – (ethylthio) propyl] – 3 – hydroxy – 2 – cyclohexen – 1 – one: It is commercially available as POAST. It controls annuals and perennial grasses in several broadleaf crops. As POAST it is safely used in alfalfa, cotton, sunflower *etc.* and for pre-plant burndown effect in soybean. At higher rates, it can be applied POST in ornamental trees, tree crops and non-crop areas for control of shrubs and hard-to-control weeds. It is rapidly degraded in soil and safe for succeeding crops.

(iv) Tralkoxydim: 2 – [1 – ethoxy imino) propyl – 3 – hydroxy – 5 – mesity cyclohexen – 1 – one: It is very effective against several annual and perennial grasses when used as POST herbicide. It is safely used in cotton, soybean, wheat, tea, coffee and other plantation crops, fruit and tree crops. It gets rapidly degraded in soil hence not harmful to crop plants.

12. Dinitroanilines

Trifluralin is the first and most prominent member of this group. Dinitroanilines are orange-yellow coloured compounds with low water solubility. They are volatile, hence need to be incorporated into soil. Selectively used as PPI and PRE to control a wide spectrum of grasses and broad leaf weeds. The susceptible weeds fail to emerge as they inhibit coleoptile growth or hypocotyl unhooking. However, seed germination is not inhibited. Root growth inhibition is a prominent symptom on emerged seedlings and established plants. Root appear stubby, tips become thickened. The base of grass shoots swell as bulbous and hypocotyls may swell in broad leaf weeds. Shoots become deformed and brittle.

(i) Benfin: N – butyl – N – ethyl – 2, 6 – dinitro – 4 – (trifluoromethyl) benzenamine: It is safely applied as PPI in groundnut, tobacco, lettuce and turf to control annual grasses such as *Echinocloa cursgalli, Brachiaria* spp. *Digitaria* spp. *Panicum dichotomiflorum, Setaria* spp. and *Cenchrus* spp. and certain annual broadleaf weeds including *Chenopodium album* and *Amaranthus* spp. It is strongly adsorbed by soil and controls weeds for a full season. The crops in rotation are allowed to express their potential productivity.

(ii) Ethalfluralin: N – ethyl – N – (2 – methyl – 2 – propenyl) – 2, 6 – dinitro – 4 – (trifluoromethyl) : It is applied PPI as effective herbicide in soybean, groundnut, sunflower vegetables, and also applied in standing wheat. It primarily controls annual grasses such as *Setaria* spp. *Echinochloa crusgalli, Panicum dichotomiflorum, Digitaria* spp. *etc.* and broad leaf weeds such as *Amaranthus retroflexus, Kochia scoparia* and *Solanum nigrum.* It injures sensitive crops grown even after one year of its application. But a deep and mould board ploughing before planting minimize the injury to sensitive crop and improve productivity.

(iii) Fluchloralin: N – (2 – chloroethyl) 2, 6 – dinitro – N – propyl – 4 – (trifluoromethyl) aniline: It is a PRE herbicide safely used in cereal crops particularly wheat, rice, groundnut, pulse crops and many vegetable crops. It selectively controls many annual broad leaf weeds and grasses. It affects germination and seedling emergence of susceptible crops.

(iv) Pendimethalin: N – (1 – ethyl propyl –) – 3, 4 – dimethyl – 2, 6 – dinitro benzenamine : It can be safely used as PRE or E-POST in cotton, maize, tobacco, sorghum, wheat, peanut, sunflower, rice, sugarcane, potato, onion, vegetables, fruit crops *etc.* It is very effective against annual grasses, broad leaf weeds that are controlled by ethalfluralin. It is also effective against *Sorghum halepense, Brachiaria* spp. *Abutilon theophrasti etc.*

(v) Prodiamine : 2, 4 – dinitro – N^3, N^3 – dipropyl – 6 – (trifluoromethyl) – 1, 3 – benzenediamine. Prodiamine inhibits root and shoot growth of susceptible plants like dinitroaniline herbicides. It is applied PRE in turf and ornamentals. It effectively controls *Sorghum halepense, Eleusine indica* and *Euphorbia humistrata.*

(vi) Trifluralin: 2, 6 – dinitro – N, N – dipropyl – 4 (trifluoromethyl) benzenamine: It is a widely used selective herbicide used in maize, pulse crops, cotton, soybean, safflower, castor, sunflower, vegetable and horticultural crops. It controls several annual grasses, small seeded broad leaf weeds.

It is applied along with liquid fertilizers or impregnated on dry fertilizers in crops like maize, soybean, wheat and sorghum. As POST herbicide it is also effective in onion, potato, sugarbeets *etc.* It is compatably mixed with several herbicides. Hence, it is a safely used herbicide in a broad spectrum of crops against weed flora.

13. Diphenylethers

Diphenylethers including nitrogen are being used in various crops *viz.,* wheat, maize, rice, soybean, cotton and groundnut. Many of them are applied at PRE. They function as contact herbicides. In susceptible plants, leaves become chlorotic, desiccated and necrotic within a day or two whereas in tolerant crops youngest expanded leaves as in soybean and groundnut. Chlorosis and necrosis is shown at higher doses sub-lethal doses produce foliar 'bronzing' on expanded leaves. Bleached spots or flecks are developed when droplets of these herbicides drift on leaves.Some of the herbicides of this group are:

(i) Acifluorfen: 5 – [2 – chloro – 4 – (trifluoromethyl) phenoxy] – 2 – nitro – benzoic acid: This is a sodium salt applied as POST, is effective in groundnut and soybean. It controls many annual broad leaf weeds including *Ipomoca* spp., *Solanum nigrum, Xanthium strumarium, Ambrosia artemisifolia, Datura stamonium, Sesbania exaltata, Chenapodium album, Amaranthus* spp., *Polygonium* spp., *Brassica kaber, Setaria* spp., *Sorghum halepense.* Its residues do not persist in the soil, hence safe to succeeding crops.

(ii) Bifenox: methyl 5 – (2, 4 – dichloro phenoxy) – 2 – nitrobenzoate: It is marketed as MODOWN. It is used as POST in grain crops and PRE or E-POST in transplanted rice. It is applied in soybean and sunflower as PRE. Bifenox controls many annual broad leaf weeds like *Kochi scoparia, Amaranthus* spp., *Datura stramonium, Polygonium* spp., *Brassica Kaber, Datura carota, Veronica* spp. *etc* and grass weeds like, *Echinocloa crusgalli* and annual sedges. The residues do not injure susceptible crops planted even after one year of application.

(iii) Fluoroglycofen: Carboxymethyl 5 – [2 – chloro – 4 – (trifluoromethyl) phenoxy] – 2 – nitro benzoate: It is also called as benzofluorfen, available as ethyl ester as COMPETE. It is effective as POST in cereal crops for control of several broad leaf weeds.

(iv) Fomesafen: 5 – [2 – chloro – 4 – (trifluoromethyl) phenoxy] – N – (methyl sulfonyl) – 2 – nitrobenzamide: Specifically fomesafen a sodium salt as REFLEX is applied POST in soybean. It controls many annual broad leaf weeds including *Ipomoea* spp., *Amaranthus* spp., *Datura stramonium, Brassica Kaber, Solanum nigrum* and *Ambrosia artemisifolia.* It persists long in the soil and injures susceptible crops like sugar beet, sunflower, sorghum.

(v) Lactofen: (±) – 2 – ethoxy – 1 – methyl –2- oxoethyl 5 – [2 – chloro – 4 (trifluoromethyl) phenoxy] – 2 – nitrobenzoate: Lactofen is being traded as COBRA. Applied as POST, is effective in soybean and cotton. It is used to control many annual broad leaf weeds such as *Datura stramonium, Brassica kaber, Solanum* spp., *Ambrosia* spp., *Amaranthus* spp. and *Xanthium* spp. The residues do not injure rotational crops. It is compatable to many herbicides. Hence, it is widely used covering many crops.

(vi) Oxyfluorfen: 2 – chloro – 1 – (3 – ethoxy – 4 – nitrophenoxy) – 4 – (trifluoromethyl) benzene : This is traded as GOAL. It is used effectively in soybean, groundnut, cassava, leguminous crops, and plantation crops such as tea, rubber and oil palm. It is also a useful herbicide in transplanted rice, vegetables and vineries.

It is also safely used as a POST – directed spray in maize, cotton, papaya, soybean, fruit trees and plantation crops. Oxyfluorfen controls many annual broad leaf weeds like *Amaranthus* spp., *Ageratum conyzoides, Borreria hispida, Commelina benghalensis, Portulaca oleracea, Chenopodium album* etc. and annual grasses such as *Digitaria sanguinalis, Echinocloa crusgalli, Eleusine indica, Paspalum conjugatum, Axonopus compressus etc.* It is moderately persistent in soil. Hence, safe for succeeding crops.

14. Imidazolidinones

Buthidazole: 3 – [5 – (1, 1 – dimethylethyl) – 1, 3, 4 – thiadazol – 2 – yl] – 4 – hydroxy – 1 – methyl – 2 – imidazolidinone: It is applied PRE and POST in maize, wheat, alfalfa and other crops. It is effective against numerous broad leaf weeds and a grass including *Elytrigia repens, Teraxacum officinale, Echinocloa crusgalli, Setaria viridis, Abutilon theophrasti, Chinopodium album, Amaranthus* spp. *etc.*

15. Imidazolinone

Imidazolinones are a new group of herbicides. Growth inhibition by the herbicide begins within a few hours after application and injury symptoms usually appear by 1-2 weeks. Meristematic areas become chlorotic followed by a slow chlorosis and necrosis. This group includes:

(i) Imazapyr: 2 – [4, 5 – dihydro – 4 – methyl – 4 – (1 – methyl ethyl) – 5 – oxo – 1 H – imidazol – 2 – yl] – 3 – pyridinecarboxylic acid: Imazapyr is traded as ARSENAL. It is formulated as isopropylamine salt. It is applied PRE and POST in sugarcane and plantation crops to control several annual and perennial grasses and broad leaf weeds. It has long persistence and hence effective weed control in tolerant crops. It thus avoids repeated sprayings and increases cost-benefit ratio.

(ii) Imazaquin: 2 – [4, 5 – dihydro – 4 – methyl – 4 – (1 – methyl ethyl) – 5 – oxo – 1 H – imidazol – 2 – yl] – 3 – quinolinecarboxylic acid: Imazaquin was registered as IMAGE and SCECTER. It is the first imidazolinone herbicide widely used. It is available as ammonium salt and applied as PRE and POST in soybean, maize, groundnut, pluses *etc.* It effectively checks a large range of annual broad leaf weeds *viz., Amaranthus* spp. *Xanthium strumarium, Datura stramonium, Kochia scoparia, Chenopodium album, Ipomoea* spp. *Sorghum nigrum, Ambrosia* spp. *Euphorbia* spp. *Anoda cistata, Abutilon theophrasti* and annual grasses such as *Digitaria* spp., *Setaria* spp., *Echinocloa crusgalli, Panicum dichotomiflorum, Sorghum halepense, Sorghum bicolor* and *Nutsedge* spp. It has long persistence. Hence, the next season crops need to be delayed.

(iii) Imazethapyr: 2 – [4, 5 – dihydro – 4 – methyl – 4 – (1- methyl ethyl) – 5 – oxo – 1 H – imidazol – 2 – yl] – 5 – ethyl – 3 – pyridine carboxylic acid: It was registered for field use as HAMMER and PURSUIT. It can be safely used in soybean, groundnut, peas, edible beans and imidazolinone – tolerant maize. It can be applied as POST and E-PPL in soybean, maize, groundnut and legumes. It controls a wide spectrum of annual broad leaf and grass weeds.

(iv) Imazamethabenz: (±) – 2 – [4, 5 – dihydro – 4 – methyl – 4 – (1 – methyl ethyl) – 5 – oxo – 1H – imidazol – 2 – yl] – 4 (and 5) – methyl benzoic acid: This was registered as ASSERT and applied as POST effectively in wheat, barley and sunflower. It specifically controlls *Avena fatua, Brassica kaber, Descurainia sophia* (flex weed), *Sisymbrium* spp., *Polygonum convolvulus* (wildbuck wheat), *Thlaspi arvense* (field pennycress), *Asperugo procumbens* (catch weed) and *Poa* spp.

16. Imines

CGA – 248757: methyl {[2 – chloro – 4 – fluoro – 5 – [(tetrahydro – 3 – oxo – 1H, 3H – [1, 3, 4] thiadiazolo [3, 4 – a] pyridazin – 1 – ylidene) amino] phenyl] thio]} acetate: CGA – 248757 referred to as KIH – 9201, was developed for use as POST in maize and soybean. It can control *Abutilon theophrasti* effectively and other broad leaf weeds. It has very short persistence in soil with 1 – 2 d half life. Hence, it is a very safe herbicide for succeeding crops besides its weed control efficacy.

17. Isoxazolidinones

Clomazone: 2 – [(2 – chlorophenyl) methyl] – 4, 4 – dimethyl – 3 – isoxazolidinone: Clomazone was developed and commercialized as COMMAND. The susceptible plants to clomazone showed foliar bleaching in later growth stages, if applied POST. It is also applied as PRE and PPI in soybean and vegetable crops. The weed species sensitive to clomazone are *Abutilon theoprasti, Ambrosia artemisiifolia, Chenopodium album, Echinocloa cusgalli, Digitaria* spp. It is less persistant in sandy loam than in silt loam or clay loam soils.

18. Nitriles

(i) Bromoxynil: 3, 5 – dibromo – 4 – hydroxy, benzonitrile: It is a benzonitrile herbicide. When applied POST, it is effective in wheat, barley, oats, rye, triticale, maize, sorghum *etc*. It is effective in controlling broadleaf weed such as *Chorispora tenella, Chenopodium album, Solanum* spp., *Polygonum convolvulus, Brassica kaber, Kochia scoparia, Xanthium strumarium, Abutilon theophrasti, Capsella bursa-pastoris etc.*

(ii) Dichlobenil: 2, 6 – dichlorobenzonitrile: It is a powerful inhibitor of germination. It is safely used in fruit and orchard crops, vineries, forest nurseries *etc*. It is used to control effectively *Cyperus rotundus* and other perennial weeds that propagate through underground rhizomes, tubers, and stolons. Young seedlings of both monocot and dicot species are killed. It effectively controls many annual and perennial broadleaf and grass weeds such as *Poa annua, Digitaria* spp., *Portulaca oleracea, Panicum* spp., *Daucus carota, Taraxacum officinale, Cirsium arvense, Euphorbia esula, Elytrigia repens etc.*

(iii) Ioxynil: 4 – hydroxy – 3, 5 – diidobenzonitrile: It is available as a sodium salt and oil soluble ammonium salt. Ioxynil was applied POST for effective control for *Labiatae Caryophyllaceae, Fumariacae, Violaceae* and *Scrophulariaceae* families specifically in crops like wheat, barley and rice.

19. Oxadiazoles

Oxadiazon: 3 – [2, 4 – dichloro – 5 – (1 – methyl ethoxy) phenyl] 5 – (1, 1 – dimethyl ethyl) – 1, 3, 4 – oxadiazol – 2 – (3H) – one : It is applied as PRE or POST in a wide spectrum of crops such as rice, cotton, sugarcane, soybean, groundnut, onion, garlic, potato, sunflower, transplanted vegetables, tea rubber and banana. It is particularly useful to control weeds that grow from bulbs, rhizomes and other deep rooted plant propagules, including *Cynodon dactylon, Lolium perenne etc*. It is also effective against many annual broad leaf and grass weeds such as *Oxalis* spp., *Amaranthus* spp., *Richardia scabra, Urtica* spp., *Poa annua, Digitaria* spp., *Mollugo vertcillata, Cenchrus incertus, Eleusine indica, Setaria viridis etc*. It affects weeds at their early growth stage. Besides wide spectrum of weed control it has moderate to long persistence.

20. Oxadiazolidines

(i) Methazole: 2 – (3, 4 – dichlorophenyl) – 4 - methyl – 1, 2, 4 – oxadiazolidine – 3, 5 – dione: It is traded as PROBE and could be applied as PPI, PRE as well as POST herbicide against a wide range of weed species including annuals and perennials. It can be used in crops like cotton, sorghum, maize, wheat, oats, soybean, onion, grapes and several fruit crops. When applied to foliage of susceptible weed plants, it causes veinal chlorosis followed by general chlorosis and desiccation. The residues, however do not accumulate in soil and so do not injure rotational crops in the following season, thus enhancing productive potential of crop plants.

21. Phenols

These organic chemicals were first used as herbicides when 4, 6, dinitro – O – cresol (2 methyl – 4,6 – dinitrophenol) was introduced. They were widely used as DNOC in Europe and USA under the name of SINOX. The dinitorphenols are widely used as uncouplers of oxidative phosphorylation. These are active by contact action.

***Dinoseb: 2- sec-butyl – 4,6 – dinitrophenol*:** Dinoseb is a most toxic substituted dinitrophenols of sodium salt applied as POST for selective weed control in potato, beans, peas, orchard crops, plantation crops and vineyards. The alkanolamine salt (PREMERGE) is used as PRE to kill germinating weeds in the top soil layer.

22. Phenoxy Alkanoic Acids

These acids and their derivatives are a major group of organic herbicides by virtue of their selectivity and outstanding ability to be translocated within the plant.

a) Phenoxy acetics: These include 2, 4 – D, MCPA and 2, 4, 5 – T. These herbicides have hormonal activity at low rates. At higher rates they exhibit herbicidal properties. They show selectivity between the susceptible broadleaf weeds and the tolerant grasses. This is the reason for their wide use in many cereals and grass crops. The susceptible plants to these herbicides show epinastic bending and twisting of stems and petioles, stem swelling and elongation and leaf curling and cupping. This is followed by chlorosis at the growing points, growth inhibition, wilting and necrosis. The affected plants die slowly. At low concentrations, young leaves may appear puckered and the tips of new leaves may develop into narrow extensions of the midrib.

(i) 2, 4 – D: (2, 4 – dichlorophenoxy) acetic acid: The potency of 2, 4 – D as plant hormone was reported in 1942 by Zimmerman and Hitch cock and herbicidal utility was reported by Krause, Marth and Mitchell; Slade, Templeman and Secton; Nutman, Thornton and Quaster[186] 2, 4 – D is a systemic applied as POST for control of many annual weeds in cereal crops, sugarcane and plantation crops. Most of the dicotyledonous crops are sensitive to 2, 4 – D. Effective when applied as PRE as it is absorbed by emerging weed seedlings. It is available in ester formulations as ethyl ester, butoxy ethyl ester, isopropyl ester, isooctyl ester *etc*. They are more volatile than their sodium and amine salts. 2, 4 – D is potentially mobile but rapid degradation in soil and removal from soil by plant uptake minimizes leaching. It is compatible with most herbicides. Hence it's a widely used herbicide. Its field half-life is 10 d only, hence not harmful for subsequent crops.

(ii) MCPA: (4 – chloro – 2 methyl phenoxy) acetic acid: It is similar to 2, 4 – D and is a selective POST herbicide against broad leaf weeds. It is more selective than 2, 4 – D at equal rates. It is widely used in cereals, legumes and aquatic weeds. It is available as ester, amines and sodium or potassium salts. The amines and salts are water soluble. It is widely used for making herbicide mixtures.

b) Phenoxy butyrics

***(i) 2, 4 – DB: 4 (2, 4 –dichlorophenoxy) butanoic acid*:**Applied as POST, 2,4 – DB is selective to control several broad leaf weeds in forage legume crops. It can also be applied as PRE and PPL. It is available as dimethyl amine salt, isooctyl ester and butoxyethyl ester.The visual injury symptoms are similar to those exhibited by 2,4– D and MCPA. 2, 4 – DB can be applied through sprinkler irrigation system also. Its field half-life is 5 d (acid), 7 d (butoxy methyl) and 10 d (dimethyl amine salt). Thus it has short residuality in soils and so safe to the crop plants to express their production potential in weed-free condition.

c) Aryloxyphenoxy propionics

The aryloxyphenoxy propionic (APP) acids and derivatives are given to the appearance of several heterocyclic oxypropionic acids. These are highly active selective herbicides and control grass weeds mainly in broad-leaf crops. These herbicides when applied POST are very toxic to graminaceous plants but are tolerated by dicotyledonous and monocotyledonous plants. The susceptible plants show chlorosis in newly formed leaves and young and actively growing tissues within 2-4 days. Progressive necrosis of the meristematic tissue in the nodes and buds develop within 1-3 weeks. The oldest leaves show senescent pigment changes. Growth of roots is strongly inhibited damaging the root tips. Mottling, scattered chlorotic appearance develop in susceptible species resembling the burning effect of contact herbicides.

(i) Dichlorprop: 2 – (2, 4 – dichlorophenoxy) propanoic acid: Dichlorprop is commercially formulated as butoxyethyl, isooctyl ester and dimethylamine salt. It is used to control weeds mainly in non-agricultural areas as POST. It also controls many hardwood brush weeds. It can be used by mixing with mecroprop, MCPA, 2, 4 – D and other herbicides.

(ii) Diclofop: 2 – [4 – (2, 4 – dichlorophenoxy) phenoxyl] propanoic acid: The diclofop methyl ester formulation and diclofop acid are herbicidally active. During metabolism diclofop methyl is hydrolysed to diclofop in plants. With the former causing leaf membrane damage and the latter causing inhibition of meristematic activity. Diclofop is applied POST in wheat and barley. It is effective in controlling annual grasses such as *Avena fatua, Setaria* spp., *Echinocloa crusgalli, Lolium* spp. *etc.* At higher rates it gives good control of *Cynodon dactylon.* Most broad leaf crops are tolerant to diclofop.

(iii) Fenoxaprop: 2 – {4 – [(6 – chloro – benzoxazolyl) oxy] phenoxy} propanoic acid: It is available as ethyl ester. It is applied as POST in soybean and turf. Fenoxaprop alone was reported to be toxic to wheat and other crops but when mixed with 2, 4 – D, MCPA, thifensulfuron and tribenuron which antagonize its activity, shows selectivity. It effectively controls many annual and some perennial grass weeds.

(iv) Fluazifop-P: (R) – 2 – [4 – [[5 – (trifluoromethyl – 2 – pyridinyl] oxy] phenoxy] propanoic acid : Fluazifop-P, is commercially available as butyl ester formulation and marketed as FUSILADE. It is applied as POST in cotton, soybean, wheat, tea, coffee, tree crops. It controls several annual and perennial grasses including *Echinocloa crus-galli, Digitaria* spp., *Bromus tectorum, Panicum* spp., *Setaria* spp., *Cynodon dactylon, Imperata cylindrica, Sorghum bicolor, Elytrigia repens, Sorghum halepense etc.* It cannot control broad-leaf weed spp. The wide spectrum of weed control by fluazifop-P provides congenial atmosphere for exhibiting productive protential of crop plants.

(v) Quizalofop-P: (R) – 2 – {4 – [(6 – chloro – 2 – quinoxalinyl) oxy] – phenoxy} propanoic acid: It was developed for use in soybean and cotton. It is commercially available as ethyl ester formulation, ASSURE. As POST it controls annual and perennial grass weeds in soybean and non-crop areas. It is effective against grass weeds including *Sorghum halepense, Cynodon dactylon, Elytrigia repense, Muhlenbergia frondosa etc.* quizalofop suppress grass weeds germinating after a POST application but its efficacy depends on rate of herbicide, soil type and environmental conditions.

23. N - phenylphthalimides

***Flumiclorac: [chloro – 4 – fluoro – 5(1, 3, 4, 5, 6, 7 – hexahydro – 1, 3, - diaxo – 2H – isoindol – 2 yl) phenoxy] acetic acid*:** Flumiclorac is commercially available as phenyl ester formulation under its trade names RESOURCE and SUMIVERDE. It causes wilting and bleaching within a day to susceptible plants under bright sunlight. The leaves become brown, desiccated and necrotic. It can be applied POST in soybean and maize. It is used to control *Abutilon theophrasti, Chenopodium album, Ambrosia artemisi folia, Amaranthus* spp. *Euphorbia maculata, etc.* It is extremely non-persistant with no effect on rotational crops.

24. Phenyl pyridazines

Pyridate: ***0*** **– (6 – chloro – 3 – phenyl – 4 pyridazinyl)** ***S*** **– octyl carbonothioate:** This herbicide marketed as TOUGH could be applied POST in a range of crops *viz.* groundnut, maize, rice, onion, alfalfa, glassland, rapeseed, orchards, vineyards and some vegetables. The susceptible plants show withering at leaf edges and the effect was complete within 5-10 days. It is non-persistant in soil. Thus, with weed control efficacy over a range of crops coupled with non-persistant residuality, provides possibility for augmenting crop productivity.

25. Phenyl triazinones (Aryl Triazinones)

Sulfentrazone: N – {2, 4 – dichloro – 5 – [4 – (difluoromethyl) – 4, 5 – dihydro – 3 – methyl – 5 – oxo – 1H – 1, 2, 4 – triazol – 1 – yl] phenyl} methane sulfonamide: Sulfentrozone, a new phenyl triazinone herbicide was registered for soybean. It could be applied PPI and PRE for control of several broad-leaf weeds including *Ipomoea hederacea* (morning glory), *Cassia occidentalis* (coffee senna), *Sida spinosa* (prickly sida), *Cucumis melo* (small melon), *Acalypha ostryifolia etc.* The crops exhibit tolerance to sulfentrazone includes tobacco, sugarcane and some species of turf grass. It is also applied POST to control a broad spectrum of weeds. The sensitive weeds for its POST activity enhanced by a surfactant include several weeds such as *Cyperus esculentus, Abutilon theophrasti etc.*

26. Phthalamates

Naptalam: 2 – [1 – naphthalenyl amino) carbonyl] benzoic acid: Naptalam is applied as PRE and POST to control many annual broad-leaf weeds such as *Chenopodium album, Amaranthus* spp., *Ambrosia artemisiifolia, Xanthium strumarium, Solanum nigrum etc.* and certain annual grasses including *Setaria faberi,* Susceptible plants show strong epinasty and antigeotropic responses. It abolishes normal curvature of root towards the ground and of shoots towards light. The tolerant crop plants include melon, cucumber, watermelon, groundnut, soybean *etc.* The residues do not injure crops planted in the following season.

27. Pyrazoliums

Difenzoquat: 1, 2 – dimethyl – 3, 5 – diphenyl – 1H – pyrazolium: The herbicidal activity of difenzoquat methyl sulphate was used to control wild oats by applying POST. In case of susceptible plants, it ceases plant growth soon after application but symptoms of injury appear after 3-7 days. The meristematic regions become chlorotic leading to necrosis.

28. Pyridazinones

Pyridazinones or fluorinated pyridazinone herbicides are applied mostly PRE which include norflurazon and pyrazon. The susceptible plants show interveinal whitening of leaf and stem tissues. Though growth continued for several days but seedlings lack green photosynthetic tissue, and eventually turn necrotic and die.

(i) Norflurazon: 4 – chloro – 5 – (methylamino) – 2 – [3 – (trifluoromethyl) phenyl] – 3 – (2H) – pyridazinone: Norflurazon (SAN 9789H) is a PRE herbicide in tree crops, vineries, ornamentals, orchard and plantation crops besides non-crop areas. The weed flora susceptible to this herbicide includes many grasses such as *Setaria* spp., *Eleusine indica, Panicum dichotomiflorum, Digitaria sanguinalis, Sorghum halepense etc.* broad-leaf weeds are: *Amaranthus* spp., *Ipomoea* spp., *Portulaca* spp., *Cirsium arvense* and many sedges like *Cyperus* spp. It has moderate to long persistence.

(ii) Pyrazon: 5 – amino – 4 – chloro – 2 phenyl – 3(2H) – pyridazinone: It is also known as chloridazon commercially available as PYRAMIN. It is applied as PRE or early POST in sugar beets and onion. Like norflurazon, it also controls many annual grasses and broad-leaf weeds. Its activity is temperature sensitive, resulting in crop damage under very unusual temperature extremes.

It is used in sprinkler or drip irrigation system. It has short to moderate soil persistence.

29. Pyridine carboxylic acid

(i) Clopyralid: 3, 6 – dichloro – 2 – pyridine carboxylic acid: It was marketed under trade names CURTAIL, STINGER and CONFRONT in different countries. It is applied POST in sugarbeets and maize specially and in pastures and rangelands at higher rates. It controls a different weed flora like many annual and perennial broadleaf weeds such as *Cirsium arvense, polygonum* spp. *Xanthium strumarium, Datura stramonium, Ambrosia artemisiifolia, Iva xanthiofolia etc.* It is mostly residual and discriminately injures certain crops like potato, peas, pulses *etc.* if planted even after one year after application.

(ii) Picloram: 4 – amino – 3, 5, 6 – trichloro – 2 – pyridine carboxylic acid: Picloram (TORDON) available as isooctyl ester and triisopropanolamine salt. Picloram injures like other auxin – type herbicides in forest plantation, non crop areas, pastures and range lands. It can be applied on cut/surfaces to control woody species. It controls many annual and perennial broadleaf weeds, vines and woody plants.

***(iii) Triclopyr: [(3, 5, 6 – trichloro – 2 pyridinyl) oxy] acetic acid*:** Triclopyr, available as a triethylamine salts and butoxy ethyl ester, marketed as GARLON. It is applied POST in non-crop areas. It can be applied to freshly cut tree stumps and mixed with oil for bark treatment on young trees. It also controls many annual and perennial broad leaf weeds, vines and woody plants at higher doses.

30. Pyridines

These are mostly soil-applied, herbicides. They cause root growth inhibition and swelling in meristematic regions such as root tips. Susceptible plants show thickened or swollen hypocotyls or internodes, but seed germination is not inhibited. These include, dithiopyr and thiazopyr:

(i) Dithiopyr: S – S – dimethyl 2 – (difluoromethyl) – 4 – (2 – methyl propyl) – 6 – (trifluoromethyl) – 3, 5 – pyridinedicarbothioate: It is introduced as DIMENSION and STAKEOUT. It is available in liquid and granular formulations. It is applied PRE or

POST, in direct seeded and transplanted rice, established turf, ornamentals and orchard and other perennial crops. It controls specifically *Echinochloa crusgalli, Digitaria* spp., *Eleusine indica, Oxalis* spp., *Euphorbia* spp. It has moderate soil persistence.

***(ii) Thiazopyr: methyl 2 – (difluoromethyl) – 5 – (4, 5 – dihydro – 2 – thiazolyl) – 4 – (2 – methyl propyl) – 6 – trifluoromethyl) – 3 – pyridine carboxylate*:** Thiazopyr commercially available as VISOR and SPINDLE in emulsifying concentrate and granular formulations. It is soil-applied and tolerable to several crops including maize, groundnut, soybean, vineries, sugarcane, sunflower, potato, vegetables, alfalfa and tree crops. It is effective in controlling several annual and perennial grasses and broad leaf weeds. But particularly wheat and sorghum are susceptible to soil residues.

31. Pyridinones

Fluridone: 1 – methyl – 3 – phenyl – 5 – [3 – (trifluoromethyl) phenyl] – 4 (1H) – pyridinone: It is an aquatic herbicide used in lakes, reservoirs, drainage canals, irrigation canals and rivers. It can be applied on water surface or subsurface and as bottom application. It controls most submerged and emerged aquatic weed species including *Utricularia* spp., *Ceratophyllum demersum, Elodea* spp., *Myriophyllum* spp., *Najas* spp., *Potamogeton* spp., *Hydrilla verticillata, Brachiaria mutica etc*. The affected weeds turn white, pink and chlorotic. Fluridone adheres to sediments, gradually desorbs into water and then is subjected to photodegradation. Hence, it is safe to the crops of semihydrophytic or hydrophytic nature.

32. Pyrimidinyl thio-benzoates (Benzoates)

These are new class of herbicides also referred as benzoates. Pyrithiobac is the first herbicide of this group.

Pyrithiobac: 2 – chloro – 6 – [(4, 6 – dimethoxy – 2 – pyrimidinyl) thio] benzoic acid, sodium salt: It is commercially available as sodium salt. It is applied POST particularly in cotton which is very tolerant to control of several broad leaf weeds such as *Xanthium strumarium, Ipomea* spp., *Abutilon theophrasti, Sesbenia exaltata, Salvia reflexa, Sida spinosa, Amaranthus* spp. *etc*. Wheat, soybean and grain sorghum can be rotated with cotton with no crop toxicity. It is used to control C*assia occidentalis*, a troublesome broad leaf weed.

33. Quinoline carboxylic Acids

(i) Quinclorac: 3, 7 – dichloro – 8 – quinoline carboxylic acid: Quinclorac is an auxin – type herbicide. It is used for direct seeded and transplanted rice to control important annual grass weeds. It is particularly very effective against *Echinochloa crusgalli, Setaria* spp. It is applied PRE, L-PRE or E-POST to control annual and perennial broad leaf weeds like *Convolvulus arvensis*. Its phytotoxicity is characterized by inhibition of shoot growth and chlorosis and necrosis of the entire shoot. It is persistant in soil and injures subsequent crops planted even after one year after application.

34. Sulfonylureas

This type of herbicides show high herbicidal potency at very low rates and hence is environmentally safer. They include widely used herbicides in agriculture now-a-days. The affected plants show growth inhibition soon after application. Chlorosis follows necrosis of the growing point. Certain affected species remain green but stunted and not competitive with the crop.

(i) Bensulfuron: methyl 2 – [[[[(4, 6 – dimethoxy – 2 – pyrimidinyl) amino] carbonyl] amino] sulfonyl] methyl] benzoate: The herbicidal activity of bensulfuron is marketed as LONDAX. It is applied at PRE or POST in irrigated rice or POST in dryland rice. It controls many annual grasses (*Echinochloa crusgalli* and *E. colona*), sedges (*Cyperus rotundus* and members of cyperaceae and broad leaf weeds such as *Commelina* spp., *Heteranthera limosa, Elipta alba, Sphenoclea zeylandica, Pontederia cordata, Ammania* spp., *Plantago* spp. *etc.* It has short field-half life and does not move deep into soil than 5-7 cm hence safer to crop plants.

(ii) Chlorimuron: ethyl 2 – [[[[(4 – chloro – 6 – methoxy – 2 – pyrimidinyl) amino] carbonyl] amino] sulfonyl] benzoate : Chlorimuron was marketed as CLASSIC and applied POST and PRE particularly in soybean, groundnut and non-crop areas. It controls specially, many annual broad leaf weeds *viz. Xanthium strumarium, Datura stramonium, Ipomoea* spp., *Ambrosia* spp., *Amaranthus* spp., *Polygonium* spp. *etc.* Chlorimuron persists longer in high pH soils, hence longer weed control efficacy and improved crop productivity.

(iii) Chlorsulfuron: 2 – chloro – N – [[(4 – methoxy – 6 – methyl – 1, 3, 5 – triazine – 2 – yl) amino] carbonyl] benzenesulfon amide : Chlorosulfuron can be applied PRE or POST in wheat specifically. It is used to control many broad leaf weeds including *Kochia scoparia, Cirsium arvense, Brassica* spp., *Amaranthus* spp., *Chenopodium album etc.* It inhibits the growth of the treated plants within a few hours after application, with symptoms appearing 1-2 weeks later. Meristematic areas become chlorotic and necrotic followed by a general foliar chlorosis and necrosis. It is toxic to susceptible crops upto 3 or 4 years after application in high pH soils.

(iv) Halosulfuron: methyl 5 – [[(4, 6 – dimethoxy – 2 – pyrimidinyl) amino] carbonyl aminosulfonyl] – 3 – chloro – 1 – methyl – 1 – H - pyrazole – 4 – carboxylate: It is introduced as MON 1200, and used as a POST herbicide in maize, sorghum, sugarcane and turf. It controls *Abutilon theophrasti, Xanthium strumarium* and several other broad leaf weeds and *Cyperus* spp. It has short to moderate persistence in soil depending on soil organic matter, rainfall and type of soil.

(v) Metsulfuron: methyl 2 – [[[[(4 – methoxy – 6 – methyl – 1, 3, 5 triazin – 2 – yl) amino] carbonyl] amino] sulfonyl] benzoate : This herbicide is marketed as ALLY and ESCORT. It is applied POST in wheat for control of broad leaf weeds such as *Kochia scoparia, Brassica* spp., *Cirsium arvense, Chenopodium album etc.*

(vi) Nicosulfuron: 2 – [[[[(4, 6 – dimethoxy – 2 – pyrimidinyl) amino] carbonyl] amino] sulfonyl] – N – N – dimethyl – 3 – pyridine carboxamide : It is marketed as ACCENT.

It is applied POST in maize for controlling most of annual and some perennial grasses including *Setaria* spp., *Sorghum bicolor, Eriochloa villosa, Panicum miliaceum, sorghum halepense* and *Elytrigia repens* as well as certain broad leaf weeds.

(vii) Primisulfuron: methyl 2 – [[[[4, 6 – bis (difluoromethoxy) 2 – pyrimidinyl] amino] carbonyl] amino] sulfonyl] benzoate : Primisulfuron is commercially used in maize to control certain annual and perennial grass weeds such as *Panicum dichotomiflorum, Sorghum bicolor, Sorghum halapense* and *Elytrigia repens* and annual broad leaf weeds such as *Amaranthus* spp., *Abutilon theophrasti, Polygonium* spp., *Ambrosia artemisiifolia, Xanthium strumarium, Solanum nigrum etc.* Its residues persist long to injure susceptible crops in the following year.

(viii) Prosulfuron: 1 – (4 – methoxy – 6 – methyl – triazin – 2 – yl) – 3 [2 – (3, 3, 3 – trifluoropropyl) – phenylsulfonyl] – urea: Prosulfuron can be applied POST in maize, sorghum, wheat, barley, sugarcane and other graminaceous crops. It effectively kills weeds such as *Xanthium strumarium, Kochia scoparia, Chenopodium album, Amaranthus* spp., *Ambrosia artemisiifolia, Abutilon theophasti.* Thus, the variaion in chemical structure results variation in function against plant species.

(ix) Sulfometuron: methyl 2 – [[[[(4, 6 – dimethyl – 2 – pyrimidinyl) amino] carbonyl] amino] sulfonyl] benzoate: Sulfometuron can be applied PRE and POST in forest areas, road sides, non-crop areas *etc* to control annual and perennial broad leaf and grass weeds including *Sonchus arvensis, Malva* spp., *Rumex crispus, Ambrosia tenuifolia, Echinochloa crusgalli, Setaria* spp., *Aegilops cylindrica etc.*

(x) Thifensulfuron: methyl 3 – [[[[(4 – methoxy – 6 – methyl – 1, 3, 5 – triazine – 2 – yl) amino] carbonyl] amino] sulfonyl] – 2 – thiophene carboxylate: This herbicide is commercially marketed as HARMONY. It can be applied POST in wheat and soybean. It is very effective to control many annual broad leaf weeds such as *Brassica kaber, Amaranthus* spp., *Kochia scoparia.* It has shorter residual effect. Hence, safer for succeeding crop plants and thus facilitate intensive cropping system.

(xi) Triasulfuron: 2 – (2 – chloroethoxy) – N – [[(4 – methoxy – 6 – methyl – 1, 3, 5 – triazin – 2 – yl) amino] carbonyl] benzene – sulfonamide: Triasulfuron is commercially available as AMBER and is selectively tolerable to wheat. It can be applied POST or PRE for control of many annual broad leaf weeds such as *Kochia scoparia, Amaranthus* spp., *Ambrosia artemisiifolia etc.* It is persistant for long enough to injure certain dicot crops even after 1 – 3 years after application.

(xii) Tribenuron: methyl 2 – [[[[(4 – methoxy – 6 – methyl – 1, 3, 5 – triazin – 2 – yl) methylamino] carbonyl] amino] sulfonyl] benzoate: It is marketed as CHEYENNE and EXPESS. It can be applied POST in wheat and barley. It effectively kills many annual broad leaf weeds including *Brassica* spp., *Chenopodium album, Lamium amplexicaule, Lactuca serriola* and *Kochia scoparia.* It has short residual effect. Hence, rotational crops can be planted safely for their maximum productivity.

35. Tetrahydropyrimidinones

These are also called as cyclic ureas (1, 3 – diaryl – 2 – pyrimidinones). These are photo-bleaching herbicides. Fluridone and fluormeturon are the parent structures of tetrahydropyrimidinones (THPS). They are specifically used in rice. They are active against a broad spectrum of monocot and dicot weeds and sedges including *Echinochloa crusgalli, Cyperus difformis, Scirpus juncodes, Monochoria vaginalis, Cyperus serotinus, Eleocharis acicularis, Saggitaria pygmaea etc.* without affecting rice. Thus, the chemical structural configuration imparts differential tolerance among crop plants while simultaneous susceptibility among the weed flora, providing less weed competition. This situation permits crop plants to explore their maximum production potential.

36. Thiocarbamates

One oxygen in the carbonic acid molecule when substituted by sulphur, thiocarbamic acid is formed. Similarly, two sulphur substitutions give rise to dithiocarbamic acids. These thiocarbamate herbicides are toxic to germinating weeds following their soil incorporation before sowing. The susceptible grass and broad leaf weeds fail to emerge. Maize is the sensitive crop to thiocarbamate herbicides. It manifests malformed and twisted seedlings in maize; leaves tightly rolled in the whorl and may not unroll normally. Cupped or crinkled leaves with a thick, leathery texture develop in case of injured dicots. These are volatile, hence need soil in corporation.

(i) Butylate: S – ethyl bis (2 – methyl propyl) carbamothioate : Butylate is applied PPI in maize for control of many annual grasses such as *Setaria* spp., *Echinochloa crusgalli, Panicum dichotomiflorum, Sorghum halepense, Sorghum bicolor, Cyperus esculentus* and *Cyperus rotundus* and a few broad leaf weeds. It does not injure succeeding crops.

(ii) Diallate: S – (2, 3 – dichloro – 2 – propenyl) bis (1 – methyl ethyl) carbamothioate: Diallate is traded as AVADEX. It is applied pre or post-plant incorporated. It is tolerated by wheat, maize, soybean, sugarbeets, oil seed crops, ornamentals, root crops and fruit trees. It is very effective against wild oats.

(iii) EPTC: S – ethyl dipropyl carbamothioate: EPTC is the first thiocarbamate herbicide developed for effective control of many annual grasses and broad leaf weeds. As it is volatile, it needs to be soil incorporated. Many crops tolerate EPTC, such as potato, safflower, vegetables, maize, cotton and orchard crops. It is effective against perennial grasses such as *Elytregia repens, Sorghum halepense etc.* sedges such as *Cyperus* spp. and many annual grasses. It is very effective against broad leaf weeds such as *Ipomoea purpurea, Solanum villosum, Solanum nigrum, Chenopodium album, Portulaca oleracea, Amaranthus* spp. *etc.* It has very short persistence.

(iv) Molinate: S – ethyl hexahydro – 1H – azepine – 1 – carbothioate : Molinate is applied in irrigated rice after transplantion before weeds emerge. It can be applied PPI in dryland rice or injected into irrigation water in direct seeded irrigated rice. Molinate controls several annual grasses including *Echinochloa crus-galli, Ceptochloa* spp., *Brachiaria platyphylla.* Its residues do not persist long enough to injure susceptible crops planted one year after application.

(v) Pebulate: -S – propyl butyl ethyl carbamothioate : Pebulate is a selective PPI and PRE herbicide like EPTC. Pebulate is specific in controlling *Cyperus* spp., *Digitaria sanguinalis, Setaria* spp., *Echinochloa crusgalli, Avena fatua* and many broad leaf weeds. It is safer to tobacco, sugar beets, and tomato.

(vi) Thiobencarb: - S – [(4 – chlorophenyl) methyl] diethyl carbamothioate : Thiobencarb also known as benthiocarb, is applied PPI or PRE in rice nursaries and PRE in irrigated rice. It is very selective for rice. It controls many annual grasses including *Echinochloa crusgalli, Echinochloa colonum, Leptochloa* spp., *Brachiaria platyphylla, Aeschynomene virginiea, Commelina* spp. and sedges of the Cyperaceae.

(vii) Triallate: S – (2, 3, 3 – trichloro – 2 – propenyl) bis (1 – methyl ethyl) carbamothioate : Triallate is marketed as AVADEX BW. Like diallate, it has two chlorines on the terminal allyl carbon. It is applied PPI in wheat, maize, soybean, barley, forage legumes, fruit crops and ornamentals. It is active against wild oats and on a wide spectrum of weeds.

37. Triazines

These are symmetrical triazines (S – triazines). They are the most prominent groups of herbicides. They are exclusively used for selective weed control. Atrazine, simazine, Ametryn and prometryn are the triazine compounds that gained major importance in agriculture or non-agricultural systems. Various substitutions on the triazine nucleus yield compounds of widely different chemical and biological properties. If the 2 – position of the triazine ring in substituted by chlorine, it is termed as chlorotriazine. If it is substituted by the methoxy group, it is called methoxy triazine, but if it is replaced by methylthio (methyl mercapto) group, it is known as methyl thiotriazine.

The typical triazine – injury symptoms begin with interveinal chlorosis of leaves and yellowing of their margins. These are followed by further chlorosis and necrosis. Older leaves are more damaged than the new growth. Browning of leaf tips can occur, while root growth is not affected. Triazines do not affect seed germination.

(i) Ametryn: N – ethyl – N^1 – (1 – methyl ethyl) – 6 (methyl thio) – 1, 3, 5 – triazone – 2, 4 – diamine: Ametryn is the methyl thio analogue of atrazine. It is selective PRE and POST herbicide depending on the crop for control braod leaf and grass weeds in sugarcane, banana, pineapple and citrus. It has contact activity also – hence its POST application must be directed on weeds. It has residual activity in tropics for several months, thus giving effective weed control for better crop yields.

(ii) Atrazine: 6 – chloro – N – ethyl – N^1 – (1 – methyl ethyl) – 1, 3, 5 – triazine – 2, 4 – diamine: Atrazine is one of the most widely used herbicides in world agriculture. It is a selective PRE herbicide for the control of many broad leaf weeds and grasses in maize, sorghum, sugarcane, pineapple, turf and orchards. Atrazine controls broad leaf weeds including *Amaranthus* spp., *Ipomoea* spp., *Datura stramonium, Convolvulus, Ambrosia* spp., *Polygonum* spp., *Abutilon theophrasti, Xanthium strumarium, Chenopodium album etc.*

and annual grasses like *Echinochloa crusgalli, Setaria* spp., *Panicum dichotomiflorum.* It is effective particularly in maize and sorghum. Many vegetable crops are sensitive to atrazine.

(iii) Cyanazine: 2 – ethyl – N[1] (1 – methyl ethyl) – 6 – (methyl thio) – 1, 3, 5 – triazine – 2, 4 – diamine : It is applied E – PPL, PPI, PRE, E – POST in maize, cotton, beans *etc.* It controls many broad leaf weeds *viz, Ipomoea* spp., *Portulaca oleracea, Ambrosia artemisiifolia, Brassica kaber, Kochia scoparia* spp., *Echinochloa crusgalli, Panicum capillare etc.* It has short residual activity.

(iv) Hexazinone: 3 – cyclohexyl – 6 – (dimethylamino) – 1 – methyl – 1, 3, 5 – triazine 2, 4 (1H, 3H) – dione : Hexazinone is a safe herbicide for the crops like sugarcane, pineapple and alfalfa. It controls many annual and perennial broad leaf and grass weeds as well as many brush species. It is effective against many woody plants and aquatic weeds. It has a moderately long residual activity.

(v) Prometryn: N – N[1] – bis (1 – methyl ethyl) – 6 (methyl thio) – 1, 3, 5 triazine – 2, 4 – diamine: Prometryn is the methyl thio analogue of propazine. It is applied PPI, PRE and directed POST in cotton. It is also safely used in soybean and non-crop areas. It has considerable contact activity. Prometryn sensitive weeds include *Physalis* spp., *Chenopodium album, Ipomoea* spp., *Amaranthus* spp., *Sida spinosa, Setaria* spp., *Eleusine indica etc.*

(vi) Simazine: 6 – chloro – N, N[1] – diethyl – 1, 3, 5 – triazine – 2, 4 – diamine : Simazine is one of the first triazine herbicides widely used as a selective PRE herbicide to control many annual broad leaf weeds and some grasses in maize, sugarcane, plantation crops like tea, coffee and rubber and many orchard crops. It is also used as non-selective herbicide for vegetation control on roadsides, banks of irrigation canals and drainage channels *etc.* It has got long residual persistence depending on soil type.

38. Triazinones

Metribuzin: 4 – amino – 6 – (1, 1 – dimethyl ethyl) – 3 – (methyl thio) – 1, 2, 4 – triazin – 5 (4 H) – one: The triazinone herbicide family is used as E-PPS, PPI, PRE or POST in sugarcane, maize, soybean, potato, peas, beans. Metribuzin is very active against annual broad leaf weeds such as *Chenopodium album, Amaranthus* spp., *Abutilon theophrasti, Datura stramonium, Brassica* spp. and *Ambrosia artemisiifolia* along with certain annual grasses. The residues of metribuzin are not injurious to susceptible crops planted one year after application. But root crops such as onion, potato and sugarbeets may show susceptibility.

39. Triazoles

Amitrole, Amitrole – T: 1 H – 1, 2, 4 – triazol – 3 – amine: Primary toxic symptom of amitrole is bleaching in leaves and shoots and it is most evident in meristems and developing leaves. Bleached leaves eventually wilt and become necrotic. Amitrole is a POST translocated herbicide. It controls many annual and perennial grass and broad leaf weeds. Amitrole – T is a mixture of amitrole and ammonium thiocyanate. It is more active than amitrole.

40. *Triazolopyrimidine Sulfon anilides*

Flumetsulam : N – (2, 6 – difluorophenyl) – 5 – methyl [1, 2, 4] triazole [1, 5 – a] pyrimidine – 2 – sulfonamide : Flumetsulam is available as BROAD STRIKE. The weed spp. most sensitive to this are killed before emergence but some weeds may die after emergence under dry conditions. The emerged sensitive plants exhibit stunting, interveinal chlorosis, veinal discoloration (purpling) and necrosis. It is safely used for maize and soybean. The sensitive weed flora include many annual broad leaf weeds such as *Brassica kaber, Amaranthus* spp., *Kochia scoparia, Chenopodium album, Polygonam* spp., *Solanum nigurm, Abutilon theophrasti etc*. but it has little activity against grasses. Residues of flumestsulam do not injure soybean, maize alfalfa, groundnut, potato, rice, sorghum and tobacco planted one year after application. It is obvious that structure of the chemical attributes for differential response among plant species.

41. Uracils

Substituted uracils like bromacil and terbacil are the two uracil herbicides used in agriculture and in non-agricultural areas. The susceptible plants show characteristic foliar chlorisis, necrosis and inhibition of root and shoot growth.

(i)Bromocil: 5 – bromo – 6 – methyl – 3- (1- methylproply) – 2,4 – (1H, 3H) pyrimidinedoine: Bromacil (Lithium salt) can be applied PRE in citurs, pine apple and on crop areas. It is good for brush control. It controls many annual and perennial grasses, sedges and broad leaf weeds including *Echinochloa crus-galli, Digitaria* spp., *Sorghum halepense, Tribulus terestris, Cenchrus* and *Chinopodium album*. It is used for selective weed control and hence offering effective weed control for higher crop productivity.

(ii)Terbacil: 5 – chloro – 3 – (1, 1 – dimethyl ethyl) – 6 – methyl – 2, 4 – (1H, 3H) – pyrimidinedione: Terbacil is a PRE and POST herbicide for control of many annual and some perennial weeds in tolerant crops like sugarcane, alfalfa, orchard crops such as apples, peaches, and citrus. The susceptible weed flora to terbacil includes *Stellariau media, Lamium amploxicaule, Chenopodium album, Descurainia* spp., *Lactuca serriola, Lolium* spp. and *Echinochloa crusgalli* and partial control of cyperus spp. It has medium to long persistence in soil.

42. Ureas

Use of substituted ureas led to the discovery of monuron and subsequently other phenyl substituted ureas including *Diuron, Linuron, Fenuron, Chloroxuron, Fluometuron, Chlorbromuron, Fluometuron, Chlorbromuron, Siduron, Tebuthiuron etc.* are developed. Besides these, hetero cyclic substituted ureas like noruron, methabenz and thiazuron were also developed.

The susceptible species following soil application emerge but become chlorotic within a few days followed by complete necrosis. Foliar application causes interveinal chlorosis of the leaves and yellowing at the leaf margins. The older leaves are more damaged than the new growth.

Among numerous substituted ureas developed, the more prominent ones are:

(i) Diuron: N[1] – (3, 4 – dichlorophenyl) – N, N – dimethylurea: It is applied PRE for controlling emerging grass and broad leaf weed seedlings in tolerant crops like tea, cotton, coffee, grapes, pineapple, apples, pears citrus and other plantation and orchard crops. It can also be applied as POST in maize, sorghum *etc.* The phytotoxic residues dissipate within a season. Hence, safe for next season crop plants.

(ii) Fluometuron: N, N – dimethyl (- N[1] – [3 – trifluoromethyl) phenyl] urea: A CH_3 substitution on the meta position of the phenyl ring of urea structure gives fluometuron. As a result its selectively increased. It is applied PPI, PRE and E-POST in its tolerant crops like cotton, sugarcane, sorghum, citrus and ornamental crops. It controls a variety of weed flora of broad leaf as well as grasses. *viz, Echinochloa crusgalli, Digitaria* spp., *Panicum dichotomiflorum, Setaria* spp., *etc*. The residues often dissipate to non-detectable levels within four months. Thus provides a moderately long weed control.

(iii) Linuron: N[1] (3, 4 – dichlorophenyl) – N – methoxy – N – methylurea: Linuron differs from diuron by having a methoxy group in place of one of the methyl substituents. It is applied as PRE or E-POST in soybean, cotton, maize, sorghum, wheat, potato and vegetable crops. It controls effectively annual broad leaf weeds including *Brassica* spp., *Amaranthus* spp., *Portulaca oleracea, Polygonum* convolvulus *etc* and annual grasses such as *Echinochloa crusgalli* and *Setaria* spp. Foliar application is effective on a wider spectrum of weeds. Its residues do not injure cover crops planted in the following season.

(iv) Tebuthiuron: N – [5 – (1, 1 – dimethyl ethyl) – 1, 3 – thiadiazol – 2 – yl] – N, N – dimethyl urea: Tebuthiuron can be applied as PRE or POST in pastures, rangelands and non-crop areas. It controls certain broad leaf weeds and woody brush species and several broad leaf grass and brush spp. Its residuals effect is more on low rainfall areas and in muck soils with high organic matter content.

43. Unclassified Herbicides

(i) Bensulide: O, O – bis (1 – methyl ethyl) S – [2 [(Phenylsulfonyl) amino] ethyl] phosphorodithioate: Bensulide is applied as PPI and PRE to control its susceptible annual grass and broad leaf weeds such as *Digitaria sanguinalis, Echinochloa crusgalli, Eleusine indica, Amaranthus* spp., *Chenopodium album, Caspella* spp., *Lamium* spp. *etc.* The tolerant crops are vegetable crops like cucumber, squash, pumpkin, melon, onion, cabbage, carrot, cauliflower and chillies. Its long term activity requires irrigation immediately after its application. It inhibits root and shoot growth.

(ii) Ethofumesate: (±) ***– 2 – ethoxy – 2, 3 – dihydro – 3, 3 – dimethyl (- 5 – benzafuranyl methanesulfonate:*** The herbicidal properties of these herbicides are available as PROGRASS. It can be applied before or after weed emergence in tolerant crops like sugarbeet. The susceptible flora include *Solanum nigrum, Stellaria media, Chenopodium album, Kochia scoparia, Amaranthus retroflexus, Cirsium arvense, Polygonum* spp., *Echinochloa crusgalli, Digitaria sanguinalis, Setaria* spp., *Bromus tectorum etc.* Thus special herbicides could be synthesized in view of controlling certain noxious weeds.

(iii) Fosamine: ethyl hydrogen (amino carbonyl) phosphonate: The fosamine is available as ammonium salt under the name KRENITE. Leaf and bud development is severely inhibited, the following spring *i.e.* after leaf senescence. The leaves that appear are abnormally small and spindly. Pines and bind weed show response soon after application. In moderately susceptible to resistant species, suppression of terminal growth may occur.

(iv) Glufosinate: 2 – amino – 4 – (hydroxy methyl phosphinyl) butanoic acid: It is a phosphorylated amino acid. The injured plants show chlorosis and wilting within 3-5 days. after application followed by necrosis. The toxic symptoms of this herbicide get enhanced by bright sunlight, high humidity and moist soil. The seedlings are not injured before emergence. It is applied POST in plantation crops like tea, coffee *etc*. The susceptible annual or perennial weed flora includes *Panicum maximum, Paspalum conjugatum, Axonopus compressus, Sataria* spp., *Imperata cylindrica, Digitaria sanguinalis etc.*

(v) Glyphosate: N – (phosphonomethyl) glycine: The herbicidal activity of glyphosate is introduced as isopropyl ammonium salt. It is commercially available as ROUNDUP and under several other trade names. The weeds susceptible to this show growth inhibition followed by general foliar chlorosis and necrosis. Chlorosis appears first more pronouncedly on immature leaves and growing points. Foliage may turn reddish purple in certain species. Regrowth of treated perennial and woody species shows often, deformed with whitish markings or striations. It is a broad spectrum selective POST herbicide used for effective control of rhizomatous and deep-rooted weeds including *Elytrigia repens, Imperata cylindrica, Cynodon dactylon, Paspalum conjugatum, Setaria palmifolia etc.* It is extensively used in plantation crops such as tea, coffee, rubber, oil palm *etc*. orchard crops, vineyards, pineapple, sugarcane and non-crop areas. It has moderate persistence in soil.

SPECIES-SPECIFIC INJURY OF HERBICIDES IN CROP PLANTS

In modern agriculture enhancing productivity of crop plants is the main and the only motto. In order to maximize crop productivity potential, the counter productive weed competition must be removed besides other pest control. In this context, herbicides occupy fore front especially those injure weeds that greatly infest agriculture crops. At the same time, it is worth to know crop species that are specially injured by herbicides. Usually a crop plant may be tolerant to a particular herbicide while it may be injured by some other herbicide. This response speaks of species - specificity to herbicides. This knowledge is particularly essential to improve the crop productivity at large. Owing to selective wetting, certain contact herbicide like iron sulphate, copper sulphate, sulphuric acid, potassium cyanide, dinitrophenyl compound or 2,4-D when applied resulted in killing of many broad leaved species and leave several grasses uninjured. If 2,4-D is removed from the above list, onions, garlic, peas *etc*., could be safely cultivated[55]. Dalapon, amitrole, endothal, dicryl *etc*., showed injury to certain crops and weeds. Although annual weeds were controlled by selective sprays of contact toxicants, entry of 2, 4-D revolutionised weed control practice. Most common weeds of grain crops are susceptible to 2, 4-D. The α- phenoxy butyric acids, 2,4-DB, MCPB and 2,4,5-TB were toxic to test species except Brassica genus. Both crop plants and weeds vary in their susceptibility to 2, 4-D. Members of grass family are resistant while most broad-leaved weeds are susceptible. Seedlings of annual and perennial grasses are commonly killed by deformation of leaves and inflorescences of cereal crops. Injury results from application to very young cereal seedlings or from root absorption when rains followed soon after treatment. Oats are most susceptible followed by barley and wheat. Deformation of roots, stems and inflorescence of corn, sorghum and rice are also observed. Rossman and Staniforth[192] showed that some corn hybrids are more susceptible than others and 6-8 leaved stage young plants are most seriously affected. Slife[213] though reported strong epinasty and some formative defects in soybean, could not record reduction in yields. Broad-leaved crop plants are extremely sensitive to effects of 2, 4-D, *viz.,* cotton, tomato, black-eyed peas, sweet potatoes and melons and grapes. Shaw *et al*[210] reported susceptible growth of wheat, barley and oats to 2, 4-D injury. Yield was reduced when applied at susceptible stages – such as pre-tillering and rapid floret development. Germination was lower and a number of morphogenic effects such as leaf rolling, late emergence of heads, reduction in tillering and delayed maturity were produced. Such effects obviously hinder the yield formation and its potential. Rojas-Gardeiduenas and Kommendahl[190] reported treatment of radicals of soybean with 2, 4-D caused inhibition of cell elongation by 75% and mitosis 25%. Disappearance of chloroplast in the cortex of the stem, lateral root initiation from the endodermis and proliferation of phloem parenchyma were reported on soybean. Muni[162] found phloem crushing and disruption of vascular tissues and roots of four species of varying susceptibility. Baskin[15] and Baskin and Walker[16] pointed out the hazards to crop were due to the use of volatile esters of 2, 4-D and its analogues. Temperature and exposure time variously effected tomato treated with 2, 4-D and 2, 4, 5-T esters. Rao[186] also reviewed that 2, 4-D, MCPA, 2, 4, 5-T *etc.,* exhibit symptoms typical of their auxin type herbicides including epinastic bending and twisting of stems and petioles, stem swelling and elongation and leaf curling and cupping. These injuries culminate into chlorosis at the growing points, growth inhibition, wilting and necrosis and death.

Sheets[211] found out plants were extremely sensitive to simazine which reduced dryweight by 27% by the presence of 1.3 mg of simazine in their tops. Crop plants were shown to vary widely in their susceptibility to simazine, *viz.,* corn, sorghum, flax, soybeans, rye, wheat and barley. Dalapon is a growth regulator and phytotoxicant that produces formative effects on grass plants by producing fusion, buckling, tubular leaf formation, stunting, excessive tillering, slow lingering death. Roots as well as foliage inhibited. These symptoms were carried through three generations of wheat. 2, 4-D and similar compounds when applied on to leaves of actively growing young plants, exhibited most drastic toxic action. Arsenicals caused root plasmolysis and leaf wilting followed by discoloration and necrosis of leaf tips and margins and shoot meristematic tissues were the major site of action of cacodylic acid and MSMAC[195]. Isoxobens are reported to move symplastically and interfere with cell wall biosynthesis and inhibits the synthesis of cellulose[52]. Benzoics inhibited elongation and growth of roots and shoots of susceptible species. Benzoics exhibit growth proliferation resulting in stem swelling[124]. Carbamates appear to inhibit cell divison and expansion of plant meristems by interfering microtubule assembly or function[244]. Cineoles inhibited mitosis in meristematic regions of shoot and roots while dinitroanilines inhibited polymerization of tubulin, the major constituent of microtubules[156,228] into microtubules and no spindle formation and kinetochore microtubules which are responsible for chromosome movement during mitosis[243]. Phenoxyalkanoic acids (2,4-D) involved in cell wall plasticity and nucleic acid metabolism and produced characteristic epinastic sympotoms[193].

The aryloxy phenoxy propionics (APPs) injure by inhibiting the growth of apical, internodal and root meristem followed by chlorosis, meristematic necrosis, membrane disruption in chloroplasts and mitochondria. The diphenyl either herbicides become phytoxic strictly depending on light. Their herbicide injury appears as water soaked lesions on treated tissues causing cell membrane lysis, destruction of lipids and chlorophylls. Derrick and Cobb *et al*[73] suggested membranes may contain primary target of diphenylethers, particularly tetrapyrole in the formation of protoporphyrinogen. Soybean treated with aciflourofin accumulated isoflavone and pterocarpans, chalcone synthase and phenylalanine, ammonia lyase and phytoalexins and stress metabolites in various crops. Lehnen *et al*[138] found that DPE herbicides inhibit the enzyme protoporphyrinogen oxidase (protox) which in healthy tissues prevents the accumulation of protodestructive protophorphyrin IX (PP IX). Matsumoto *et al*[159] reported that the protox from rice and barnyard grass was more susceptible than that from radish, cucumber and buckwheat.

Thiobencarb of thiocarbamate group caused adnation of leaves, leaf deformities, separation of laminal joints shorter and narrower leaves, shorter panicles with malformed spikelets in *Echinochloa crugalli*[186]. If, by any means, the rate of thiobencarb causes toxicity on sensitive crop lands, the yield forming components are affected leading to decreased productivity.

Thus, herbicides affect plant parts, organs, cell wall integrity, cell organelles and enzymes of formative process in susceptible crop plants and hinder their productivity. However, the discovery of selective herbicides for specific species of crop plants augmented their productivity is reviewed on page 260.

MECHANISM OF HERBICIDE ACTION IN PLANTS

(a) Morphological, Physiological and Biochemical Basis of Herbicide Action

Herbicides particularly in susceptible plants or in tolerant plants when applied at higher rates cause a number of anatomical changes, disturb physiological and biochemical processes as well as aberrations in genetic structure. These changes may differ depending on the type of herbicide and plant species. The rate of appearance of these effects varies with the characteristic actions of the herbicide and depends in the degree of tolerance or susceptibility of the crop plant species.

These changes can be understood in the following categories:

(a) Morphological (anatomical) (b) Physiological processes and (c) Biochemical reactions

Morphological changes

Both crop plants and weeds vary widely in their susceptibility to 2,4-D sprays. Members of the grass family are resistant, while most broad leaved plants are susceptible to 2,4-D. Deformation of leaves and inflorescenses of cereal crops follow the selective spraying, though not a common injury. The injury is conspicuous, if applied to very young cereal seedlings. Oats, barley and wheat are most susceptible. Deformation of roots, stems and inflorescence occur in corn, sorghum and rice[55] Rossman and Staniforth[192] reported that some corn hybrids were more susceptible at 6-8 leaf stage. Soybeans showed strong epinasty and some formative effects to 2,4-D[213]. Shaw *et al* [210] showed pre-tillering and rapid floret development stages were sensitive to 2,4-D in wheat, barley and oats. When applied at these sensitive stages, many morphogenetic effects were produced such as rolling, lateral emergence of heads, reduction in tillering and delayed maturity. Soybean when sprayed with 2,4-D resulted in loss of chloroplasts from the stem, initiated lateral roots from endodermis and proliferation of phloem parenchyma. Muni[162] found phloem – crushing and disruption of vascular tissues in roots in susceptible species.

Alachlor and propachlor inhibited cell division and cell enlargement[69,78,79], while metolachlor inhibited seed germination and early seedling growth besides cell division and cell enlargement. Propanil inhibited the growth of tomato radicles and auxin – induced growth of Avena coleoptiles[116]. Eepicuticular wax on the leaves was decreased or altered its form[188]. Arsenical (DSMA, MSMA) caused cessation of growth and gradual browning. Benzamides in susceptible plants caused reduced root growth, root hair distortion and root clubbing. Carbamates[186] inhibited cell division in barley roots at 1 ppm. Barban (Carbyne) affected growings points by swelling and distorted meristems and axillary buds (tillers) failed to develop. Chloracetamide herbicides, (CDAA), randox, reduced the rate of cell division in barely roots but had little effect on symptoms similar to auxinic herbicides on susceptible plants such as twisting and curling of stems and petioles (epinasty), stem swelling (at nodes) and elongation and leaf cupping. Flumetsulam caused stunting, interveinal chlorsis, veinal discoloration (purpling) and necrosis in sensitive plants.

(ii) Physiological and bio-chemical processes

Understanding the mechanism of each herbicide is essential because of two reasons:

(a) For effective weed elimination from admist of crop plants without injuring crop plants is a pre-requisite to enhance crop productivity;

(b) For herbicides affecting formative structures of crop yield forming components, fixing of a safe and economic dose is essential for enhancing crop productivity;

(c) For synthesizing new generation herbicides of plant process target *vis-a-vis* herbicidal efficacy.

Different groups of herbicides have different characteristic mechanism of herbicide action on plants. These mechanisms may be modified in the tolerant plant spices by detoxifying or degrading the herbicide molecule through various bio-chemical reactions.

Herbicides cause various physiological and biochemical effects and anatomical changes resulting in killing of susceptible weed plants. The physiological and biochemical effects are followed by various types of visual injury symptoms on susceptible plants. They are chlorosis, defoliation, stunting necrosis, stand reduction, epinasty, morphological aberration, growth stimulation, cupping of leaves, marginal failure of seedling emergence *etc*. The injurious symptoms may appear on any part of the plant *viz.*, roots, flowers, fruits, foliage *etc*. After reaching the site of action, the herbicide affects the living processes of the plant. Herbicides have more than one site of action. The secondary and tertiary sites of action become involved herbicide concentration builtup in the tissue.

The various physiological and biochemical processes that form the mechanism of herbicides that affect susceptible plant species are:[195, 186]

1. Photosynthesis, 2. Mitochondrial activities, 3. Protein and nucleic acid biosynthesis 4. Pigment biosynthesis 5. Fatty acid Biosynthesis 6. Branched – chain amino acid biosynthesis 7. Aromatic compound biosynthesis, 8. Ethylene biosynthesis, 9. Glutamine biosynthesis, 10. Hydrolytic enzyme activities. Most of the herbicides affect either one or more of these processes.

1. Herbicides affecting photosynthesis mechanism

Photosynthesis, a vital physiological process, energizes plants by offering (a) The phosphoryl group-transfer potential of ATP and (b) the reducing power of NADH and NADPH. Further ATP and NADH or NADPH are produced by the oxidation of carbohydrates via the glycolysis and citric acid cycle and oxidative phosphorylation which yield more ATP. The ultimate carbohydrate, the source of chemical energy is produced by photosynthetic process. Hence, some herbicides affect the process of photosynthesis for blocking carbohydrate synthesis. Thus weakening the susceptible plants succumb to stages of photosynthesis process as their target cites to inhibit photosynthesis carbohydrate synthesis. These sites are :

(a) Electron transport inhibitors (b) Uncouplers (c) Energy transfer (d) Inhibitory un-couplers and (e) Electron acceptors.

The only two of these processes that are demonstrated as the primary site of action on photosynthesis are :

(a) Inhibition of PS II electron transport and

(b) Diversion of electron transfer through PS I

The PS II electron transport inhibitors belong to a variety of chemical groups. These are phenols, nitriles, pyridazinones *etc*. Besides inhibiting electron transport, nitorphenols and phenols possess uncoupling activity, pyridazinones inhibit lipid and carotenoid biosynthesis. The PS II herbicides block the flow of electrons through PS II and thus indirectly block the transfer of excitation energy from chlorophyll molecules to the PS II reaction centre. The triplet chlorophyll (^{3}Chl) reacts with molecular oxygen to form singlet oxygen ($^1O_2$)

$$\text{Chl} + \text{PS II (hv)} \xrightarrow{\text{isc}} 1\text{Chl} - {}^{3}\text{Chl}$$

$$^{3}\text{Chl} + O_2 \longrightarrow {}^{1}O_2 + \text{Chl}$$

Reaction involved in the generation of ^{3}Chl (triplet chlorophyll) and $^{1}O_2$ (siglet oxygen) by photosystem II transport inhibitory by intersystem crossing (isc)[186]

Both triplet chlorophyll and singlet oxygen can abstract hydrogen from unsaturated lipids, producing a lipid radical and initiating a chain of lipid peroxidation. Lipids and proteins are then oxidized resulting in the loss of chlorophyll and carotenoids and in leaky membranes, which allow cells and cell organelles to dry and disintegrate rapidly[186].

Arsenics may bring about uncoupling of oxidative phosphorylation by poisoning the triose phosphate dedydrogenase system. Triose phosphate is oxidised without production of high energy phosphate bonds. So catabolic respiration processes may go on until the substrate supply is exausted.

2. Herbicides Affecting Mitrochondrial Activities

Mitochondria carryout the processes of cellular respiration responsible for the controlled oxidation of the respiratory substrates (organic acids) and the conservation of the energy thus released in forms usable for the energy requiring functions of the cell. The processes involve : (a) synthesis of ATP (adenosine tri phosphate) and (b) transport of electrons and protons from the substrate to oxygen. During the electron transport of oxygen, ADP (adenosine diphosphate) is phosphorylated to ATP in a process known as oxidative phosphorylation. The Krebs citric acid cycle and the fatty acid oxidation cycles are the preparatory processes for the aerobic regeneration of ATP from ADP and phosphate (Pi) by energy coupling in the respiratory chain.

Herbicides affect the mitochondrial activities by uncoupling the reactions responsible for ATP synthesis or interfere with electron transport and energy transfer. The uncouplers act on membranes of the mitochondria in which phosphorylation takes place. Electrons leak through the membranes so that the charges that they normally separate are lost. As a result, energy is not accumulated for ATP synthesis. Besides, uncouplers also stimulate respiration of mitochondria, promote hydrolysis of ATP or inhibit exchange reactions catalyzed by mitochondria in the absence of inorganic phosphate, ADP, ATP and H_2O.

3. Herbicides Affecting Protein and Nucleic Acid Biosynthesis

Protein synthesis is a pre-requisite process, along with RNA synthesis, essential for cell elongation. The synthesized proteins are highly organized molecules composed of a specific sequence of amino acids joined by peptide bonds. The protein synthesis is an elaborate process requiring ribosomes, transfer RNA (tRNA) or soluble RNA (sRNA).

4. Herbicides Affecting Pigment Biosynthesis

Many herbicides cause either white or yellow chlorisis of leaves as a consequence of the total or partial absence of the normal chloroplast pigments, *viz.* chlorophylls and carotenoids. Chlorosis results from the inhibition of pigment biosynthesis or from destruction of the existing pigment. Pigment biosynthesis inhibitors cause chlorosis only in newly developed leaves and they are most effective when herbicides are applied pre-emergence on the other hand, existing chlorplast. Pigments are destroyed by photosynthesis and photosynthetic electron transport inhibitor herbicides *e.g.* Atrazine, phenols, nitriles, pyridazinones *etc*.

a) Chlorophyll Biosynthesis

The precursor for chlorophyll (chl) is α-aminolevulinic acid (ALA). In plants, ALA is formed by utilizing the intact carbon skeleton of glustamic acid through the catalytical role of ALA synthetase[36]. Condensation of groups of two ALA molecules is catalyzed by ALA dehydratase resulting in the formation of porphobilinogen (PBG), the first pyrolic intermediate in the chlorophyll biosynthesis pathway.

b) Carotenoid Biosynthesis

Carotenoid biosynthesis is an integral of the construction of the pigment the formation of other components such as chlorophyll, proteins and lipids. Carotenoid biosynthesis begins with head-to-tail condensation of two molecules of geranylgeranyl pyrophosphate to pre-phytoene pyrophosphate, which after dephosphorylation, is converted to 15-Cis phytoene[197]. The subsequent pathway involves four desaturation steps (removal of 2H each time) and two cyclization steps to form β-carotene.

The target sites for herbicidal activity during carotenoid biosynthesis are:

(a) Phytoene synthesis (b) phytoene desaturases (c) carotenoid cyclization (d) caroteine hydroxylation.

The herbicides that inhibit pigment biosynthesis at target sites are :

(a) Carotenoid biosynthesis- by pyridazinones (norflurazon);
(b) Carotenoid biosynthesis – phytoene desaturases- by Tetrahydro pyrimidinones;
(c) Carotenoid biosynthesis, desaturation, cyclization-by pyridinones (fluridene). Hydroxylation - by nitriles (bronoxynil, ioxynil);
(d) Protoporphyrin IX (PP IX)-by Diphenyl Ethers (bifenox, protporphyrinogen oxidase (protex), acifluorfen, fomesafen, lactofen oxyfuorfen). Phenyl triazinones (Aryl triazinones, Sulfentrazone) N-phenylphthalamides (flumiclorac), Imines (CGA 248757) Oxadiazole (oxadiazon);
(e) Bleaching;
(i) Caratenoid Biosynthesis inhibition (unknown target) by Triazole (amitrole, amitrole-T);
(ii) Diterpenes inhibition - by Isoxazolidionones (Clomazone)

5. Herbicides Affecting Fatty acid (lipid) Biosynthesis

Plastids contain acetyl.CoA synthetase, acetyl CoA carboxylase and fatty acid synthetase, which are the key components of fatty acids in plants[134]. Young developing leaves and meristematic tissues which depend on the efficient supply of fatty acids are the most affected by the inhibitors of these enzymes which block glycerolipid and phospholipid synthesis which results in the inhibition of membrane formation[134].

Acetyl – CoA carboxylase (ACCase) is the major target of inhibitors affecting fatty acid biosynthesis. The enzyme produces malonyl CoA which is a key intermediate in both fatty acid and favinoid (gibberellins, abscisic acid, carotenoids and other isoprenoids) biosynthesis[108]. Inhibition of ACCase would deprive the plant of a key intermediate (malonyl CoA) essential to both lipid and flavonoid biosynthesis and would lead to phyotoxic effect[134].

Four classes of herbicides have a specific effect on fatty acid synthesis: Pyridazinones, Cyclohexanediones, aryloxyphenoxy propionics and thiocarbamates. Of these, cyclohexanediones and aryloxyphenoxypropionics inhibit acetyl – CoA carboxylase by binding to the same region of the target enzyme but occupy different binding sites. These compounds inhibit lipid and / or flavonoid biosynthesis in susceptible species. This suggests ACCase is playing a key role in thse processes.

Herbicides inhibiting fatty acid (lipid) biosynthesis on the basis of primary site of biochemical action[186].

Site of action	Herbicide family (herbicides)
(i) Acdtyl CoA carboxylase (ACCase inhibition)	Aryloxyphenoxypropionics (diclofop, fenoxaprop, fluazifop-P, haloxyfop, quizalofop-P)
(ii) Acetyl elongase (ACEase) (very long chain fatty acids inhibition)	Thiocarbamates (butylate, EPTC, mobinate, pebulate, thiobacarb, triallate, vernolate) Acetamides (acetochlor, alacholor butachlor, metalachlor, pronamide, propachlor, dimethenamide).

6. Herbicides Affecting Branched-Chain Amino Acid Biosynthesis

Valine, leucine and isoleusine[120,128,129] the branched-chain amino acid in the growing tissues are essential for survival of the plants under starvation. These amino acids on addition reverse the toxic effects of certain herbicides. Inhibition of acetolactate synthase, a prominent enzyme in the biosynthesis of branched chain amino acids causes disruption of protein synthesis and cell division as well as increase in translocation of photosynthates to the growing points.

Acetolactate synthase (ALS) also referred to as acetohydroxy acid synthase (AHAS) is the first common enzyme in the combined pathway responsible for the biosynthesis of valine, leucine and isoleucine[235, 236]. The four enzymes involved in the three biosynthetic pathways are: acetolactate synthase, acetolactate reducto isomerase, dihydroxyacid dehydratase and branched-chain amino (amino acid) transferase. Besides, isoleucine requires one additional enzyme, threonine deaminase and leucine requires three additional enzymes, 2-isopropylmalate synthase, 3-isopropylmalate dehydratase and 3-isopropylmalate dehydrogenase[236]. The enzymes of branched chain amino acids biosynthesis reside in the chloroplast. The pathway could be regulated by inhibition of threonine dehydratase by isoleucine inhibition of 2-isopropylmalate synthase by leucine and inhibition of ALS by all the amino acids.

The herbicides inhibiting the branched chain amino acid biosynthesis are:

Site of action	Herbicide families (herbicides)
(i) Acetolactate synthase (ALS) (also called Aceto hydroxy acid synthese AHAS) inhibition.	Sulfonyl ureas (benesulfuron, chlorimuron, chlorsulfuron, primisulfuron, rimislfuron, sulfometuron, triasulfuron, tribenuron, thifenulfuron, triflusulfuron, halosulfuron methyl MON, 1200; MON 37500) Imidazolinones (imazapyr, imazamethabenz, imazaquin, imazethpyr, AC 263, AC 922, AC 299, AC 263 Triazolopyrimidine sulfoanilides flumet sulam, pyrimidinylthio benzoates pyrithiobac)

7. Herbicides Affecting Aromatic Compund Biosynthesis

Aromatic amino acids, phenyl alanine, tyrosine, tryptophan *etc* and p-aminobenzoic acid, ubiquinone, folic acid, Vitamins E and K, lignin, flavonoids *etc* account for 35% by dry weight of higher plants. Thus herbicides which inhibit biosynthesis of these aromatic compounds effectively block the growth and development of plant tissues. Biosynthesis of these aromatic compounds proceeds by way of the shikimate pathway, condensation of phosphoenol pyruvate (PEP) with erythrose 4-phosphate. Inhibition of EPSP (5-enolpyruvyl-shikimate-3-phosphate) synthase, involved in the synthesis of 5-enolpyruvyl-sikimate 3-phosphate inhibits the formation of chorismate. Chorismate, the end product of this pathway is the common precursor of all aromatic aminoacids, other aromatic compounds and secondary metabolites. Phenyl alanine leads into the production of phenolic compounds via phenylalanine ammonia lyase (PAL) to produce a diverse array of phenolic compounds such as lignin precursors, flavonoids and tannins.

Chorismate also gives rise directly to a number of phenolic compounds.

Herbicides inhibiting aromatic compound biosynthesis

Site of action	Herbicide families (herbicides)
(i) EPSP synthase (inhibition)	Glyphosate
(ii) Glutamine synthetase (inhibition)	Glufosinate
(iii) Aluxin-type action	Phenoxys (2, 4-D, MCPA, dichlorprop, MCPA, MCPB)

8. Herbicides Affecting Ethylene Biosynthesis

Ethylene exists in the gaseous state under normal physiological conditions. It regulates many aspects of plant growth, development and senescence. It is actively produced from all parts of higher plants. Ethylene production is induced during germination, ripening of fruits, abscission of leaves and senescence of flowers. Ethylene production can also be induced by external factors such as mechanical wounding, environmental stresses and certain chemicals, including auxin and other growth regulators. Increased ethylene production can also bring out many important physiological consequences.

Methionine is an effective major precursor of ethylene. Methionine conversion requires ATP and oxygen and is inhibited by an uncoupler of oxidative phosphorylation like DNP. The S-adenosylmethionine (SAM), the first intermediate in the presence of oxygen is converted through the catalytic enzymatic action of ACC synthase to 1-aminocyclopropane-1-carboxylic acid (ACC), is later oxidized to form ethylene. Hence inhibition of ACC synthase prevents formation of ACC (from SAM) leading to inhibition of ethylene biosynthesis.

Herbicides inhibiting ethylene biosynthesis are :

Site of action	Herbicide families (herbicides)
Ethylene biosynthesis ACC - synthase (inhibition)	Pyridine carboxylic acids (clopyralid, fluroxypyr, picloram, triclopyr) Benzoic acids (dicamba) quinoline carboxylic acid (quinclorac)

9. Herbicides affecting Glutamine Biosynthesis

Glutamine synthetase (G.S) is the first enzyme involved in the assimilation of ammonia *i.e.* the coverion of inorganic nitrogen to an organic form. The glutamine synthetase catalyzes the conversion of glutamate to glutamine. The glumine is transferred to 2-amino position of glutamate by the enzyme glutamate synthase. The glutamine synthetase (GS) and glutamate synthase act in conjunction to form glutamate synthase cycle[140]. In plants, chloroplast is the major site of ammonia assimilation.

The rapid turnover of ammonia in leaves in light made the GS activity an important target for herbicide action. The glutamine formed by chloroplast GS is used for the biosynthesis of amino acids and nucleotides[140]. Abell[1] observed partial inhibition of GS is required for phytotoxicity. Inhibition of GS produces not only an increase in ammonia levels, but also a decrease in the rate of CO_2 fixation. Reduction of GS activity also affects electron transport and CO_2 fixation under photorespiratory conditions. Hence inhibition of GS leads to multiple deleterious effects, making it an exquisitely toxic site[1]. The herbicide inhibiting glutamine synthetase is glufosinate.

10. Herbicides Affecting Hydrolytic Enzyme Activities

Many enzymes are activated, synthesized or stored during seed germination in order to degrade or solubulize stored insoluble foods and to translocate soluble foods or to mobilize nutrients to synthetic reactions related to growth[3]. The major hydrolytic enzymes include amylases, proteases, lipases, phosphotases, esterases *etc.* These hydrolytic enzymes are under the control of gibberellins in aleurone layers of cereal grains. Thus the production of hydrolytic enzymes requires the synthesis and the presence of proteins, polyribosomes, nucleic acids. Herbicide affecting anyone or more of these events would affect the other events affecting eventual germination of the seed.

Herbicides inhibiting metabolic processes of seed germination include pre-emergence herbicides, like acetamides (acetanilides), thiocarbamates and others.

Table : 2. Herbicide classification based on the primary site of biochemical action

WSSA Group No.	Site of Action	Herbicide Families / Herbicides
	PHOTOSYNTHESIS (Inhibition)	
22	PHOTOSYSTEM I (electron diversion)	Bypiridiliums (diquat, paraquat)
	PHOTOSYSTEM II	
5	Same site, binding behaviour I	Triazines (ametryn, atrazine, cyanazine, prometon, prometryn, propazine, simazine) Triazinones (metribuzin, hexazinone) Uracils, (bromacil, terbacil) Pyridazinones (pyrazon)
7	Same site, binding behaviour II	Phenylcarbamates (desmedipham, phenmedipham)
		Ureas (diuron, fluometuron, linuron, metobromuron, monuron, siduron, tebuthiuron) acetamides (propanil)
6	Same site binding behaviour III	Nitriles (bromoxynil, ioxynil) Benzothiadaziole (bentazon) Imidazolidinones (buthidazole) Phenyl pyridazines (pyridate)
	PIGMENT BIOSYNTHESIS (Inhibition)	
12	CAROTENOID BIOSYNTHESIS phytoene de3saturase desaturation, cyclization hydroxylation	Pyridazionones (norflurazon) Tetrahydropyrimidinones Pyridinones (fluridone) Nitriles (bromoxynil, ioxynil)
14	PROTOPORPHYRIN IX (PP IX) Protoporphyrinogen oxidase (protox)	Diphenyl Ethers (bifenox, acifluorfen, formesafen, lactofen, oxyfuorfen)
		Phenyl triazinones (Aryl triazinones) (sulfentrazone) N-phenylphthalamides (flumiclorac) Oxidiazole (oxadiazon) limines (CGA - 248757)
11	BLEACHING : carotenoid biosynthesis inhibition (unknown target)	Triazole (amitrole, amitrole - T)

Contd....

Table : 2 *Contd.*

WSSA Group No.	Site of Action	Herbicide Families / Herbicides
13	BLEACHING : Diterpenes (Inhibition)	Isoxazolidinones (clomazone)
	FATTY ACID (LIPID) BIOSYSTHESIS	
1	ACETYL CoA CARBOXYLASE (ACCase) (Inhibition)	Aryloxyphenoxypropionics (diclofop, fenoxaprop, fluazifop-P, haloxyfop, quizalofop-P)
		Cyclohexanediones (clethodim, cycloxidim, sethoxydim, tralkoxydim)
	ACETYL ELONGASE (ACEase) (Very long Chain Fatty Acids) (Inhibition)	Thicarbamates (butylate, EPTC melinate, pubulate, thiobacarb, triallate, vernola Contd.
		Acetamides (acetochlor, alachlor, butachlor, metolachlor, pronamide, propachlor, dimethenamid)
	BRANCHED-CHAIN AMINO ACID BIOSYNTHESIS	
2	ACETOLACTATE SYNTHASE (ALS) (also called acetohydroxy-acid synthase : AHAS) (Inhibition)	Sulfonylureas (bensulfuron, chlorimuron, chlorsulfuron, ethametsulfuron, metsulfuron, nicosulfuron, primisulfuron, rimisulfuron, sulfometuron, triansulfuron, tribenuron, thifensulfuron, triflusulfuron, halosulfuron methyl MON 1200, MON 37500)
		Imidazolinones (imazapyr, imazamethabenz, imazaquin, imazethapyr, AC 263, 922, AC 299, 263
		Triazolopyrimidine sulf anilides flumetsulam
		Pyrimidinythio benzoates pyrithiobac)
	AROMATIC COMPOUND BIOSYNTHESIS	
9	EPSP SYNTHASE (Inhibition)	Glphosate
10	GLUTAMINE SYNTHETASE (Inhibition)	Glufosinate
4	AUXIN-type action	Phenoxys (2, 4-D, MCPA, dichlorprop, MCPA, MCPB)

Contd....

Table : 2 *Contd.*

WSSA Group No.	Site of Action	Herbicide Families / Herbicides
	Ethylene biosynthesis - ACC Synthase	Pyridinecarboxylic acids (Clopyralid, fluroxypyr, picloram, triclopyr)
		Benzoic acids (dicamba)
20		Quinolinecarboxylic acids (quinclorac)
		Quinolinecarboxylic acids (quinclorac)
21	CELL WALL synthesis (site B) (Inhibition)	Benzamides (isoxaben)
19	AUXIN - type action (inhibition)	Phthalamates (naptalam)
23	MITOSIS (Cell division inhibition)	Carbamates (chlorpropham, propham)
		Cineoles (Cinmethylin)
		Bensulide
3	MICROTUBULE ASSEMBLY (Inhibition)	Dinitroanilines (benefin, ethalfluralin, oryzalin, pendimethalin, trifluralin)
18	7, 8-DIHYDROPTEROATE (DHP) SYNTHASE (Inhibition)	Carbamates (aulam)
22	UNCOUPLERS (membrane disruptors) (Inhibition of oxidative phosphorylation)	Phenols (dinoseb), Arsenicals (DSMA, MSMA)
I	PROTEIN SYNTHESIS (Inhibition)	Acetamides (acetachlor, alachlor, butachlor, metolachlor, pronamide, propachlor, dimethenamid, napropamide)
		Thicarbamates (butylate, ETCm molinate, pebulate, thiobencarb, triallate, vernolate)
II	PROTEIN BISOSYNTHESIS and/or RNA BIOSYNTHESIS (RNA polymerase) (Stimulation)	Pyridine carboxylic acids (cloyralid, fluroxypyr, picloram, triclopyr)
		Phenoxys (2, 4-D, MCPA, dichlorprop, MCPA, MCPB) Benzoic acids (dicamba)
III	RNA BIOSYNTHESIS (Inhibition)	Pyrazoliums (difenzoquat)

*I, II and III are not WSSA groups. These have been designated by the author[2,186].

(b) Mechanism of action of herbicides (groups) based on their structure- function specificity:

The ever increasing strategies of modern agriculture, particularly with an emphasis on intensive agriculture for augmenting food production and production of cost-effective commercial agricultural products to meet global quality requirements, forced weed control strategies to change its facetes. Consequently scores of herbicides are developed as detailed elsewhere in this review. Above all, the weed control efficacy of herbicides has been increased enormously due to continued research and development. In achieving this success of weed control efficacy much of the credit vowed to the knowledge of development of herbicides with specific mechanism of action. The discovery of varied mechanisms of herbicidal action in relation to their structural specificity, made it easy to descriminate 'susceptible' or 'tolerant' weed species or crop plants. Developing a particular mode of action is a tricky job, as the same chemical has to kill the unwanted weed while saving the wanted crop plant. Development of a herbicide with a specific mechanism requires thorough knowledge of physiology of response of the crop plant and the structure- function relationship of the herbicide molecule. The intermediates of the herbicide metabolized should get ameliorated with crop plants metabolism to render the herbicide safer for crops and at the same time retaining its phytotoxicity against a broad spectrum of weeds or a specific weed. The literature on the efficacy weed control by present generation herbicides offer considerable possibility for improved agricultural productivity. This could be certainly attributed to the substantial progress in the development of mechanism of action safer to crop plants at the same time toxic to weed species. A variety of important groups of herbicides with varied types of mechanisms of action are reviewed hereunder in brief and presented in (Table 2). For detailed information the reader may consult Rao[186] and anonymous[5].

Acetamides (Amides)

This group of herbicides (Alacholor, Prapachlor, Pyrana chlor) inhibits seed germination by interfering with the related metabolic activities and suppress seedlings emergence. Alachlor and propachlor inhibited GA_3 induced α - amylase production in seeds of barley, the susceptible species during germination[77, 120]. This group also severly inhibited protease synthesis by barley seeds[184] and suggestively these were acting as repressors of gene action preventing normal expression of the hormonal effects of GA through the synthesis of DNA dependant RNA. But GA_3 at higher level though overcome alachlor inhibition by removing the repressor effect[184], didn't result in the removal of inhibition on seedling growth. The results indicate that acetamides are possibly acting on biosynthesis of hydrolytic enzymes and that their action on α - amylase and protease was secondary. The primary effect of propachlor was on protein biosynthesis while it was secondary on nucleic acid biosynthesis[81, 83, 84, 85]. Similar reports of inhibition of protein and RNA synthesis besides GA induced a - amylases were also obtained.

These herbicides inhibited the formation of mRNA-ribosome complex (polyribosome). It was concluded that the chain initiation process was the primary target of these herbicides during protein synthesis[184]. The propachlor formation of amino acetyl-tRNA required for polyribosome complex[78, 79] increased peptide chain and that a-chloroacetamides undergo nucleophilic displacement with methionyl-tRNA[120]. It prevents occurance of peptide chain linkage[139]. Thus, these acetanilide herbicides prevent : (a) formation of mRNA- ribosome complex, (b) activation of amino acid to form aminoacyl tRNA or, (c) linkage of the growing peptide chain (polysome-tRNA-AA(n)) and (d) the incoming amino acyl-tRNA thereby inhibit chain initiation and eventually protein synthesis.

Metolachlor: Structurally similar to alachlor metolachlor, inhibits seed germination and early seedling growth by inhibiting protein synthesis[70] and root growth. Truelove *et al* [239] reported that it inhibited choline incorporation into phosphotidyl choline during cotton seed germination. Hence, protein synthesis was the primary mechanism of metolachlor action in susceptible species.

Butachlor CDAA: They have strong inhibitory effect on the incorporation of ^{14}C-acetate into lipids in susceptible speices. Chloroacetamides interfere with fatty acid metabolism by alkylating key enzymes of fatty acid biosynthesis interfering with CoA metabolism[94].

Napropamide: It reduced bud break, terminal growth, root development in cranberry cuttings[76] and retarded root growth[33]. It blocked the progression of dividing cells through cell cycle into mitosis. It reduced cell division and DNA synthesis by inhibiting cell cycle-specific proteins. Mercado[154] found it inhibiting germination of *cyperus rotundus* tubers and shoot development by inhibiting α - amylase activity.

Pronamide: Pronamide inhibits mitosis by binding to tubulin and prevents their assembly into microtubules. So the cells unable to form spindle fibres, resulting in their inability to separate chromosomes to the poles of the cell. Pronamide treated plants are thus devoid of cortical microtubles, prevent isodiametric cell expansion thus had club-shaped root tips[215].

Propanil: It inhibits radicle elongation in coleoptiles[116] and destroys permeability of root membrane[186]. It strongly inhibits RNA and protein synthesis as well as α -amylase synthesis. Propanil also inhibited oxygen uptake and phosphate esterification in soybean mitochondria[116] while Good[96] and Moreland and Hill[186] found that propanil inhibited the Hill reaction of photosynthesis by reducing cytochrome by PS II[170]. Propanil destroys chloroplast membranes and permeability of red beet membranes[116]. According to Moreland and Huber[156], it interfered ATP generation and electron transport in mung bean mitochondria. At lower concentrations it acted as uncoupler.

Aliphatics

Acrolein: It is an aquatic herbicide and kills the plant through its sulphahydryl reactivity which destroys vital enzyme systems in plant cell.

Dalapon: It affects meristematic activity of the root tip and arrests mitotic activity at prophase. Dalapon has no precise mechanism or one site of action but affects many metabolic processes of the plant. Jain *et al*[119] reported that it affected glucose utilization. Glycolytic pathway inhibited initially in Krebs cycle. It affected nitrogen metabolism by degradation of protein to ammonium compounds to amides and ammonia in susceptible setaria and tolerant sugarbeets. It also reduces surface wax on leaves of various plants indicating that it might be disturbing lipid metabolism.

Arsenicals

DSMA, MSMA: The first arsenical toxicity symptom is cessation of growth and gradual browning. Arsenicals cause root plasmolysis and leaf wilting followed by discoloration and necrosis of leaf tips and margins. Shoot meristematic tissues are the major site of action of cacodylic acid and MSMA[195]. They are uncouplers of oxidative phosphorylation and inhibit transformation of energy to ATP. During uncoupling, energy is dissipated as heat and become unavailable for cellular processes. They denature protoplasm and bind organic sulphahydryl groups, inhibit pyruvate oxidases and phosphatases so that respiration is inhibited.

Benzamides

Isoxaben: It interferes with cell wall biosynthesis in susceptible weeds by inhibiting synthesis of cellulase [52].

Benzoics

***Chloramben**:* Inhibits elongation and growth of roots and shoots of susceptible species. It promotes cell division in the presence of kinetin as well as cell enlargement indicating their auxin like properties[124]. It also causes growth proliferation resulting in swelling.

***Dicamba**:* Inhibits oxygen uptake in *Cyperus rotundus* leaves[146] and in isolated mitochondria[92]. Moreland *et al*[157] showed dicamba inhibition of GA_3 induced α - amylase synthesis in barley seeds but did not affect RNA and protein synthesis. Watson *et al*[249] reported of reduced ATPase activity. Arnold and Nalwaja[8] suggested that dicamba increased RNA and protein levels in susceptible plants by directly affecting the removal of histones from the DNA template. Dicamba influences DNA precipitating properties of histones and affected normal functioning of genetic mechanism[180].

Benzothiadiazole

Bentazon: It inhibits photosynthesis by inhibiting the PQ_B – binding niche on the D protein of the PS II complex in chloroplast thylakoid membranes. Thus, it blocks electron transport from PQ_A to PQ_B. This stops CO_2 fixation and production of ATP and $NADPH^{++}$; But plants die by other processes in many cases[186]. The inability to reoxidize PQ_A promotes formation of triplet state chlorophyll (^3Chl) which interacts with ground state oxygen to form singlet oxygen ($^1O_2$). Both ^{3} Chl and $^1O_2$ can abstract a hydrogen from unsaturated lipids, producing a lipid radical and initiating a chain reaction of lipid peroxidation. Obviously lipids and proteins are attacked and oxidized resulting in loss of chlorophylls and carotenoids, leaky membranes consequently cell organelles dry and disintegrate rapidly.

Bipyridiliums

Diquat, Paraquat: These herbicides cause wilting and rapid desiccation of foliage as they are contact poisions. Their phytotoxicity is enhanced in the presence of light. Diquat injury requires both chlorphyll and light. Paraquat disrupts membrane integrity to possibly cause wilting, necrosis and eventual death of foliage[18]. Bipyridiliums form toxic free radicles to yield H_2O_2 and OH in the presence of chloroplasts, oxygen and light. They inhibit reduction of NADP to NADPH by removing electrons from the electron transport system of PS II[258]. Thus, diquat can shut electrons from ferredoxin to form diquat-free radicals and prevent reduction of NADP to NADPH. Diquat and Paraquat exert their herbicidal effects more rapidly in light than in darkness by generatings oxygen readicles[29]. Reoxidation of paraquat leads to the generation of superoxide (O_2^-).

Carbamates

Asulam: It produces a toxic effect in the frond buds situated in the rhizome branches of bracken fern (*Pteridium aquilinum*). Results of Veerasekharan *et al*[244] showed that they interfere with RNA and protein sysnthesis at the metabolically active sinks (rhizome buds) as a mojor site of action of asulam.

Asulam appears to inhibit cell division and expansion of plant meristems by interfering with microtubule assembly or function. It inhibits 7,8 dihydropteroate synthetase, an enzyme involved in folic acid synthesis which is required for purine nucleotide biosynthesis[186].

Desmedipham, Phenmedipham

Bethlenfalvay and Norris[20] found that desmedipham causes photosynthetic inhibition in sugarbeets when tempeature was raised from 10 to 35°C. High light intensity following foliar application greatly inhibited photosynthesis than low light intensity. Both of them contain the peptide bond structure common to many photosynthetic inhibitors[201]. Phenmedipham inhibits CO_2 assimilation while both reduce photosynthesis by inhibiting Hill reaction.

Cineoles

Cinmethylin, appears to inhibit mitosis in meristematic regions (shoots and roots) of susceptible plants.

Cyclohexanediones (cyclohexenones)

Clethodin, cycloxidim, sethoxydim, tralkoxydim

These herbicides interfere with lipid metabolism. Burgstahler and Lichthenthaler[38] and Ishihara *et al*[118] reported reduced glycolipid and phospholipid content of maize seedlings with sethoxidim affecting lipid metabolism at early stage. Further, Burton *et al*[39] reported that sethoxydim inhibited acetyl-CoA synthesis in chloroplasts of sensitive maize, but not in chloroplasts of tolerant pea.

Dinitroanilines

Benefin, Dinitramine, Fluchloralin, Isopropalin, Pendimethalin, Trifluralin.

The striking symptoms of these herbicides are similar to those seedlings treated with colchicines[243]. Dinitroaniline herbicides inhibit polymerization of tubulin into microtubules[156, 228].

When polymerization is blocked by these herbicides, depolymerization process shortens the microtubules continuously until they eventually become undetectable resulting in arrest of cell division, formation of polynucleate cells and inhibition of root and plant growth.

Diphenylethers

Aciflurofen, Bifenox, Fomesafen, Flurodifen, Fluoroglycofen, lactofen, oxyfluorfen.

The phytotoxicity of these herbicides is strictly light dependent. Lehnen *et al.,* [138] found that DPE herbicides inhibit protoporphyrinogen oxidase (Protox) which oxidizes protoporphyrinogen (PPG IX) to protoporphyrin IX (PP IX) a tetrapyrrole intermediate. Inhibition of protox by herbicides leads to uncontrolled accumulation of PP IX, which degrade cellular molecules by peroxidative attack[200]. Light absorption by exicited PP IX produces triplet state PP IX which interacts with ground state oxygen to form singlet oxygen. Both triplet chlorophyll and singlet oxygen can abstract hydrogen from unasaturated lipids (fatty acids) producing lipid radical and initiating a chain reaction of lipid peroxidation. Lipids and proteins then are attacked and oxidized resulting in loss of chlorophyll, carotenoids and in leaky membranes eventually leading to rapid disintegration and death of cell and cell organelles.

Imidazolidinones

Buthidazole : Buthidazole disrupts mesophyll chloroplasts and reduces starch as shown by Hatzois and Penner[110]. It inhibits respiration and photosynthesis. York *et al*[256] found no significant inhibition of PS I – mediated electron transport while York and Amtzen[255] reported inhibition of the whole electron transport from water to methyl viologen, PS II – dependent dichloro phenolindophenol reduction, non-cyclic photophosphorylation and silicomolybdate photoreduction. The major effect of buthidazole, according to York *et al*[257] was inhibition of electron transport on the reducing side of PS II at the site of diuron and atrazine inhibition, while the oxidizing side of PS II was the secondary site.

Imidazolinones

Imazamethabenz, Imazapyr, Imazaquin, Imazethapyr : Shaner and Reider[204] reported that DNA synthesis was not the primary site target for imazapyr but 1.3 to 20 fold increased levels of most aminoacids. But markedly reduced levels of valine and leucine were found by Shaner *et al*[208]. Valine, leucine and isoleucine since are linked by a common biosynthetic pathway, imazapyr likely inhibits acetolactate synthase (ALS), the first enzyme common to all these aminoacids. Several reports[160, 207, 226] also stressed that imazapyr treated plants had low levels of extractable ALS, also called as acetohydroxy acid synthase (AHAS). Sulfonylureas also inhibit ALS *in vitro*, but only imidazolinones decrease the level of extractable ALS[225].

Imines

CGA – 248757

CGA – 248757, like diphenyl ether herbicides, stimulates accumulation of protoporphyrins (PP IX) by inhibiting protoporphyrinogen oxidase (Protox) which oxidizes protoporphyrinogen (PPG IX) to PP IX. Accumulation of PP IX leads to peroxidation of membrane lipids in the presence of light and oxygen and irreversible damage to cell membranes and cell function[186].

Nitriles

Bromoxynil, Ioxynil

Both of these affect a number of essential physiological and biochemical processes in higher plants. Hill reaction was inhibited by Ioxynil and uncoupled oxidative photophosphorylation[245]. Its toxicity was greater in light than in dark Paton and Smith[178] reported that ioxynil inhibited electron transport, non cyclic photophoshorylation and CO_2 fixation in leaf chloroplasts. Uptake of inorganic phosphate (Pi) and oxygen were depressed by ioxynil resulting in a low P/O ratio.

Since electron transport is essential for photosynthesis, Smith *et al*[186] suggested that blockage of this transport by ioxynil could stop synthesis of ATP resulting in death of the plant. The actual site of action of ioxynil was near plastoquinone (PQ). Recently, Rao[186] reviewed that both these nitriles inhibit photosynthesis by binding to the PQ_B – binding niche on the D_1 protein of the PS II complex in chloroplast thylakoid memebranes and block electron transport from PQ_A to Q_B. This stops CO_2 fixation and production of ATP and $NADPH^{++}$ which are required for plant growth. The inability to reoxidize PQ_A promotes formation of ^3Chl (triplet chlorophyll) which interacts with ground state oxygen to form singlet oxygen (O_2). Both ^3Chl and $^1O_2$ abstract hydrogen from unsaturated fatty acids producing a lipid radical and initiating a chain reaction of lipid peroxidation. Lipids and proteins are attacked and oxidized resulting in loss of chlorophyll and carotenoids as well as leaky membranes. Consequently, cells and cell organelles dry and disintegrate rapidly.

Oxadiazoles

Oxadiazon: Oxadiazon causes rapid loss of pigmentation and desideration as of those herbicides that kill by causing rapid peroxidative damage. Oxadiazon affected protein synthesis, CO_2 fixation, nucleic acid synthesis and lipid synthesis. According to Hatzios,[109] it specifically inhibits chlorophyll biosynthesis prior to lipid peroxidation. It caused accumulation of protoporphyrin IX and reduced protochlorophyllide accumulation of protoporphyrin IX. This is a single oxygen generating pigment[186].

Phenols

(i) Dinoseb: Dinoseb kills the plants rapidly by contact. The primary mechanism of action of phenols is inhibition of oxidative phosphorylation. They act as an uncoupling agent and inhibitor of oxidative phosphorylation. They penetrate the membrane of the cell organelles and destroy its permeability by interfering with oxidative phosphorylation. Dinoseb releases respiration from its dependence on a phosphate acceptor. Consequent stimulation of respiration and failure of ATP formation rapidly affects the living plant[186].

Phenoxyalkanoic Acids

Phenoxyacetics (2,4-D, MCPA); Phenoxybutyrics (2,4-DB, MCPB)

The chlorophenoxy acids also known as auxin-type herbicides cause characteristics phytotoxic symptoms such as epinasty, swelling, twisting and bending of the treated areas and distant plants. The plants stop growing, meristematic cells stop dividing, young leaves stop expanding, mesophyll tissue ceases development. Roots loose ability to absorb water and mineral nutrients and leave loose the photosynthetic ability. When the tip of the shoot is killed, some lateral buds may begin to grow because of release of apical dominance.

At herbicidal concentrations these phenoxyalkanoic acids promote doubling of RNA content of 2,4-D treated soybean hypocotyls with over half of the increase appearing in microsomal fraction and one quarter in the soluble fraction[48]. 2,4-D induced aberrations in the nucleic acid metabolism of sensitive plants and synthesis of RNA polymerase in soybean hypocotyls[41, 125, 171]. This could lead to read out template which is not read out by endogenous polymerases. This causes a change in DNA sites available for transcription, there by exerting influence at the gene level. 2,4-D affects all types of RNAs as well as DNA and ribonuclease (RNase). Thus, various reactions related to gene action seems to be the primary mechanism of 2,4-D. Recent review of Rao[186] portrays that the primary action of auxinic herbicides appears to be cell wall plasticity and nucleic acid metabolism. 2,4-D acidifies cell wall by stimulating the activity of a membrane-bound ATPase driven proton pump. Cell wall loosening enzymes are induced by reduced apoplasmic pH which causes cell-elongation. Ethylene production is stimuated by 2,4-D and other auxin – type herbicides, so characteristic epinastic sympotoms are developed with these herbicides.

N- Phenylphthalamides

Flumiclorac

The primary site of action of flumiclorac, appears to be protoporhyrinogen oxidase (Protox) an enzyme of chlorophyll and heme biosynthesis, catalyses the oxidation of protoporphyrinogen (PPG IX) to protoporphyrin IX (PP IX). Inhibition of protox leads to accumulation of PP IX, the first light absorbing precursor of chlorophyll. The unprocessed PPG IX that overflows the normal localization in the thylakoid membrane is oxidized by molecular oxygen to PP IX which is catalyzed by a plasma lemma enzyme that has Protox activity. It is insensitive to diphenylethers. PP IX is separated from Mg chelatase and produces triplet state PP IX. The latter then interacts with ground state oxygen to form singlet oxygen ($^1O_2$). Both, triplet PP IX and singlet oxygen, abstract hydrogen from unsaturated lipids, producing a lipid radical and initiating a chain reaction of lipid peroxidation. Lipids and proteins are attacked and oxidized resulting in loss of chlorophyll and caroteinods as well as leaky membranes. This results in dry cell and cell organelles and rapid disintegration[186].

Phenylpyridazines

Pyridate: Pyridate, upon entering the plant, is hydrolysed to 3-phenyl-4-hydroxy-6 chloropyridazine, which then inhibits PS II electron transport[186].

Phenyl Triazinones (Aryl triazinones)

Sulfentrazone

Nandihalli and Duke[165, 166] found that sulfentrazone inhibited protoporphyrinogen oxidase (Protox) in barley, the susceptible species. Like other photobleaching herbicides, sulfentrazone also causes accumulation of PP IX which reacts with oxygen to form highly destructive oxygen radicals. This is because like most of the photobleaching herbicides, sulfentrazone also mimics half of the tetrapyrrole ring of protoporphyrinogen and compete for the catalytic site on the enzyme[85, 165,166].

Phthalamates

Naptalam

Naptalam appears to have both herbicidal and plant growth regulator activity. It is an auxin inhibitor. It appears to attach to a phytotropin binding site and inhibits auxin efflux from the basal end of cells, there by blocking basipetal auxin transport[186].

Pyrazoliums

Difenzoquat

It inhibits nucleic acid biosyntheis, photosynthesis, ATP production, K^+ absorption and P incorporation into phospholipids and DNA.

Pyridazinones

Metflurazon (SAN 6706) Norflurazon, Pyrazon, SAN 9785. Substituted pyradazinone herbicides inhibit one or more of processes: (a) photosynthesis (b) carotenoid biosynthesis and (c) fatty acid desaturation.

Norflurazon causes bleaching in susceptible species[107, 117, 196, 197]. It inhibits carotenoid biosynthesis and photoxidizes chlorophyll. Based on current evidence, norflurazon blocks carotenoids biosynthesis by inhibiting phytoene desaturase, in chloroplast membrane which catalyzes the destruction of phytoene. Mayer *et al*[151] supported this hypothesis. He found norflurazon is a reversible, non-competitive inhibitor of phytoene desaturase. Metflurazon and norflurazon inhibit the desaturation of phytoene to lycopene during carotene biosynthesis, thus causing bleaching of leaves, besides inhibition of carotenoid biosynthesis and fatty acid desaturation[107].

Pyrazon largely inhibits photosynthetic electron transport. It binds D_1 the quinone – binding protein on the reducing side of PS II[121]. Pyradazinones, except pyrazon, alter the fatty acid composition of lipids. Galactolipds are increased by blocking conversion of linoleic acid to linolenic acid[82, 107, 117].

Pyridine carboxylic acids

Clopyralid, Picloram, Triclopyr

The mechanism of action of pyridine carboxylic acids is similar to that of exogenous auxin (IAA) and auxin type herbicides. The primary action involves cell wall plasticity and nucleic acid metabolism. Clopyralid, picloram and triclopyr acidify the cell wall by stimulating activity of the membrane bound ATPase proton pump. The reduced apoplasmic pH induces cell elongation by increasing the activity of cell wall loosening enzymes. Low concentrations of these herbicides stimulate RNA polymerase, increase in DNA, RNA and protein biosynthesis. Abnormal increase in these processes lead to uncontrolled cell division and growth causing destruction of vascular tissue. In contrast, high concentration of pyridine carboxylic acid like other auxin herbicides inhibits cell division and growth in meristematic tissue that accumulate herbicides from phloem[186].

Pyridines

Dithiopyr, Thiazopyr

These herbicides disrupt cell division by inhibiting mitosis in late metaphase and causing multipolar mitosis. This does not bind to tubulin and mother protein a 65-58 kDa protein, which may be a microtubule associated protein (MAP). MAPs stabilize growing microtubules. The pyridine herbicides shorten microtubules such that they cannot form spindle fibres normally responsible for migration of chromosomes to the poles of the cell during mitosis.

Pyridinones

Furidone: Furidone blocks carotenoid biosynthesis by inhibiting phytoene desaturase. They allow $^1O_2$ and ^3Chl to abstract a hydrogen from an unsaturated lipid, producing a lipid radical. The lipid radical interacts with O_2, yielding a peroxidized lipid and another lipid radical. Lipid peroxidation leads to destruction of chlorophylls and membrane lipids. Proteins are destroyed by $^1O_2$ which leads to leaky membranes and rapid tissue desiccation[186].

Pyrimidinyl Thio-Benzoates (Benzoates)

Pyrithiobac

Like sulfonylureas, imidazolinones, pyrimidinylthiobenzoate herbicides represented by pyrithiobac, inhibit acetolactate synthase (ALS) an enzyme in the synthesis of branced – chain amino acids valine, leucine and isoleucine. ALS inhibition causes plants to appear stunted and chlorotic[186].

Quinoline Carboxylic Acids

Quinclorac

Quinclorac, a new auxin type herbicide inhibits shoot growth with chlorosis and subsequent necrosis of the entire tissue. With simultaneous stimulation of ethylene production by activating the synthesis of 1-aminocyclopropane-1-carboxylic acid (ACC) synthetase, in the root tissue the ACC and its conjugate (N-malonyl-1-aminocyclopropane-1-carboxylic acid, (MACC) accumulate predominantly in shoot suggesting transfer of ACC into the shoots. But ethylene is inherently not lethal even at non-physiological high concentrations. The phytotoxicity then is due to the formation of cyanide as a co-product during quinclorac – stimulated ethylene biosynthesis[104].

Sulfonylureas

Bunsulfuron, chlorimuron, chlorsulfuron, Halosulfuron, Metsulfuron nicosulfuron, Primisulfuron, Prosulfuron, Sulfometuron, Triasulfuron, Thifensulfuron, Triflusulfuron, Tribenuron.

Sulfonylurea herbicides first affect the meristematic tissues where growth ceases soon after treatment. Chlorosis and the necrosis soon follow with die-back to the mature parts of the plant.

La Rossa and Schloss[130] provided evidence that sulfonyl ureas inhibit the biosynthesis of branched chain amino acids. Chlorsulfuron was an extremely potent inhibitor of acetolactate synthase (ALS)[139] and it forms the target of sulfonylurea herbicides.

Inhibition of ALS causes increase in biosynthesis of fatty acids via the pyruvate dehydrogenase complex due to an increased pool of pyruvate.

Tetrahydropyrionidinones

They are also called as cyclic ureas, a new class of photo-bleaching herbicides. Chlorophyll bleaching was demonstrated to be a secondary photo-oxidative event caused by the loss of carotenoids. They also showed that phytoene desaturase step is the primary molecular target of these herbicides[186].

Thiocarbamates

Butylate, EPTC, Diallate, Pebulate, Triallate

The evidence accumulated indicates that the inhibitory effect of thiocarbamates on the synthesis of surface lipids such as waxes, cutin and suberin is due to their ability to inhibit the biosynthesis of very long- chain fatty acids (VLCFA). Preventation of acyl-CoA elongases catalyzes the synthesis of VLCFA. This effect could be due to selective effect of thiocarbonate on the in vivo synthesis of VLCFA. Inhibition of VLCFAs leads to reduction of epiculticular wax formation on the plant foliage. This increases leaf wettability, allowing more affective absorption of herbicide for effective killing of the plant. Thiobencarb caused aberrations similar to growth regulating properties of hormones, although it is a non-hormonal herbicide. It effects protein and RNA metabolism and causes morphological changes.

Triazines

Atrazine, Ametryn, Propazine, Prometon, Prometryn, Simazine, Terbutryn

Triazines are known to inhibit growth of emerged seedlings. Triazines characteristically inhibit photosynthesis. The specific site of action is the oxygen evolution or photolysis in photosynthesis. It is also well known that they also inhibit ATP formation via photophosphorylation in the chloroplasts when flavin mononucleotide (FMN) is used as an electron acceptor. Bishop[25] found that Simazine inhibited FMN catalysed photophosphorylation but not when catalyzed by vitamin K_3. The reducing power required for CO_2 reduction to carbohydrates will not be possible when oxygen evolution and light reactions are blocked. All triazine herbicides inhibited state 3 respiration and cyclic photophosphorylation[233,234] the effect of triazines was similar to that of oligomycin[152]. Copping and Davis[51] reported that 3 triazine herbicides inhibited non cyclic photophosphorylation and were toxic only in light. Madhulety[143] and Madhulety and Chandra Singh[144] also demonstrated that Atrazine inhibited photosynthesis by blocking carbohydrate synthesis and trapping photoliberated electrons using sucrose feeding and riboflavin spray to the treated plants. They also demonstrated that atrazine was phytotoxic to maize only in the presence of light[143]. It is further emphasized by Brewer *et al*[31] that atrazine inhibited electron transport, but not uncouple photophosphorylation. The primary site of action was on reducing side of PS II between primary acceptor (Q) and the plastoquionine pool of the electron transport chain (for atrazine)

According to recent review by Rao[186] triazine herbicides inhibit photosynthesis by binding to the PQ_B – binding niche on the D1 protein of the PS II complex in chloroplast thylakoid membranes. Thus, blocking electron transport from PQ_A + PQ_B. This stops CO_2 fixation and production of ATP and NADPH, which are needed for plant growth. The inability to reoxidize PQ_A promotes ^3Chl interact with O_2 to form $^1O_2$. Both ^3Chl and $^1O_2$ then lead to lipid peroxidation resulting in leaky membranes, loss of chlorophyll, carotenoids consequently the cell and cell organelles dry and degintegrate rapidly.

Triazoles

Amitrole, Amitrole – T

The development of albino leaves and shoots is the distinct feature of amitrole toxicity. Amitrole causes destruction of chloroplyll and impairment of chlorophyll development in tissues prior to the protochlorophyll stage. Thus, the developing leaves show yellowing and browning followed by death. Amitrole induced albinism is due to the destruction of chloroplast pigments or an inhibition of their synthesis. Bartels[14] reported that amitrole blocked light induced plastid development as evidenced by the absence of normal lamellar system and membranal organization, hence affecting photosynthesis. Bartles and Hyde[13] further suggested that amitrole brings metabolic change in the developing plastids by destroying chloroplast DNA or blocking synthesis of chloroplast enzymes or structural components necessary for chloroplast synthesis.

The recent review by Rao[186] states that aminotriazole inhibits carotenoid biosynthesis by inhibing the phytoene desaturation step, resulting in accumulation of phytofluene and g-carotene. It causes accumulation of lycopene by inhibiting the cyclization process.

Triazolopyrimidine Sulfon Anilides

Flumetsulam

It inhibits acetolactate synthase (ALS), a key enzyme in the biosynthesis of branched chain amino acids-valine, leucine and isoleucine. The plant death results from events that inhibit ALS.

Uracils

Bromacil, Terbacil, UCC –C4243

Bromacil and terbacil inhibit photosynthesis by binding to the PQ_B -binding niche on the D1 protein of the PS II complex in chloroplast thylakoid membranes, thereby blocking the electron transport from PQ_A to PQ_B. This inhibits CO_2 fixation and production of ATP and $NADPH^{++}$. The inability to reoxidize PQ_A promotes the formation of ^3Chl which interacts with O_2 to form $^1O_2$. Both ^3Chl and $^1O_2$ then cause lipid peroxidation, resulting in loss of chlorophyll, carotenoids and in leaky membranes.

Ureas

Diruron, Fluometuron, Isoproturon, Linuron, Methabenzthiazuron, Monuron, Siduron

The substituted urea herbicides inhibit the oxygen evolution step or photolysis in photosynthesis. The primary site of action of urea herbicides like triazine herbicides is located in PS II involving the oxygen evolution step. They interfere with the reducing site of PS II[181]. Ureas also inhibit non –cyclic electron transport as well as cyclic electron transport. Giannpolitis and Ayers[186] provided evidence that monuron accelerated photo-oxidations. (bleaching of chlorophyll and lipid oxidation) in chloroplasts thus lending support to the hypothesis of depletion of reducing potential (NADPH) which is responsible for chloroplast oxidations and death of plant.

The latest review of Rao[186] narrates that ureic herbicides inhibit photosynthesis by binding PQ_B -binding niche on the D1 protein of the PS II complex in chloroplast thylakoid membranes, thus blocking electron transport from PQ_A to PQ_B following similar reactions as described for triazines.

Unclassified herbicides

Bensulide: It inhibis root growth Cutter *et al* [57] reported curved root, root hair and elongated epidermal cells at the root tip. It caused binucleated cells in epidermal and cortical tissues of roots suggesting direct inhibition of mitosis. Ashton *et al* [10] found that bensulide inhibited proteolytic activity in squash seedlings.

Ethofumesate: It inhibits epicuticular wax on the surface, suggesting that it affects lipid biosynthesis. The epicuticular wax of cabbage was totally eliminated by ethofumesate resulting in increased epicuticular transpiration [140].

Glufosinate: It is also referred as phosphinothricin, an analogue of glutamate. Glufosinate inhibits glutamine synthase (GS)[1] which converts glutamate to glutamine in chloroplasts. The inhibition is a two step process : (a) a reverisible inhibition of GS (glufosinate being competitive with glutamate) and (b) an irreversible inhibition of phosphorylation.

Kocher *et al*[128] found spraying of glufosinate increased ammonia level in sorghum. Build up of toxic ammonia level could cause both inhibition of photosynthesis and the death of plants. It also stimulated leakage of 'K' ions. Lea *et al*[136] found a decrease in free glutamine with increased ammonia levels due to glufosinate which enters cell and inhibits GS activity. Glufosinate inhibits rate of photosynthetic CO_2 assimilation affecting Calvin cycle by insufficiency of amino donors under high ammonia levels. This cannot dissipate energy in the formation of ATP and $NADPH^{++}$ resulting in photo-inhibition. Extensive photo-inhibition leads to the formation of ^3Chl and $^1O_2$ and hydroxy radical (OH)[126].

Higher levels of ammonia can cause the uncoupling of photosynthetic phosphorylation. The unprotonated NH_3 diffuses freely through the chloroplast and thylakoid membrane, takes up H^+ to form NH_4^+. Influx of anions towards ionic balance, results in osmotic uptake of water. This dramatic swelling of chloroplast and thylakoids cause protrusions in the affected leaves. Higher ammonia can also bind to the water splitting, manganese containing site in PS II and thus inhibit O_2 evolution in electron transport. However, all these above inhibit O_2 evolution in metabolic processes form secondary, while inhibition of GS forms the primary site of action of glufosinate. Inhibition of GS leads to multiple deleterious effects making it an exquisitely toxic site for glufosinate[1].

Glyphosate

Glyphosate is a highly specific competitive inhibitor of the enzyme 5- enolpyruvyl shikimate 3-phosphate (EPSP) synthase. It is involved in the skimate pathway of biosynthesis of aromatic compounds (aminoacids) including tryptophan, tyrosine and phenylalanine, all needed for protein synthesis or for biosynthetic pathways leading to growth. Increase in shikimate -3- phosphate due to inhibition of EPSP synthase glyphosate, affects chorismate, the precursor for the biosynthesis of aromatic compounds. This indicates that EPSP synthase inhibition is the primary site of action of glyphosate. The secondary effects of glyphosate responsible for causing cell death in plants are the arrest of both protein synthesis and diverse phenolic compound formation attributable to a deficit of aromatic aminoacids and destruction of indole acetic acid and inhibition of porphyrin synthesis. Recent studies showed that glyphosate may be trapped on the EPSP synthase enzyme in the presence of EPSP[186] but Abell[1] suggested that glyphosate binds near, but not at the site of EPSP synthase.

EFFECT OF HERBICIDES ON FORMATION OF YIELD COMPONENTS OF CROP PLANTS

It is well known that some of the herbicides are phytotoxic to crop plants also due to lack of universal tolerance to an array of herbicides. Consequently, crop plants need to be screened for the given herbicide. This phytotoxicity if developed at formative stage of either at plant growth and development or at the formation of yield components, crop yields will be drastically affected. The extent of yield loss thus depends on the target tissue of the plant organ.

As the herbicides affect the formative tissues at the very early stages, development of plant organs especially yield contributing panicles, number of grains or fruits drastically lower the yields per unit area, which run into considerable economic losses. Cinosulfuron was phytotoxic to rice since early stage onwards resulting in low tillering and poor growth of the crop. This may lead to ill-formed yield components (panicles). Since toxicity was at early stages, crop could recover later and the adverse effects may not last long. On the other hand many formative effects leading to enhancement of yield components were also reported resulting in improved yields. Significantly higher rice grain yield due to cinosulfuron was reported[253]. Similarly, such formative effects on yield components obviously enhanced yield as it is evident with pretilachlor which improved panicle number (530) in rice compared to unweeded control (283) or hand weeding (441). This resulted in the grain yield of 4100 kg ha^{-1} against unweeded control (1577 kg ha^{-1}).

QUANTITATIVE AND QUALITATIVE EFFECTS OF HERBICIDES ON CROP YIELDS

Herbicides increase yields in field crops considerably over farmer's practices. Pyrazosulfuron could cause an increase in rice yield by 10%, pretichlor about 13% while butachlor by 11% over hand weeding. Estimating yield losses in field crops due to farmer's methods, herbicides could increase about 14 – 20% yield in sprouted rice over farmer's practice while it was 29% by oxyfluorofen in transplanted onion and 11% by pendimethalin in carrot[182]. Nagaraju[164] reported about 28% increased yield of potato while Srinivasa Murthy[222] demonstrated about 14% in potato by oxyfluorofen. Sannappa[198] showed enhanced grain yield of 25% in drilled finger millet by isoproturon over farmer's practice. Demonstrating the effect of organic matter along with herbicide on groundnut pod yield, higher pod yields were obtained with sole herbicides *viz.,* metachlor and pendimethalin. Pendimethalin treatment recorded higher groundnut pod yield than hand weeding as it could effectively control grass weeds[253]. Better yields of chick pea with herbicides were obtained[12]. Maximum net profit was obtained with fluchloralin in chick pea[240]. Nadannassababady and Kandasamy[163] reported highest seed cotton yield by glyphosate, which was comparable with weed free treatment which gave maximum net returns as also reported by Panwar *et al*[175]. Paraquat similarly gave highest seed cotton on par with glyphosate. Yaduraj and Devender Reddy[252] reported higher rice grain yield with cinosulfuron and butachlor compared to control. Compared to farmer's practice, rice grain yield was higher with 2, 4-D + butachlor in kharif followed by pretilachlor in rabi[252]. Lange *et al*[132] showed the improvements in yield, ranging from 13% in cotton, to 24% in rice by using herbicides in comparison with normal local weeding practices. The over all improvement in yields for seven different annual crops was 19.3%. The potential for yield improvement by the use of herbicides was probably more than 20% as estimated by Parker and Fryer[177]. While evaluating herbicides for weed control and economics in onion and cold desert region, Rameshwar *et al*[183] reported that pendimethalin at 1.5 kg ha^{-1} supplemented with one hand weeding gave highest net return 43.4% more than farmer's practice and a benefit : cost ratio of 54. The pursuit plus and alachlor resulted in significantly higher grain yield of soybean by 12.2% increase compared to weedy plot and therewas no toxicity carried into succeeding rabi season potato[187]. Moorty and Saha[186] reported that butachlor + safener, produced comparable grain yield in direct seeded puddle rice. Clodinafop and fenoxaprop – ethyl increased grain yield by 68.7. and 60.4% over control and 34.1 and 29.8% over two hand hoeings respesectively[246]. This also clearly demonstrates that herbicides could increase crop yields. Similar increase in wheat was reported with tralkoxydim (by <45%) followed by clodninfop (45%) and diclofop methyl (45%)

HERBICIDES IN THE IMPROVEMENT OF CROP YIELDS

Improvement in crop productivity so far possible could be conveniently ascribed to the following four areas:

(a) Herbicide use management
(b) Herbicide resistant crops
(c) Traditional plant breeding methods
(d) Biotechnological improvement

a. Herbicide Use Management

A given crop productivity obviously depends on, whether the given herbicide exhibits its full potential of killing weed flora. This, to a large extent depends on management of herbicide use, however effective and specific it is to the weed flora. Hence, management of herbicide use considers proper method of use of the herbicide cognisant to its physical, chemical properties and its relationship with function as well as plant parts to which it is applied.

Herbicide will be effective on the physiological and biochemical processes of plant only when herbicide is sufficiently absorbed and translocated (except for contact herbicides) to the site of action. Differential absorption and translocation, which forms the basis for herbicide selectivity, determine the tolerance and susceptibility of plant species to a particular herbicide.

Herbicides are applied either to the soil or plant foliage. Hence, herbicide absorption depends upon the method of application and the plant part. Herbicides enter the roots in both passive and active mechanisms and move with the water through out the plant in the apoplast system. Besides roots, the soil applied herbicides are also absorbed by the developing shoots and coleoptiles. Shoot entry is the primary site of entry for some herbicides. For example exposure of roots of barnyard grass.

Ehinochloa crusgalli to EPTC had little or no effect while exposure of the shoots results in severe injury[186]. Further, uptake of EPTC, diallate and propham through coleoptile was more toxic to oat seedlings than uptake through the roots[7]. Parkar[176] suggested that volatility might be a factor in shoot uptake of herbicides such as thiocarbamates and for lethal effect of trifluralin on *Setaria italica* and *Panicum miliaceum* when shoots but not roots were exposed to the herbicide. Vapours of trifluralin injured the shoots of maize drastically but not roots as found by Negi and Funderburk[167]. Triazines and urea herbicides kill the plants via the shoot. They allow plants to emerge and effect shoot growth by inhibiting Hill reaction of photosynthesis. Seeds play an important role in the uptake of herbicides.

The aqueous solutions of soil applied herbicides move into the seeds by diffusion. Phillips *et al*[179] showed that absorption of chloropropham, atrazine, linuron, chloramben and 2,4-D by seeds of 11 soybean strains differed in concentration and total quantity and that latter affected total oil content and percentage oil in seeds. Rubin and Eshel[193] found susceptible snapbean absorbed larger amounts of terbutryn and fluometuron from soil than the seeds of resistant cotton.

Thus root, shoot and seed are the major sites of uptake of soil applied herbicides and the combination of the three obvisouly serve effective absorption and killing of weed resulting in better crop productivity.

When herbicides are applied to soil, placement of the herbicides in the soil enhances the efficacy, selectivity and reproducibility. Surface or shallow application of herbicide ensures greater selectivity when shallow rooted weedsa are to be removed from the deep rooted crops. The herbicide treatment will be ineffective if herbicide is not concentrated in the zone in the upper 2-8 cm of soil where weeds grow[186].

When herbicides are applied post-emergence, the spray is directed at the weed plant and then most of the chemical reaching the foliage is absorbed. In such cases, the herbicide has to pass through a wax layer, the cuticle, which protects the inner cells. The cuticle is made up of cutin, cutin wax, pectin and cellulose. The structural properties of these compounds decide the rate and quantity of herbicide to be penetrated. For effective penetrance, of herbicide formulation should match the structural make up of the cuticle. Thus, foliage applied herbicides exist in polar form or non-polar form which are electrically positive as well as negative. For example, 2,4-D salts, PCP salts, TCA, thiocyanates *etc* are soluble in polar solvents and insoluble in non-polar solvents. In contrast non-polar ions, oils, waxes, 2,4-D acid, 2,4-D esters and most organic substances dissolve in non-polar solvents such as oil, but not in water. Foliage applied herbicides enter the living cells either by a lipoidal route or an aqueous route or both. Herbicides such as dinitroaniline phenoxy compounds penetrate the cuticle by the lipodial route in the non-polar undissociated form. Polar herbicides take aqueous pathway and penetration is greatly assisted by humidity and saturated atmosphere around the leaves. Water, polar solutes and polar herbicides readily penetrate the protein and cellulose portion of the cuticle of cell wall.

Trichomes and leaf hairs contain living protoplasm provide micro climate and alter the drying time of aqueous sprays and hence improves penetration of herbicide.

The leaf stomata, allow gaseous exchange when they are open but acqueous solutions do not penetrate stomata. Hence, addition of a wetting agent which lower surface tension is required to improve spreading of a herbicide solution on the leaf surface, along the inter cellular spaces, enhance the wetting of the walls of the mesophyll and bundle sheath cells and increase penetration between wax particles of the cuticle. The greater permeability of the lower leaf surface is due to greater number of stomata, trichomes, morphology and chemistry of the cuticle and amount of waxes present - all together influence the herbicide penetration.

The stems of plants though offer limited target area, have an extensive and highly developed transport system. The degree of penetration through stem depends on the growth and stage of the development of the plant. Penetration through the bark of woody plants is very difficult but it is normally done through mechanical cuts or by injections.

The herbicide once absorbed into the plant system moves either apoplastically or symplastically. Apoplastic movement takes place predominantly through xylem and inter cellular spaces. The symplast refers to the system of interconnection from cell to cell by means of plasmodesmata, excluding vacuoles. Any chemical, which penerates the symplast, first penerates the apoplast.

Herbicide that enters the symplast migrate to the phloem and translocate into the lumina of the sieve tubes in the assimilation stream[54]. Crafts and Yamaguchi[56] reported that compounds such as 2,4-D get accumulated in living cells during migration across the mesophyll along the sieve tube conduits. Others such as dalapon 2,3,6-TIBA, picloram, amitrole *etc.* are not accumulated but move freely across the mesophyll along the phloem and into the living growing tips. All herbicides can enter the symplast and are potentially phloem transported. Some are predominantly phloem-mobile, while others are almost exclusively xylem mobile. Maleic hydrazide and amitrole transfer from the phloem to the xylem before being circulated in plants. Glyphosate is translocated apoplastically following phloem translocation out of a leaf. The root applied picloram first moves to the shoots in the xylem and is then translocated into phloem before accumulating in young leaves. Imazaquin following root application and translocation to the shoot in the xylem is translocated into phloem of *Xanthium strumarium*[94]. Dicamba is continously remobilized, first accumulating in young leaves and translocated when they have expanded. Thus, most herbicides are ambimobile – transported in either system, depending on conditions within the plant. Some of ambimobile herbicides are first phloem mobile but its metabolites may be xylem mobile and translocated apoplastically. Hence, a herbicide translocation is phloem, depends on the production, remobilization and loading of assimilates into phloem translocation from the source to sink and unloading at the sink[74]. So any herbicidal action that interferes with these processes can disrupt phloem transport of assimilates, hence translocation of the herbicides. A herbicide applied in mixture can reduce the translocation of another herbicide than that applied alone; contrarily, translocation of one herbicide may stimulate the mobility of another herbicide.

Thus, efficacy of a herbicide could be improved if herbicide use is properly managed keeping in view of the chemical structure, function of the herbicide molecule and composition of the plant cell surfaces.

The modified herbicide usage includes rotation of herbicides with different modes of action and use of combination of herbicides with different modes of action. Herbicide rotation makes weeds exposed to different classes of herbicides. This decreases the frequency of resistance alleles. The rotation of herbicides to be effective, it must involve alternate generations of the weed species and cost of the rotational compound[98] Gould[98] suggested difficulty to implement herbicide rotations.

Herbicide mixtures would substantially delay or preclude evolution of resistance to the more vulnerable or at risk herbicides. Herbicide mixtures are of two types:

Broad spectrum mixtures (BSM)-each member of the mixture affects different weed spectra. But increases more weed control and more open niche for resistant biotypes[250]. The second type is target weed mixtures (TWM) : Both herbicides aim to kill same weed species and have same persistence[250].

b. Herbicide Resistant Crop

Improvement in crops productivity could also be attributed substantially in some crops due to herbicide resistant crops. Development of herbicide resistant crops (HRC) provides efficient, safe and economical production of crops[34, 83]. They lead to use less persistant herbicides in the environment, less expensive production costs, increased production options to growers by providing more weed control strategies.

S-triazine herbicides resistance (psb A gene Mutant) has been used to incorporate into several crops *i.e.* from weed birds rape (*B compestris*)[24] to rutabaga and rape seed[23] and from weedy green foxtail (*S. viridis*) to foxtail millet (*S. viridis*)[61]. By using tissue culture selection, resistant tobacco (mutant D-1) and potato (mutant D-1) were produced[214]. Protoplast fusion transfer of the psb A gene to potato was the source for resistance[186]. Similarly, S-triazine resistance in rape seed and in several other *Brassica* sp including broccoli (*B. oleraceae var italica)*were used as source of resistance. Three cvs, OAC Triton, OAC Tribute and OAC Triumph canola cultivars resistant to S-triazine were released. The sulfonyl ureas, imidazolinones resistant crops have been produced. Because of the plasticity of the enzyme, selection at the seed or whole plant[202] organ[186] and tissue or cellular level[105] has been a very simple and successful strategy to produce such crops. Sulfoxylurea herbicide resistance from a resistant weed Prickly lettuce, has been transferred to lettuce crop by conventional breeding methods[149]. Mutant selection has been created sulfonyl urea herbicide resistant lines of barley, tobacco[44], canola[236], sugarbeet[105, 199], soybean[202], rice[231] as well as some horticultural crops. The imidazolinone resistant crops are reviewed previously,[169,206]. Imidazolinine resistant maize was produced from tissue culture[169] and by pollen mutagenesis[206] . Imidazolinone resistant wheat[169] and canola[230] were also developed but the yield was same as that of susceptible varieties. Crop resistance to acetyl – CoA carboxylase inhibitors, which include aryloxy phenoxypropionates and cyclohexanediones is due to insensitive ACCase[39, 75]. Wheat is resistant to these herbicides as it metabolically degrade them rapidly[212]. Red fescue is naturally resistant by virtue of a herbicide resistant ACCase[227]. Maize resistant to both aryloxyphenoxy propionates and cyclohexanediones has been produced by selection from mutations conferring resistance in tissue culture and then regenerating the plant[150, 219]. With the same approach herbicide resistant wheat and Kentucky bluegrass were developed[219]. Several types of mutations varying degrees of cross resistance to ACCase inhibitors were produced. Maize mutants Acc 1-S1, Acc 1-S2 and Acc-S3 are resistant to only haloxyfop and Acc 1-H2 mutant is very resistant to haloxyfop, but partially resistant to sethoxydim[150]. Herbicide treatment had no adverse effects on grain yield or quality. The germ plasm for resistance to ACCase inhibiting herbicides was transferred to commercial maize breeding companies in 1990.

Production of glyphosate resistant crops was in focus in recent past. It is readily translocable to meristem, hence was difficult in developing glyphosate resistant crops. The resulting resistant lines from tissue or cell culture and those regenerated had more EPSPS resistance than the wild type due to gene amplification[95, 203]. Selection with glyphosate in cell cultures resulted in resistance due to a glyphosate resistant EPSPS. Forlani *et al*[186] selected maize cell cultures with glyphosate and produced a cell line with a glyphosate resistant EPSPS. Madsen and Jensen[145] found glyphosate resistant sugarbeets. Plant enzymes do not degrade glyphosate. The leading transgenic glyphosate resistant soybean line expresses the CP4 EPSPS[174] and there was no yield penalty. It conferred high level of glyphosate resistance.

Promoxynil resistant tobacco and cotton have been generated by transformation with a plasmid gene from the bacterium *Klebsiella ozaenae* that encodes a bromoxynil nitrilase enzyme[223]. The resultant plants were resistant upto 10 fold the recommended dose of bromoxynil. In cotton, the gene was introduced via *Agrobacterium*. Bromoxynil resistant cotton was the first transgenic herbicide resistant crop introduced into the market.

Cotton and tobacco have been made resistant to 2,4-D by genetic transformation with the tfdA gene[19, 142]. The resultant transformants were resistant to more than three times the highest recommended doses of 2, 4-D for wheat, corn, sorghum or pasture. In cotton this represents a 50 to 100 fold increase in tolerance to 2,4-D compared to untransformed controls[186]. In Australia, the 2,4-D tolerance gene is currently transferred to the best commercial cotton varieties released as 2,4-D drift tolerant cultivars[142]. To develop glufosinate resistant crops, two genes that encode enzymes that metabolically inactivate glufosinate were used to produce resistant plants.

Using bar gene from *Streptomyces hygroscopicus* and/or the pat (phosphino thricin-acetyl transferase) gene from *S. viridochromogenes* about 20 crops have been transformed to glufosinate resistant crops, including carrot[80], oats[220], beet[28], oilseed rape[67], barley[247], tomato[67], alfalfa[59], rice[63], potato[66], sorghum[42], wheat[242], tobacco[80], sugarbeet[58] and maize[133]. Sugarcane has been transformed with pat gene in callus culture, but no plant regenerated[47]. It was found apparently no crop yield or quality sacrifice for either of these genes. The herbicide use was substantially reduced by using glufosinate resistant crops[161]. Tobacco has been made resistant to cyanamide by transformation with a gene from the fungus *Myrothecium verrucaria* that encodes cyanamide hydratase[147]. This enzyme converts the herbicide to urea.

Dalapan resistant tobacco (*Nicotiana plumbaginifolia*) has been produced by transgenic methods[37].

Because herbicides are not commonly applied to susceptible crop species, there is a less opportunity for tolerant mutants of crops. Use of paraquat for nine successive years to control plants of *Lolium perenne* lead to find alleles of paraquat tolerance which contributed to develop paraquat tolerant lines of *Lolium perenne*[141]. Variation in tolerance of 2,4-D among cultivated stocks of turf grass, *Agrostis stolonifera* L. has been recognized[2]. Exploitable degrees of tolerance to 2,4-D were found in 12 winter wheat cultivars to chlorsulfuron and isoproturon[237]. Significant differences between 35 cultivars of *Lolium perenne* were found and concluded that dalapon had selective grass swords when sown with more tolerant cultivars of ryegrass. Fisher[91] found marked differences between 12 cultivars of *Festuce rubra* to paraquat and glyphosate tolerance. The hexaploid cultivars of *F.rubra* were relatively tolerant. The most tolerant cultivar to both herbcides was Dawson which was choosen for a selection on experimentation.

Necessity for herbicide resistant crops

The research conducted in recent years on S-triazine tolerance led to a much greater understanding of photosynthesis and its inhibition and of herbicide-binding sites in the thylakoid membranes of chlorplasts. The HRCs have a great potential as tools for future studies on herbicide mechanism of action, plant biochemical processes, screening for new herbicides and other research in theoretical genetics and applied breeding. Another potential benefit from herbicide resistant plants is in the possible transfer of resistance from one plant (*eg.* weed) to another plant (*eg.* crop)[132]. The herbicide resistant crops developed are furnished in Table: 3.

Table : 3. Herbicides resistant crops developed.

Herbicide group / herbicide	Resistant crops
ALS are inhibitors - sulfonylureas	Barely, cotton, maize, rapeseed, rice, soybean, gygarbeet, tobacco
Imidazolinenes accase inhibitors	Maize, rapeseed, wheat
Bipyricliliums	Potato, tobacco
DHPS inhibitors	Tobacco
Penoxycarboxylic acids	Cotton, tobacco
Triazenes	Foxtailmillet, potato, rapeseed, tobacco
Bromaxyril	Cotton, tobacco
Cyanamide, dalapon	Tobacco
Glufosinate	Barely, sugarbeet, carrot, maize, oats potato, rapeseed, sorghum, tobacco, tomato, wheat

The HRCs developed so far through different means that lead to the improvement crop yield are reviewed here under. The mechanisms of herbicide resistantance include: Exclusionary resistance mechanisms, herbicide uptake, translocation[75], compartmentation[225], metabolic detoxification, sequestration of herbicides in vacuoles[53] - such as oxidation[214], reduction[131], hydrolysis and conjugation[108] of herbicides to glucose or glutathion[108].

c. Traditional Plant Breeding Methods

The necessity of developing herbicide resistant crop (HRC) plants has been prompted from the natural development of herbicide resistance in weed species over-time. The development of HRC plants need considerable variability in herbicide response among the crop plants /weed species. In many cases sufficient variability in herbicide response among the plant population exists from which improved lines could be selected[26, 71, 106, 238] but in other instances the amount of variability in herbicide resistance has been insufficient[126].

Hence, HRCs were developed with traditional plant breeding approaches[21, 22, 216, 217, 218, 241]. Direct selection of variants within a species with enhanced herbicide resistance has been successful in many crops[88, 122]. Traditional plant breeding methods were used the existing herbicide resistance was derived from weedy sources and produced HRCs with resistance to S-triazines in rapeseed[23] and Lettuce[149]. Triazine resistance under cytoplasmic control, discovered in wild genetype of *Brassica compestris*[134] has been transferred into cultivated varieties of *B.compestris* (Polish rape) and *B. napus* (seed rape) by back crossing. Plants selected out of more tolerant diploid cultivars were used in the development of Rathlin, a cultivar of *Lolium perenne* bred for high dalapon tolerance.

However, traditional methodologies have their intrinsic drawbacks like, cumbersome, long time taking process than the patent life of a herbicide and mostly meager natural resistance resources availability in crop plants. Given these drawbacks, the rate of development process of HRCs in crop plants by traditional methods thus set a slow pace with limited spread, even though the need of HRCs is inevitable in augumenting productivity of crop plants in the present era of intensive agriculture and diversified human resource and social aptitude.

d. Biotechnological Improvement

Eventhough there are substantial evidences that crop yields could be increased in multitude on both quantitatively and qualitatively, presently there is a live debate going on across farming world to accept biotechnological means to improve crop yields. Improvement of herbicide-resistant crops through biotechnological means offer a promise for speedy development even with meager amount of available source of resistance, either in plants, weed species or micro organisms.

In vivo or *in vitro* selection and subsequent crop transformation permit shorter time to achieve such a development.

Since the use of HRCs reduces the cost and quantity of herbicide to be used, pollution and other hazards, HRCs are in great demand globally[6]. In 2000 herbicide tolerance deployed in soybean, corn and cotton occupied 74% of the 44.2 m ha, with Bt insect resistant crops taking up 19% and stacked genes for herbicide tolerance and insect resistance in cotton and corn accounting for 7%. Use of genetically modified HRCs in case of canola reduced the application of 6000 tons of herbicide in canola alone besides direct green house gas savings of reduced field operations and reduced tillage helping to conserve soil structure[6].

BIOTECHNOLOGICAL MANIPULATIONS FOR DEVELOPMENT OF HERBICIDE RESISTANT CROPS (HRC'S)

a. Development of transgenic plants
   (i) Identification of inter-specific sources
   (ii) Identification of intra-specific sources

b. Induction of herbicide resistance
   (i) Plant sources of resistance
   (ii) Microbial sources of resistance

Today there are some 800 million people (18% of the population in the developing world) who do not have access to sufficient food to meet their needs. Malnutrition plays a significant role in half of the nearly 12 million deaths of children under five in developing countries each year. In addition to lack of food, deficiencies in micro-nutrients (especially vitmin A, iodine and iron) are wide spread. Achieving the minimum necessary growth in total production of global staple crops maize, rice, wheat, cassava, yams, sorghum, potatoes and sweet potatoes, without further increasing land under cultivation will require substantial increase in yields per acre. Increase in production is also needed for other crops, such as legumes, millet, cotton, grape, bananas and plantains. In this context, biotechnological manipulations to develop transgenic HRC are more relevant and promising. Despite past successes, the rate of increase of food crop production has been decreased from 3% per annum in 1970s to approx. 1% per annum in 1990s[186]. There are still heavy losses of crops owing to biotic (*eg.* Pests and diseases) and abiotic stresses. Transgenic plants with important traits such as pest and herbicide resistance are most necessary where no inherent resistance has been demonstrated with in the local species. The benefits from transgenic plants include increased flexibility in crop management, decreased dependency on chemical insecticides enhanced yields and harvesting higher proportions of the crop available for trading. Even when some genes for herbicide resistance are useful in different regions, they will be introduced into locally adapted cultivars by biotechnological means. For example, transgenic crops containing insect resistance genes from *Bacillus thuringiensis* could reduce significant amount of insecticide applied on cotton in US. It could reduce 5 million acre-treatment or about one million kg of chemical insecticides in 1999 compared with 1998.

Development of Transgenic Plants

Some of the existing HRCs were developed by biotechnological methods. Development of HRCs through biotechnological techniques though promises speedy developments, the techniques invoving *in vivo* and *in vitro* selection and subsequent crop tranformation further shorten the time to achieve this type of crop improvement. These methodologies include:

1. Cell and tissue culture selection; 2 Hybridization; 3. Microspore (gametophytic) and seed mutagenesis; 4. Plant transformation.

1. Cell and Tissue culture selection

By this technique, somaclonal variation in cultured plant cells has been exploited to select herbicide resistance traits for crop improvement. This provided one of the most important selection technique for the development of HRCs[43, 44, 45]. Callus for maize[4], microspores and protoplasts for rapeseed[230] and plant cell suspension cultures for maize were used for selecting herbicide resistance. These techniques were used to develop HRCs for those herbicides inhibiting the ALS site of action[168, 169], but not with complete crop safety in some cases[17].

2. Hybridization

The herbicide resistance from weedy relatives has been transferred to crops to develop HRCs. For example, protoplast fusion techniques have been used to incorporate S-triazine resistance between Solanum species[11] and into cytoplasmic male sterile Brassica variatnts. S- triazine resistance was transferred from weedy birds rape (*Brassica compestris*)[24] to several Brassica crop species, including rapeseed and rutabaga[23] and from the weedy green foxtail (*Setaria viridis* sub spp *viridis Briquet*) to the foxtail millet crop (*S. viridis*, sub spp *italica Briquet*)[61]. Sulfonylurea – resistant prickly lettuce (*Lactuca serriola*) was used as the source of ALS inhibitor resistance into a new cultivar of lettuce, ID-BRI[149].

3. Microspore (Gametophytic) and seed mutagenesis

Mutagenesis forms one of the most powerful means of deriving novel herbicide resistant variants for HRCs. Christianson[49] reviewed several of these techniques and approaches. Sebastian *et al*[202] used mass selection of mutagenized soybean seed to find herbicide resistant crop variants. Dyer *et al*[86] used similar approaches in other crops. Microspore mutagenesis was employed in developing herbicide tolerance in rapeseed[230].

4. Plant transformation

Several techniques were used to transfer genes of herbicide tolerance by different researchers. The transgenics were developed aiming target site and metabolic detoxification resistance mechanisms. The resistant transgenic tobacco plant lines[46] were developed against S-triazine by constructing with mutant resistance coding sequence, expression level control sequence and transit peptide encoding sequence. Glyphosate resistant crop development was developed using a mutant *E.colgene* fused to a EPSPS enzyme chloroplast transit sequence[72]. Crops like cotton, rapeseed, soybean, tobacco and tomato resistant to glyphosate were developed for commercial usage[86]. Crops resistant to ALS – inhibiting herbicides were developed by transfer of resistant genes between different higher plant species[112, 155]. Some crop plants are also developed for herbicide resistance by exploiting metabolic detoxification resistance from microbial speicies. For example the bromoxynil specific nitrilase gene, encoded by the bxn gene has been transferred into cotton, potato, tomato and rapeseed[86]. The Cyanamid hydratase encoding gene from the soil fungus *Myrothecium verrucaria* has conferred resistance in the transgenic tobacco[147]. Detoxification of glufosinate by acetylation is achieved by acetyl transferase encoded by the bar gene from streptomyces. This gene is fused to high expression promoters and used to produce high levels of glufosinate resistance in transformed alfalfa, poplar, rapeseed, potato, sugarbeet, tobacco and tomato crops[66]. Genes for herbicide resistance were also used as selectable marker in transformation studies[254].

(i) Identification of Inter-specific Sources

Development of HRCs using biotechnological techniques provides more ways for crop and cultivar improvement by inter specific sources or intergeneric or even distant related resources. These techniques have been reviewed[28, 99, 100, 101, 102]. Herbicide resistant mutants probably occur in populations of all plant species but frequency of their occurrence is unknown[248]. Since the enzyme that these herbicides inhibit is common to all weeds as well as crop plants, these herbicides can not be sprayed for postemergence purpose. So to use these broad spectrum herbicides selectively, herbicide resistance has to be engineered for two strageties.

1. Decreasing the sensitivity of plant to herbicide by overproduction of the target enzyme or by expression of modified target that is insensitive to the herbicide but retains enzymatic activity
2. Providing an herbicide detoxifying pathway

Examples of the first approach are engineered resistance to glyphosate and sulfonylurea compounds for inhibitor enzyme 5- Enolpyruvyl Shikimate-3- phosphate synthase (EPSPS) and acetolactate synthase[158].

Introduction of mutant EPSPS genes to encode enzymes less sensitive to glyphosate[50,89,115], or the over production of plant EPSPS[203] conferred tolerance to glyphosate in the transgenic plants. Similarly, transgenic tobbaco plants expressing mutants ALS genes were found tolerant to sulphonylurea compounds[112,137]. High levels of resistance for several herbicides were obtained with engineering for a novel detoxification pathway in plants. The herbicide detoxifying enzymes are identified in several plant species and in microorganisms. Bialaphos produced by *streptomyces hydroscopicus* is a peptide consisting of PPT (a herbicide glufosinate or 1-phosphinothricin) is commercilised as Basta (by Hoechst) and Herbiace (by Neiji Seika Ltd). A PPT – resistant gene, bar has been isolated from *S. hygroscopicus*[186] and was shown to encode the enzyme phosphinothricin acetyl transferase (PAT) which specifically converts PPT to an acetylated, non-herbicidal form[235]. The bar gene under the control of the cauliflower mosaic virus (CaMV) 35 S promoter was introduced into tobacco, tomato, potato, oilseed rape, alfalfa, sugarbeet, aspen and poplar plants via the Agrobacterium transformation system[28, 59, 65, 66, 67]. In all the cases, the bar gene conferred resistance on green house – grown transgenic plants to doses of glufosinate and bialaphos much higher than that used in agriculture. Bacterial genes encoding bromoxynil and 2,4-dichlorophenoxy acetic acid (2,4-D) detoxifying enzymes were likewise engineered in tobacco plants and conferred high levels of herbicide resistance[142, 223, 229].

(ii) Identification of Intra-specific Sources (for herbicide tolerant genes)

The main resource for obtaining intra-specific herbicide tolerance could be traditional plant breeding techniques and natural mutations or point mutations few HRCs are developed by using traditional plant breeding approaches[21, 22, 216, 217, 218, 241]. Direct herbicide selection of variants within a species with enhanced resistance has been successful in many crops[88, 122].

Various reports suggested that sufficient variability in herbicide response among plant populations exist to enable selection of improved lines[26, 71, 106, 238]. Crop plants are inherently resistant to many herbicides. So improvements can be made with proper selection efforts in many instances. The first recorded evidence of intra-specific resistance to the S-triazine herbicides was in 1970, when Ryan[194] reported that *Senecio vulgaris* could no longer be controlled in a Western Washington nursery after several years of satifactory results with simazine. Similar reports were obtained within the following few years that *Amaranthus* spp and *Chenopodium album*, first susceptible to triazinc were found resistant to it later.

Triazine resistance occurring under cytoplasmic control in a wild, *Brassica comprestris* L., has been transferred into cultivated varieties *B. Campestris* (Polish rape) and *B. napus* (seed rape) by back crossing[221]. De Gournay *et al*[68] considered the possibility of existence of herbicide tolerance in primitive wheats and related species. Since the primitive and wild varities are not broadly adapted, their isolated genes of herbicide tolerance could be used as a last resort in crops such as cereals.

b. Induction of Herbicide Resistance

(i) Plant sources of resistance

Where there is no possibility of resources for native herbicide genes, they could be induced *in situ* in crops by using laboratory techniques such as tissue culture, followed by hybridization of induced cell lines for somaclonal variation by adopting protoplasmic fusion techniques *etc*. Somaclonal variation in cultured plant cells has been exploited to select herbicide resistance traits for crop improvement. Plant cell and tissue culture has provided one of the most important selection techniques for the development of HRCs[43, 44, 45, 148].

Herbicide resistance was induced by using callus of maize[4] and microspores and protoplasts for rapeseed[230] and plant cell suspension cultures of maize. Herbicide resistance by inhibiting the ALS site of action has been developed using these techniques[168, 169]. A tolerant line of *Poa annua* has arisen under unusually strong selective pressure as a result of continued use of paraquat. The normal *P. annua* is killed by 0.1 to 0.2 kg/ha paraquat but more than 0.8 kg/ha was required to kill the tolerant strain[114] and perennial ryegrass [141].

Thomas and Pratt[232] isolated a paraquat tolerant mutant of tomato using an inter specific hybrid strain called L2. The new callus cultures initiated from the regenerated plants typically had at least a 30 fold increase in paraquat tolerance than that of normal callus of strain L2. Since most of the agronomically important crops were not amendable to transformation, the resistance to suflonyl ureas and other ALS inhibitors was generated using alternative methods. Fertile plants exhibiting greater than 100 fold increase in resistance to imidazolizonones and cross resistance to the sulfonylureas were regenerated from one line. Homozygous progeny showed more than 300 fold increase in resistance to herbicides. Thus, highly resistant plant might be achieved through induction and inbreeding to produce homozygous tolerant genotypes through hybridization to combine different tolerance alleles in a single strain, or through isolation of new and better mutants.

Thus, where there are no resistant genes for a particular herbicide, using a hetergenic population of crop plants, native genes could be forced to express for the tolerance. Such genes after isolation could be transferred into many crop plants.

(ii) Microbial sources of resistance

As a non plant alternative strategy for achieving herbicide resistance, where native plant genes are not available, a microbial model has been proved useful. In *Streptomyces griseolus*, sulfonylurea herbicides such as chlorsulfuron, sulfometuron methyl and chlorimuron ethyl are metabolized by a three step process requiring an NADP dependent **reductase and either of the two distinct, inducible cytochrome P-450 enzymes[172, 191].**

This bacterium could serve as a source for the isolation of genes encoding the metabolically active proteins. These genes could be introduced into plants by transformation to confer herbicide resistance. Another example is the expression of microbial genes encoding herbicide resistant ALS in plants. Alternatively, the bacterial genes might serve as probes for the isolation of plant genes encoding detoxifying enzymes. Certain crop plants are transformed by transferring resistant genes from various sources using several techniques as given below .

Herbicide	Source of resistant gene	Mode of resistance	Transfer in plants
Glyphosate	*Escherichia coli Agrobacterium lumefaciens*	Altered EPSPs over expression	Tobacco
Glufosinate	*Streptomyces hygroscopicus*	Expression of PAT	Soybean, rapeseed
Chlorsulfuron	*Nicotiana tabacum*	ALS mutated gene	Tobacco, tomato, potato, maize, soybean, wheat
Bromoxynil	*Klebiella ozene*	Nitrilase	Tobacco, rice
Norflurazon	*Erwinia ureadovora*	Enhanced carotenoid Biosynthesis	Cotton, clover Rapeseed
Dalapon	*Pseudomonas putid*	Dehalogenase	Tobacco
2, 4-D	*Alcaligene* eutrophus	Mono oxygenase	Tobacco, cotton
Phenaedipham	*Arthrobactor oxidans*	Carbomate hydroxylase	Tobacco

CONCLUSIONS

The analogy of the forerun paras evidently reveal that the sound built up of a herbicide needs the appropriate chemical architect and biochemical link up with physiological processes of a weed plant for its efficacy of killing. It also elucidates the irony of physiological and biochemical processes of a tolerant crop plant in degrading or metabolizing the toxic herbicide into benevolent ions or metabolites. The molecular search paves the way for rebuilding new generation plants for target processes and these efforts culminated in augmenting crop productivity, a potential already existing in the present day crop plants. The herbicides have become inevitable chemical tools for enhancing agriculture productivity in this stride fast era of intensive agriculture.

REFERENCES

Abell L M (1996) Biochemical approaches to herbicide discovery: Advances in enzyme target identification and inhibitor design. *Weed Sci*. **44**: 734-742.

Albrecht H R (1947) Strain differences in tolerance to 2, 4-D in creeping bent grasses. *J. Amer. Soc. Agron.* **39**: 163.

Amen R D (1968) A model of seed dormancy. *Bot. Rev.* **34**: 1-31.

Anderson P C and Georgeson M (1989) Herbicide tolerant mutants of corn Genome **31**: 994-999.

Anonymous (1994) *Herbicide Hand Book*. Weed Sci. Soc. of America, Lawrence, Champaign I L, USA 352.

Anonymous (2002) Biotechnology Global update, 2002 (Oct) published by Monsanto.

Appleby A P, W R Furtick and S C Fang (1965). Soil placement studies with EPTC and other carbamate herbicides on *Avena sativa*. *Weed Res.* **5**: 115-122.

Arnold W F and J D Nalwaja 1971. Effect of dicamba on RNA and protein *Weed Sci.* **19**: 301-305.

Arya M P S and R V Singh (1998). Direct and residual effect of oxadiazon and oxyflurrefen herbicides on the control of Oxalis Latiofolia in Soybean. *Indian Journal of Weed Sci.* **30(1 & 2)** : 36-38.

Ashton F M, D Penner and S Hoffman 1968. Effect of several herbicides on proteolytic activity of squash cotyledons *Weed Sci.* **16**: 169-171.

Austin S and Helgeson J P (1987) Interspecific somatic fusions between *Solanum brevidens* and *S.tuberosum*. In "Plant Molecular Biology" (D Von Wettstein Barsby, T L Kemble, R J and Yarrow (1987) Brassica cybrids and their utility in plant breeding *Plant Molecular Biology* **140**: 223-234.

Balyan R S, R S Malik and R P S Vedwan and V M Bhan (1987). Chemical Weed Control in Chickpea (CIcer arietinum). *Trop. Pest Management* **33**: 16-18.

Bartels P G and A Hyde 1970. Buoyant density studies of chloroplast and nuclear deoxyribonucleic acid from control and 3-amino-1,2,4-tria zole – treated wheat seedlings. *Plant Physiol*. **45**: 825-830.

Bartels P.G. 1965. Effect of amitrole on the ultra structure of plastids in seedlings. *Plant Cell Physiology* – **6**: 227-230.

Baskin A D (1953) Vapor hazard of 2,4-D esters. *Agr Chem* **8(8)**: 46-48

Baskin A D and E A Walker (1953). The responses of tomato plants to vapors of 2,4-D and/or 2,4,5-T formulations at normal and higher temperatures. *Weeds* **2(4)**: 280-287.

Bauman T T, Owen M D K and Liebl R A (1992) – Evaluation of Garst 85321 corn for herbicide tolerance. *Weed Socieity Am.Abstr*. 32.15.

Baur J R and J J Bowman, 1972. Effect of 4-amino 3,5,6-teichloropicolsic acid on protein systhesis. *Physiol. Plant* **27**: 354-359.

Bayley C, Trolinder N, Ray C Morgan, M, Quisenberry J E and Ow D W (1992). Engineering 2, 4-D resistance into cotton. *Theor. Appl. Genet.* **83**: 645-649.

Bethlenfalvay G and R N Norris (1975) Phytotoxic action of desmedipham : Influence of temperature and light intensity *Weed Sci* **23**: 499-503.

Beversdorf W D (1987) Classical approaches to the development of herbicide tolerance in crop cultivars *Amer Chem Soc.* Symp Ser **334**: 108-114.

Beversdorf W D and Kott K S (1987) Development of triazine resistant crops by classical plant breeding. *Weed Sci.* **35** (suppl 1): 9-11.

Beversfdorf W D and Hume D J (1984) OAC Triton spring rapeseed *Can J Plant Sci.* **64**: 1007-1009.

Beversodorf W D, Wein herman J, Erickson L R and Souza Machado V (1980), Transfer of cytoplastiyccallus inherited triazine resistance from birds rape to cultivated oil seed rape (*Brassica campestris and B napus*) *Can J. Genet. Cytol.* **22**: 167-172.

Bishop N T 1962. Inhibition of oxygen evolving system of photosynthesis by amino-triazines Biochem. *Biophys. Acta* **57**: 186-189.

Boerboom J, Ehlke N J, Wyse D L and Somess D A 1991. Recurrent selection for glyphosate tolerance in birdsfoot trefoil. *Crop Sci* **31**: 1124-1129.

Boldt P F and A R Putnam 1980. Selectivity mechanisms for foliar applications of diclofop methyl . I. Rentention, absorption, translocation and volatility. *Weed Sci* **28**: 474-477.

Botterman J and Leemans J, 1988. Engineering herbicide resistance into plants. *Trends genet.* **4**: 219-222.

Bowler C M, V Montagu and D Inze 1992. Superoxide bismutase and stress tolerance. Ann-Rev. *Plant Physiol. Biology* **43**: 83-116.

Brar S S, Sajeev Kumar, L S Brar and S S Walia (1998). Effect of crop residue management systems on the grain yield and efficacy of herbicides in rice-wheat sequence. *Ind. J. on Weed Sci.* **30(1&2)**: 36-38.

Brewer P E, C J Arntzen and F W Slife 1977. Effect of atrazine, cyanazine and procyanazine on the photo-chemical reactions of isolated chloroplasts. *Weed Sci.* **27**: 300-308.

Bridges D C (1994) Impact of Weeds in human endeavours: *Weed Tech.* **8**: 392-395

Briggs B A (1978) Manipulation of Herbicides and effect of herbicides on rooting. In combined *Proc. Inter. Ph. Propagator's Soc.* 1977, **27**: 426-467.

Bright S W J (1992) "Herbicide resistant crops Biosynthesis and molecular regulations of aminoacids in plants" (*eds*) B R Singh, H E Flores and J C Shannon pp 184-194. *Amer. Soc. Plant Physiol.* Rockville M D.

Brillie A M R, Rossnagel B G and Kartha K K (1993). *in* vitro selection for improved chlorsulfuron tolerance in barley (*Hordeum vulgare L*). *Euphyticek* **67**: 151-154.

Brittion, G P Barry and A J Young, (1989). Carotenoids and Chlorophylls Herbicidal inhibition of pigment biosynthesis. In A D Dodge (ed) *Herbicides and Plant Metabolism* Cambridge University Press, Cambridge, 51-7.

Buchaman – Wollaston V Snape A and Cannon F (1992). A plant selectable marker based on detoxification of herbicide Dalapon. *Plant Cell Rep* **11**: 627-631.

Burgstahler R J and H K Litchthenthaler (1984). Inhibition by sethoxydin of phosphor and galactolipid accumulation in maize seedlings. In P.A. Siegenthaser an dW. Eichenberger (*eds*) structure, function and metabolism of plant lipids. Elsevier, Amsterdam 619-622.

Burton J D, Gronwald J W, Somers D A, Connelly J A, Gengenbach B G and Wyse D L (1987). Inhibition of acetyl CoA carboxylase by the herbicides sethoxydim and haloxybop. *Biochem. Biophys. Res. Commun,* **148**: 1039-1044.

Burton J D, J.W. Gronwald, D.A. Somers B.G. Gengenbach and D L. Wyse (1989). Inhibition of acetyl CoA carboxylase by cyclohexanedione and aryloxyphenoxy propianate herbicide. *Pestic Biochem Physio.* **34**: 76-85.

Cardenas J F W, Slife, J.B. Hanron and J Butler (1967). Physiological changes accompanying the death of cocklebur plants treated with 2,4-D. *Weed Sci* **16**: 96-100.

Casas A M, Kononoviicz A K, Zehr V B, Tomes D T, Axtell J D, Butler L G, Bressan R A and Hasegawa P M (1993). Transgenic sorghum plants via microprojectile bombardment. *Proc. National Acade. Sci.* **90**: 11212-11216.

Chaleff R S (1988). *Herbicide resistant plants from cultured cells.* In "Applications of Plant cell Tissue Culture" pp 3-20. Ciba Foundations symposium 137 – Wiley.

Chaleff R S and T B Ray (1984) Herbicide resistant mutants from tobacco cell cultures. *Sci.* **224**: 1143-1145.

Chaleff R S and T B Ray (1984). Herbicide – resistant mutants from tobacco cell cultures. *Sci.* **223**:1148-1151.

Cheung A Y, Bogorad L, Van Montague M and Schell J (1988). Relocating a gene for herbicide tolerance. A chloroplast gene is converted into a nuclear gene Rve. *Nathl Acad.Sci.*USA **85**: 391-395.

Chowdhury M K U and Vasil I K (1992) Stably transformed herbicide resistant callus of sugarcane via micro projectile bombardment of cell suspension cultures and electroporation of protoplasts. *Plant Cell Rep* **11**: 494-498.

Chrispeels M J and J B Hanson (1962). The increase in ribonucleic acid content of cytoplasmic particulates of soybean hypocotyls induced by 2,4 dichlorophenoxyacetic acid. *Weeds* **10**: 123-125.

Christianson M L (1991) Fun with mutants. Applying genetic methods to problems of Weed Physiology: *Weed Sci.* **39**: 489-496.

Comai L Facciolli D, Hiatt W R, Thompson Y, Rose R E and Stalker D M (1985) Expression in plants of mutant A gene from Salmonella thyphimurium confers tolerance to glyphosate: *Nature* **317**: 741-744.

Copping L G and D E Davis (1972) Effects of atrazine on chlorophyll retention in corn leaf discs. *Weed Sci.* **20**: 86-89.

Corio-Costet M F J.Lherminer and R S Scalla 1991. Effects of isoxaben on sensitive and tolerant cell cultures. II. Cellulose alterations and inhibition of synthesis of acid – insoluble cell wall material. *Pestic Biochem. Physiol.* **40**: 255-265.

Coupland D (1991) The Role of compartmentation of Herbicides and their metabolites in Resistance Mechanisms In J C Caseley, G W Cussans and R K Atkins (eds), *Herbicide Resistance in weeds and crops*. Butterworth, Heinermann Ltd Oxford 263-278.

Crafts A S and C E Crisp (1971) *Phloem Transport in plants* W H Freeman and Co., San Francisco, 481.

Crafts A S and Rabbison W W (1973) Properties and Functions of Herbicides, In *"A text book and Manual of Weed Control"*. Tata McGraw Hill Publishing Company Ltd, New Delhi.

Crafts A S and S Yamaguchi 1964. The autoradiography of plant materials Agr. Extension Serv. Manual 35, Univ. of California, Berkely, California, 143.

Cutter E G , F M Ashton and D Huffstutter (1968). The effects of bensulide on the growth, morphology and anatomy of oat roots. *Weed Res* **8**: 346-352.

D'Halluin K, Bossul M, Bonne E, Mazur. B, Leemans J and Botterman J (1992). Transformation of sugarbeet (*Betavulgarii L*) and evaluation of herbicide resistance on transgenic plants. *Biotechnology* **10**: 309-314.

D'Halluin K, Botterman J and De greef W (1990) Engineering of herbicide resistant alfalfa and evaluation under field conditions. *Crop Sci.* **30**: 866-871.

Daniel Mc, J L and R E Frans 1969. Soybean mitochondrial response to prometryn and fluometeson. *Weed Sci.* **17**: 192-196.

Darmancy H and Pernes J (1989) Agronomic performance of a triazine resistant foxtail millet (*Setaria italica*, L Bauv), *Weed Res.* **29**: 147-150.

Darmancy H J P Compoint and J Gasquez (1981) La resistance auxtrianes chez *Polyqonlum lapathipolium L Acad Agric France* 231.

Datta S K, Datta K, Soltanifar N, Donn G and Potrykus I. (1992) Herbicide resistant indica rice plants from IRRI breeding line IR72 after PEG mediated transformation of protoplasts. *Plant Mol.Biol.* **20**: 619-629.

Davies A O and J L Harwood (1983) Effect of substituted pyridazinones on chloroplast structure and lipid metabolism in greening barley leaves. *J.Exp. Bot* **34**: 1089-1100.

De Block (1990) Factors influencing the tissue culture and the *Agrobacterium tumefaciens* mediated transformation of hybrid aspen and poplar clones. *Plant Physiol.* **93**: 1110-1116.

De Block M, Botterman J, Vandewiele M, Dockx J, Thoen C, Gossele V, Movva N Thompson V C, Van Montagu M and Leeman J (1987). Engineering herbicide resistance in plants by expression of detoxifying enzyme. *EMBOJ* **6**: 2513- 2518.

De Block M, De Brouwer D and Tenning P (1989) Transformation of *Brassica napus* and *Brassica oleracea* using *Agrobacterium tumefaciens* and the transgenic plants. *Plant physiology* **91**: 614-701.

De gournay, J Koller and J L Dufour (1975) Defficultes rencontrees dans la prospection de geniteurs pour la resistance a un-herbicide cas de L' atrazine at du ble d' hiver. Proc. Eur. *Weed Res. Soc. Symp. Status Biol. Control.* Gressweeds Eur., Paris, 388.

Deal L M and F D Hess 1980. An analysis of the growth inhibitory characteristics of alachlor and metalachlor. *Weed Sci.* **28**: 168-175.

Deal L M, J T Reeves, B A Larkins and F D Hess 1980. Use of an *in vitro* protein synthesizing system to test the mode of action of chloroacetamides *Weed Sci.* **28**: 334-340.

Dekker J and Burmster R (1988) Fluorometric determination of *in vivo* p-halohydroxy benzonitrite detoxification kinetics in Zea mays, *Anal. Lett.* **21**: 077-2089.

Della Cioppa G, Bauer S C, Taylor M L, Rochester, D E Klein, B K Shah, D M Fraley R T and Kishore G M (1987) Targeting a herbicide resistant enzyme from *Escherichia coli* to chloroplasts of higher plants. *Biotechnology* **5**: 579-584.

Derrick P M and A H Cobb (1987) The effect of acifluorfeon on membrane integrity in Gallium aponine leaves and protoplast. *Proc. Bro crop protect. Conf. on weeds,* **3**: 997-1004.

Devine M D 1989. Phloem translocation of herbicides. *Rev. Weed Sci.* **4**: 191-123.

Devine M D, Duke S O and Fedtke C 1993. *Physiology of Herbicide action.* Prentice Hall, Englewood Cliffs, N.J. 141-169.

Devlin R M and I E Demoranville 1974. Influence of dichlobenil and three experimented herbicides on bud break, terminal growth and root development of cranberry cuttings. A*bstr Weed Sci Soc.* Amer 14-15.

Devlin R.M. and R.P. Cunningham (1970). The inhibition of gibberlic acid induction of a - amylase activity in barley endosperm by certain herbicides. *Weed Res* **10**: 316-320.

Dhillon N S, Anderson J L 1972. Morphological, anatomical and biochemical. Effects of propachlor on seedling growth. *Weed Res* **5**: 115-122.

Dhillon N S and J.L. Anderson, (1972), Morphological anatomical and biochemical effects of propachlor on seedling growth. *Weed Res.* **12**: 182-189.

Droge W, Broer I, and Puhler A (1992). Transgenic plants containing phosphenothricin – N- acetyl transferase gene metabolise the herbicide L-phosphenothricin gene (glyphosinate) differently from transformed plants. *Planta* **187**: 142-151.

Duke S O (1988) Glyphosate In *Herbicides : Chemistry, degradation and Mode of action.* (eds) P C Bearney and D D Kaufman, **3**: 1-70 Dekker.

Duke S O and Kenyon W H 1988. Polycyclic alkanoic acids In *Herbicides : Chemistry, degradation and mode of action* (*eds*) (P.C. Kearney and D D Kaufmna **3**: 71-116. Dekker.

Duke S O, Chrity A L, Hus F D and Holt J S (1991) *Herbicide resistant crops* comments from AST 1991-1, Council for Agricultural Sci. and Technology Ames 1A.

Duke S. O (1992) *Mode of action of herbicides used in weeds of cotton : characterization and control.* (eds) C.G. Mc Whorter and J.L. Abernathy, pp 403-437. The cotton foundation, Memphis T.N.

Duke S O, U B Nandihalli and M.V. Duke (1994). Protoporphyrinogen oxidase as the optimal herbicide site. In S.O Duke and C.A. Rebeiz (eds). *Porphuric Pesticides : Chemistry, Toxicology and pharmaceutical applications.* ACS sym ser 559.

Dyer W E, Hess F D and Holt J S (1993 b) Potential benefits and risks of herbicide resistant crops produced by biotechnology. *Horti. Rev* **15**: 367-408.

Faulkner J S (1974) The effect on dalapon on 35 cultivars of *Lolium perenne. Weed Res.* **14**: 405.

Fedtke C (1991) Deamination of metribuzin in tolerant and susceptible soybean (*Glycine max.*) cultivars Pestic soc **31**: 175-183.

Fillatti, J J Kiser, J Rose R and Comai L (1987 a) Efficient transfer of a glyphosate tolerance gene into tomato using a binary *Agrobacterium tumefaciens* vector. *Biotechnology* **5**: 726-730.

Fisher R (1975) Herbicide tolerance in amenity grasses, Ph.D. thesis. The Queens University of Belfast, U K,

Fisher R and Wright C E (1980) The breeding of lines of Agrostis tenuis Sibth and *Festuca rubra,* L tolerant of grass killing herbicides. *Proc.3rd Ind. Turf grass Res.* Munich, 1977, 11.

Foy C.L. and D. Penner (1965). Effect of inhibitors and herbicides on tricarboxylic acid cycle substrate oxidation by isolated cucumber mitochondria. *Weeds* **13**: 226-231.

Fuerst E P and M A Norman, (1991). Interactions of herbicides with photosynthetic electron transport. *Weed Sci.* **39**: 458-464.

Furest E P (1987). Understanding the mode of action at chloroacetamide and thiocarbonmate herbicides. *Weed Technol.* **1**: 270-277.

Goldsbrough P B, Hatch E M; Huang B Kosihski, W G Dyer W E; Herrman K M and Weller S C (1990). Gene amplification in glyphosate tolerant tobacco cells. *Plant Sci* **72** : 53-62.

Good N E (1961). Inhibitors of Hill reaction. *Plant Physiol.* **36**: 788-803.

Goodman R M (1987) Future Potential, Problems and Practicalitiers of herbicide tolerant crops from genetic engineering. *Weed Sci.* **35** (Suppl 1): 28-31.

Gould F (1995). Comparison between resistance management strategies for insects and weeds. *Weed Technology*. **9**: 830-839.

Gressel J (1987) *Genetic manipulation for herbicide resistant crops in combating resistance to xenobiotics* (M G Ford, D W Hollomon, B P S Khambay and R M Sawicki eds) 266-280. VCH Verlags, Weinheim, Germany.

Gressel J (1989) Conferring herbicide resistance on susceptible crops. *In Herbicides and Plant Metabolism* (A D Dodge, ed) pp 237-259. Soce.Exp. Biol. Seminar Ser 38. Cambridge Press, Cambridge.

Gressel J (1992) The need for herbicide resistant crops. In Achievements and Development in combation Pest Resistance (I. Denholm, A.Devonshire, and D.Hollomon, eds) 283-294. Elsevier, London.

Gressel J (1993) Advances in achieving the needs for biotechnologically derived herbicide resistant crops. *Plant Breed Rev.* **11**: 155-198.

Gronwald J W (1991) Lipid biosynthesis inhibitors. *Weed Sci* : 39-43

Grossman K and J Kwiatkowski (1993). Selective induction of ethylene and cyanide biosynthesis appears to be involved in the selectivity of the herbicide quinclorac between rice and barnyardgrass. *J Plant Physiol.* **172**: 457.

Hart S E Saunders J W and Fenner D (1993). Semi dominant nature of marogenic sulfonyl urea herbicide resistance in sugarbeet (*Beta vulgaris*). *Weed Sci.* **41**: 317-324.

Hartwig E A 1987. Indentification and utilization of variation in herbicide tolerance in soybean (*Glycine max*) breeding *Weed Sci.* **35** (Suppl 1): 4-8.

Harwood J L, S M Ridley and K A Walker (1989) Herbicides inhibiting lipid synthesis. In A.D. Dodge (Ed) *Herbicides and Plant Metabolism*, Cambridge Univ. Press. Cambridge pp 277.

Hatzios K K and Penner, D (1982). *Metabolism of Herbicides in Higher plants*, Burgess, Minneapolis pp 264-275.

Hatzios K K (1987). Comparative effects of oxadiazon and its metabolism on biochemical process of enzymatically isolated leaf cells of soyabean. *Enzymology* **1**: 235-242.

Hatzios K K and Penner D, (1980). Some effects of buthidazole on corn (Zea mays) photosynthesis, respiration, anthocyanin formation and leaf ultrastructure. *Weed Sci.* **28**: 97-100.

Hatzois K K (1989). Biotechnology application in weed management now and in the future. *Adv Agron* **41**: 325-375.

Haughn G W, Smith J, Mazur B and Somer ville C (1988) Transformation with a mutant Arabidopsis acetolactate synthase gene renders tobacco resistance to surfonyl urea herbicides. *Mol. Gen. Genet.* **211**: 266-271.

Hawkes T R, J L Howard and S E Pontin, (1989) Herbicides that inhibit the biosynthesis of branched chain aminoacids. In A D Dodge (ed) *Herbicides and Plant Metabolism*, Cambridge University. Press Cambridge, 277.

Hawkins A F, ICI, Jealott's Hill, England, Personal Communication (1979)

Hinches M A W, Connor Ward, D V Newell C A, Mc Donnell, RE, Sato S J Gasser, C S Fischhoff D A Re D B Fraley R T and Horsch R B (1988). Production of transgenic soybean plants using *Agrobacterium* mediated DNA transfer. *Biotechnology* **6**: 915-922.

Hofstra G and C.M. Switzer 1968. The phytotoxicity of propanyl. *Weed Sci.* **16**: 23-28.

Hoppe H H (1989). Fatty acid biosynthesis a target site of herbicide action. In P Boger and G. Sandman (eds) *Target sites of Herbicide action*. CRC Press, Inc. Boca Raton, Fl USA.

Ishihara K H Mosaka M. Kubota, H. Kamimura, N. Takakusa and Y. Yasuda (1987). Effects of sethoxydin on the metabolism of excised root tips of corn. In R. Greenhalgh and T.R. Roberts (eds). *Pesticide chemistry and Technology*. Blackwell Scientific, Palo Alto, C.A. USA, 187-190.

Jain M.L, E V Kurtz and K C Hamilton (1966). Effect of Dalapon on glucose utilization in the shoot and root of barley. *Weeds* **9**: 431-442.

Jaworski E G. (1969) Analysis of mode of action of herbicidal a - chloroacetamides. *J.Agr. Food Chem.* **17**: 165-170.

John, St J B (1982) Effects of herbicides on the lipid composition of plant membranes. In D E Moreland, J B St John and F D Hess (eds) *Biochemical Responses induced by herbicides*. Amer. Chem. Soc. Washington, D C, pp 97-109.

Johnston D T and F S Faulkner (1991) Herbicide resistance in graminaceae-a Plant breeders view. In J C Caseley, G W Cussans and R K Atkin (eds) *Herbicide resistance in Weeds and Crops*. Butterworth-Heinmann Limited Oxford U K. pp 319-330.

Kandasamy O S, C Chinna Samy and T Chitdeshwari (2002). Herbicide Resistance in crop plants – A review. *Agric. Rev.* **23** (3): 208-213.

Keitt G W and R A Baker (1966). Auxin activity of substituted benzoic acid and their effect of polar auxin transport. *Plant Physiol.* **41**:1651-1569.

Key J L, C Y Lin, E M Gifford, Jr and R.Dengler (1966) Relation of 2,4-D induced growth aberrations to changes in nucleic acid metabolism in soy bean seedlings. *Bot Gaz* **127**: 87-94.

Kibite S and Harker S N (1991) Evaluation of oat germplasm for resistance to diclofop-methyl. Can *J. Plant Sci.* **71**: 491-496.

Kleczkowski L A (1994) Inhibitors of photosynthetic enzymes/carriers and metabolism. Ann Rev *Plant Physiol* **45**: 339-367

Kocher H H, M Kellner, K Lotzsch and E Dom (1982) Mode of action and metabolic fate of the herbicide fenoxyprop – ethyl, Hoe 33171. Proc Br Crop Protec Conf. on *Weeds* pp 341-347.

Kolhe S S and Tripathi R.S. (1998). Integrated Weed Management in direct seeded rice. *Ind. J. Weed Sci.* **30(1 & 2)**: 51-53.

La Rossa R A and J V Schloss (1984). The sulfonylurea herbicide sulfometuron methyl is an extremely potent and selective inhibitor of acetolactate synthase in Salmonella typhsimurium. *J Biol. Chem.* **259**: 8753-8757.

Lamourex G L and Rusness D G (1981) : Sulfur in Pesticide Action and Metabolism: (Rosen J.D. *et al* eds) *ACS symp. Ser* 158. Washington, D.C 133.

Lange A Cardenes J and Cruz R (1973) Crop losses caused by weeds. *Weeds Today* **4(1)**: 14-22.

Laursent C M, Krzyzek R A, Flick C E, Anderson P C and Spencer T M (1994). Production of fertile transgenic maize by electroporation of suspension culture cells. *Plant Mol. Biol.* **24**: 51-61.

Le Baron H M and J Gressel J (1982) *Herbicide Resistance in Plants*. John Wiley and Sons, New York.

Lea P J and Ridley S M (1989) Glutamine synthetase and its inhibition in A. D. Dodge (*ed*) *Herbicides and Plant Metabolism.* Cambridge Univ. Press, Cambridge, 277.

Lea P J, K W Joy, J L Ramos and M G Guarrero (1984). The action of 2-amino -4- (methyl phosphenyl) – butanoic acid (phosphinothricin) and its 2-oxo derivative on the metabolism of cyanobacteria and higher plants. *Phytochemistry* **23**: 1-6.

Lee K Y, Townsend J, Tepperman J, Balck M, Chui C F, Mazur B, Dunsumuir P and Bedbrook J (1988). The molecular basis of sulfonyl urea herbicide resistance in tobacco. *EMBOJ* **7**: 1241-1248.

Lehnen L.P, T D Sherman, J.E. Becerill and S.D. Duke (1990). Tissue and cellular localization of acifluorfen – induced porphyrins in cucumber cotyledons. *Pestic Biochem. Physiol* **37**: 239-248.

Leis J P and E V Keiler (1970). Protein chain initiating methionine RNA's in chloroplast and cytoplasm of wheat leaves. *Proc Natl. Acad Sci* (USA) **67**: 1593-1599.

Levitt J R C, D N Duncan, D Penner and W F Meggitt (1978) Inhibition of epicuticular wax deposition on cabbage ethofumesate. *Plant Physio.* **61**: 1034-1036.

Lovelidge B (1974) 'Rogue' grass defies paraquat. *Arable Farming*. 9 (Sept.)

Lyon B R, Liewellyn D J, Huppatz J L, Dennis E S and Peacock W J (1989) Expression of bacterial gene in transgenic tobacco plants confers resistance to herbicide 2,4-Dichlorophenoxyacetic acid. Plant *Mole. Biol.* **13**: 533-540.

Madhulety T Y (1972) *Physiological and Bio chemical aspects of atrazine and lasso on maize* (*Zea mayes L*) M.Sc. (Ag) Thesis. APAU, Rajendranagar, Hyderabad.

Madhulety T Y and Chandra Singh D J (1976) Biochemical and Physiological Studies of alachlor and atrazine in maize. *Andhra Agri* J **23** (3 and 4) 85-92.

Madsen K H and Jensen J E (1995). Weed control in herbicide tolerant sugarbeet (*Beta vulgari L*) : A comparison between glyphosate and currently used herbicides.

Magalhaes A C and F M Ashton (1969). Effect of dicamba on oxygen uptake and cell membrane permeability in leaf tissues of *Cyperus rotandus. Weed Res* **9**: 48-52.

Maier – Greiner U H, Klavs C B A, Estermaier L M and Hartman G R (1991) Herbicide-resistance in transgenic plants through degradation of the phytotoxin to urea. *Angew Chemical Int.Ed.Engl* **30**: 1314-1315.

Maliga P, Fejes E, Steinback K and Menczel L 1987. Cell culture approaches for obtaining herbicide resistance in chloroplasts of crop plants. *ACS Symp Ser* **334**: 115-124. Am Chem Soc. Washington DC.

Mallory-Smith C A, Thill D C and Dial M J (1993) ID-BRI : Sulfonylurea herbicide resistant lettuce germplasm, *Hort Sci*. **28**: 63-64.

Marshall L C, Somers D A, Dotray P D, Gengenbach B G, Wyse D L and Gronwald, J W (1992). Allelic mutations in acetyl – coenzyme A carboxylase confer herbicide tolerance in maize. *Theor. Appl. Gent*. **83**: 435-442.

Mayer M P, D L Bartlett, P Beyer and H Klienig. 1989. The *in vitro* mode of action of bleaching herbicides on the desaturation of 15-Cis-phytoene and cis-d-carotene in isolated daffodil chloroplasts. Pestic. *Biochem Physio*. **34**: 111-117.

McDaniel J L and R E Frans (1969). Soy bean mitochondrial response to prometryn and fluometuron. *Weed Sci* **17**: 192-196.

McDevitt S C, Chui R E, Hartnett C F, Knowlton M E, S Mauvaris C J, Smith J K Falco and Mazuri B J (1989). Engineering herbicide resistant acetolactate synthase. *J Indus Microbil.* 30 (suppl) **4**: 187-194.

Mercado L.R. (1975). Some effects of Amitrol on the respiratory activities of *Zea mays*. *Weeds* **8**: 29-38.

Miki B L, Labbe H, Hattori J, Ovellet T, Gabard J, Sunohara G, Charest P J and Iyer V V (1990) Transformation of *Brassica napus* canola cultivars with *Arabidopsis thaliana* acetohydroxy acid synthase genes and analysis of herbicide resistance. *Theor Appl Genet* **80**: 449-458.

Morejohn L C, T E Bureau, J. Mole – Bajer and A.S. Bajer. Plant tubulin that inhibits microtubule polymerization *in vitro. Plant* **172**: 252-264.

Moreland D E and S.C. Huber (1979). Alterations to properties and other (phenyl carbamate and phenyl amide) herbicides. *Abstr. Weed* Sci. Amer P. 104.

Moss J P (1992) Biotechnology and crop improvement in Asia: ICRISAT, India.

Motsumoto H, Lee J J and Ighizuka K (1993) A rapid and strong inhibition of protoporphyrinogen oxidase from several plant species by oxyflurfen. *Pestic. Biochem. Physiol*. **47**: 113-118.

Muhitch M.J, D L Shaner and M A Stidham 1987. Imidazolinones and acetohydroxyacid synthase from higher plants. *Plant Physiol* **83**: 451-456.

Mullner H, Eckes P and Donn G 1993. Engineering crop resistance to the naturally occurring glutamine synthetase inhibitor phosphinothricin. *ACS Symp Ser*. **524**: 38-47.

Muni A P (1960). *Investigations on the anatomy of weeds and selective weed control*. Ph.D. Dissertation, Bose Res Inst Univ of Culcutta, India.

Nadanassababdy T and O S Kandasamy (2002). Evaluation of herbicides and cultural method for weed control in cotton. *Ind J Weed Sci.* **34** (1 & 2): 143-145.

Nagaraju (2003) Annual Programme Report for the year 2002, UAS, Hebbal, Bangalore.

Nandihalli D B and S O Duke (1993). The porphyrin pathway as a herbicide target site. In S O Duke, Menn and J.R. Plimmer (Eds) *Pest Control with enhanced environmental safety* ACS Symp Ser **524**: 62-78.

Nandihalli V B and S O Duke (1994). Structure, activity relationships of 12 protoporphyrinogen oxidase – inhibiting herbicides. *In Porphyric pesticides, chemistry, toxicology and pharmaceutical applications*. ACS Symp Ser **559**: 143-146.

Negi N S and H H Funderbunk (1968). Effect of solutions and vapours of trifluralin on growth of roots and shoots. *Abstr. Weed Sci Soc. Amer*. 37-38.

Newhouse K E, Singh B, Shaner D and Stidham M (1991a) Mutations to corn (*Zea mays L*) conferring resistance to imidazolinone herbicides. *Theor. Appl. Genet*. **83**: 65-70.

Newhouse K E, Wang T and Anderson P C (1991 b)Imidazolinone tolerant crops in *The imidazolinone Herbicides. (eds)* (D L Shaner and SLO Conner pp 139-150; CRC Press Boca Raton F L.

Nischimura M and Takamiya (1966) Energy and electron transport system in algal photosynthesis. I Action of two photochemical systems in oxidation – reduction reactions of cytochrome in porphyra. *Biochem Biophys. Acta* **120**: 45-56.

O'Brien J J, B C Jarvin, J H Harvey and J B Hanson (1968) Enhancement by 2,4-D of chromatin RNA polymess in soy bean hypocotyl tissue. *Biochem Biophys Acta* **169**: 35-43.

O'Keefe D P, Romesser J A and Leto K J 1988. Identification of constitutive and herbicide inducible cytochromes. p – 450 in *Streptomyces griseolus. Arch, Microbiol*. **149**: 406-12.

Ossman E C and Staniforth D W (1949) : Effects of 2,4-D on inbred lines and a single cross maize. *Plant Physiol* **24**: 60-74.

Padgette S R, D B Barry, J F Eicholtz, D E Delanney, X. Fucsh and R L Kishore G M and Fraley R T (1994). New Weed Control Opportunities development of soy beans with a round up Ready gene. *In Herbicide – Resistant crops, Agricultural Environmental, Economic, Regulatory and Technical aspects (ed)*. S.O. Duke Lewis Publishers, Chelsca M I.

Panwar R.S, S S Rathi, R K Malik and R S Malik (1995). Effect of application time and hoeing efficiency of pendimethalin in cotton. *Ind. J. Agron*. **40**: 153-155.

Parker C (1966). The importance of shoot entry in the action of herbicides applied to the soil weeds **19**: 117-121.

Parker C and Fryer J D (1975) Weed Control problems causing major reduction in world food supplies. FAO *Plant Prot*. Bull. 23 (3/4)

Paton D and J E Smith (1965). The effect of 4-dydroxy 3,5 – idobenzonitrite in CO_2 fixation, ATP formation and NADP reduction in chloroplast of *vicia faba L. Weed Res*. **5**: 75-77.

Phillips R E, D B Egli and L Thompson Jr 1972. Absorption of herbicides by sorghum seeds and their influence on emergence and seedling growth. *Weed Sci.* **20**: 506-510.

Quimby P C (1967). Studies relating to the sensitivity of dicamba for wild buck weed (*Polygonum convolvulus L*) V S Selkrik, Wheat (*Triticum aesticum L*) and possible mode of action. Ph.D. Thesis N. Dakota State Univ. Fargo 87.

Radosevich S R, K E Steinback and C J Arntzen (1979) Effect of Phtosystem II inhibitors on thylakoid membranes of two common groundsel (*Senecio vulgaris*) biotypes. *Weed Sci.* **27**: 216-218.

Ramesh (2003). All India Coordinated Research Programme on Weed Control, Bangalore (ICAR) (eds) Yaduraju N.T, T.V. Ramachandra Prasad. Annual Programme Report for the year 2002, UAS, Hebbal, Bangalore.

Rameshwar, Sanjay Chandra, G D Sharma and Surinder Rana (2002) Evaluation of herbicides for weed control and economics in oinon (*Alium cepa*, L) under cold desert region of Himachal Pradesh. *Ind. Jour. of Weed Sci.* **34** (1&2): 68-71.

Rao Shivaji V and W B Duke (1976) Effect of Alachlor, Propachlor and Pryrochlor on GA_3 induced production of protease and a - amaylase. *Weed Sci.* **24**: 616-618.

Rao V S (2000) Introduction, In *Principles of Weed Sci*., 2nd Edn. Oxford and IBH Publishing Co Pvt. Ltd, New Delhi 1-6.

Rao V S (2000) *Principles of Weed Sci*. 2nd edn. Oxford IBH Publication Co Pvt Ltd, New Delhi.

Raskar B S and P G Bhoj (2002) Bioefficiency and phytotoxicity of pursuit plus herbicide against weeds in soybean (*Glycine max L*). *Ind. J. Weed Sci.* **34** (1 & 2): 50-52.

Rawlinson C J, Muthylu G and Turner R H (1978) Effects of herbicide on epicuticlar was of winter oilseeds Rape (*Brassica napus*) and infection by *Pyrenopeziza brassicae*. Trans. Britishmyucol. Soc. **71**: 441-451.

Ray T B (1984). Site of action of chlorsulfuron. *Plant Physiol.* **75**: 827-831.

Rojas - garciduenas M and T. Kommerndahl (1958). The effects fo 2-4-Dichlorophenoxyacetic acid on radicle development and stem anatomy of soybean. *Weeds* **6**(1): 49-51.

Romesser J A O'Keefe, D.P. (1986) Induction of cytochrome P_{450} dependent sulfonylurea metabolism in *Streptomyces griseolus*. *Biochem Biophys Res Commu* **140**: 650-59.

Rossman E C and D W Staniforth (1949) Effects of 2,4-D on inbred lines and a single cross of maize *Plant Physiol*. **24**: 60-74.

Rubin R and Y Eshel (1977) Absorption and distribution of terbatrya and Fluomenturon in cotton (*Gossypium hirsutum*) and Snap beans (*Phaseolous vulgaris*). *Weed Sci*. **25**: 499-505.

Ryan G F (1970) Resistance of Common groundel sp to simazine and atrazine, *Weed Sci*, **18**: 614-616.

Sachs R M and J L Michael (1971). Comparative phytotoxicity among four arsenical herbicides. *Weed Sci.* **17**: 550-564.

Sandman G, I E Clark, P M Bramley and A Boger (1984) Phytoene desaturase, the essential target for bleaching herbicides. *Weed Sci.* **39**: 474-490

Sandman, G, I E Clark, P M Bramley and P Boger (1984). Inhibition of phytoene desaturase mode of action of certain bleaching herbicides. Z Naturforsch. 39C : 443 : 449.

Sannappa (2003). All India Coordinated Research Programme on Weed Control. Bangalore (ICAR) (*eds*) Yaduraju, N T and T.V. Rama Chandra Prasad. Annual Programme Report 2002, UAS, Hebbal, Bangalore.

Saunders J W Acquad G Fenner K A and Doley W P (1992) – Monogenic dominant sulfonylurea resistance in sugarbeet from somatic cell selection. *Crop Sci*. **32**: 1357-1360.

Scalla R and M. Matringe, (1994). Inhibitors of prtoporphyrinogen oxidase as herbicides, diphenyl ethers and related photobleaching molecules. *Rev Weed* **6**: 103-132.

Schweizer E E and D.M. Weatherspoon (1971). Response of sugar beets and weeds to Phenedipham and two analogs. *Weed Sci*. **19**: 635-639.

Sebastian S A, Fader G M, Ulrich J F, Forney D R and Chaleff R S (1989). Semidominant soy bean mutation for resistance to sulphonylurea herbicides. *Crop Sci*. **29**: 1403-1408.

Shah D, Horsch R B, Klee H J, Kishore G M, Winter J A, Turner N E, Hironaka C M, Sanders P R, Gasser C S, Aykent S A, Siegel N R, Rogers S G and Fraley R T (1986) Engineering herbicide tolerance in transgenic plants. *Sci*. **233**: 478-481.

Shaner D L and M L Reider (1986). Physiological responses of corn (*zea mays*) to AC 243, AC 997 in combination with valine, leucine and isoleucine. *Pestic Biochem. Physiol*. **25**: 248-257.

Shaner D L and P A Robinson (1985). Absorption, translocation and metabolism of AC 252, 214 in soybean (*Glycine max*), common cocklebur (*Xanthium strumeriun*) and Velvetleaf (*Abuliton theophrasti*). *Weed Sci.* **33**: 469-471.

Shaner D L Bascomb N F and Smith W 1994. Imidazolinone resistant crops : Selection, characterization and management, *In Herbicide – Resistant crops : Agricultural, Environment Economics, Regulatory and Technical aspects.* (*ed*) (S O Duke ed), Lewis Publihers, Chelsea, M I

Shaner D L, B K Singh and M A Stidham 1990. Interaction of imidazolirones with plant acetohydroxyacid synthase ; evidience for *in vivo* binding and competition with sulfoneuron methyl. *J. Agric Food Chem*. **38**: 1279-1282.

Shaner D L, M A Stidham, M Muhitch, M. Reider, P. Robson and P. Anderson (1985). Mode of action of the imidazolinones. *Proc. British Crop protect conf.* **1**: 147-154.

Shaw N C and W A Gentner (1957). The elective herbicidal properities of several variously substituted phenoxyalkylcarboxylic acids. *Weeds* **5(2)**: 75-92.

Shaw W C, C J Willard and R L Bernard (1955). The effect of 2,4-dichlorophen oxyacetic acid. (2,4-D) on wheat, oats, barley. *Ohio Agr. Expt.Sta Res* Bull.

Sheets T J (1961b) Uptake and distribution of Simazine by oat and cotton seedlings. *Weeds* **9(1)** : 1-13.

Shimabukuro R. H (1985) In *Weed Physiology* (*ed*) S.O. Duke Vol II. CRC Press, Boca Raton, FL : 215-240.

Slife F W (1956) The effect of 2,4-D and several other herbicides on weeds and soybeans when applied as post emergence sprays. *Weeds* **4(1)**: 61-68.

Smeda R J, Haregawa P M and Weller S C (1989) Mechanisms of tolerance to Atrazine in Photoautotrophic potato cells. *Weed Sci Soc Amer. Abstr* **29**: 164.

Smith L W, R L Peterson and R.F. Worten (1971) Effects of dimethyl propynal benzamide herbicide on quack grass rhizomes. *Weed Sci*. **19**: 174-177.

Snape J W Angus W J Parker B B and Leckie D (1987) The chromosomal locations in wheat of genes conferring differential responses to the wild oat herbicide, difenzoquat. *J Agric., Sci.* **108**: 543-548.

Snape J W, Nevo E, Parker B B, Leckie D and Morgunov A (1990). Herbicide response polymorphism in wild populations of Emmer Wheat. *Heredity* **66**: 251-257.

Snape J W, Leckie D, Parker B B and Nevo E (1991). The genetical analysis and exploitation of differential responses to herbicides in crop species. *In Herbicide Resistance in Weeds and Crops* (*eds*) J.C. Carelay G.W. Currians and R.K. Atkin pp 305-317. Butterworth – Heinemann, Oxford U.K.

Somers D A (1994). Aryloxy phenoxy propionate and cyclohexanedion resistant crops. *In Herbicide – Resistant crops : Agricultural Environmental, Economic, Regulatory and Technical Aspects.* (*ed*) (S.O. Duke), Lewis Publishers, Chelsca, M I.

Somers D A, Rines H W, Gu W, Kaeppler H F and Bushnell W R (1992). Fertile, transgenic oat plants. *Biotechnology*, **10**: 1589-1594.

Sousa Machado V, J D Bandeen, G R Stephenson and P Lavigne (1978) Uniparental inheritance of chloroplast atrazine tolerance in *Brassica compestris Can J Plant Sci.* **58**: 77.

Srinivasamurthy (2003) All India Coordinated Research Programme on Weed Control, Bangalore (ICAR) *(eds)* Yaduraj, NT and T. V. Ramachandra Prasad. Annual Programme Report 2002, UAS, Hebbal, Bangalore.

Stalker D M, Mc Bride KE and Malyj L D (1988) Herbicide resistance in transgenic plants expressing a bacterial detoxification gene. *Sci*. **242**: 419 -423.

Stengink, S J and Vaughn, K.C. (1988). Norflurazon (SAN-9789) reduces abscisic acid levels in cotton seedlings. a glandless isoline is more sensitive than its glanded counterpart. *Pestic Biochem Physiology* **31**: 269-275.

Stidham M.A (1991). Herbicides that inhibit acetohydroxy acid synthase. *Weed Sci* **39**: 428-434.

Stidham M A and D L Shaner (1990). Imidazolinone inhibition of acetohydroxy acid *synthase in vitro* and *in vivo. Pestic Sci* **29**: 335-340.

Stoltenberg D E, Gronuwald J W, Wyse D L, Burton J D, Somers D A and Gengenbach B G (1989). The effect of sethoxydim and haloxyfop on acetyl – coenzyme A carboxylase activity in tolerant and susceptible Festuca Species. *Weed Sci*. **37**: 512-516.

Strachan, S D and F D Hess, (1983). The biochemical mechanism of action of dinitro aniline herbicide oryzalin. *Pestic Biochem Physiol*. **20**: 141-150.

Streber W R and Willmitzer L (1989) Transgenic tobacco plants expressing a bacterial detoxifying enzyme resistant to 2,4- Ed *Biotechnology* 7: 811-815.

Swanson E B, Herrgesell M J, Arnoldo M, Spell D W and Wang R S C (1989). Microspore mutagenesis and selection : Canola plants with field tolerance to the imidazolinones. *Theor. Appl. Genet* **78**: 525-530.

Terakawa T and Wakasa K (1992) Rice mutant resistant to the herbicide bensulfuron methyl (BSM) by *in vitro* selection. *Japan J. Breed*, **42**: 267-275.

Thompson B R and Pratt D (1982) Herbicide tolerance, In (Agricultural applications : Integrating conventional and molecular genetics (ed) Qualset *C. California Agri.* (Spl. Issue). **36** (8) : 33.

Thompson D C, B Truelove and D E Davis (1969). Effect of herbicide prometryne [2,4-bis-(*Isopropylamino*) -6- (methylthio) – s- triazine] on mitochondria. *J Agr Food Chem.* **17**: 997-999.

Thompson D C, B Truelove and D E Davis (1974) Effects of triazines on energy relations of mitochondria and chloroplasts. *Weed Sci.* **22**: 164-166.

Thompson D C, J Ray, Movva N, Tizard R, Crameri, R Davies, J E Lauwereys M and Botterman J (1987) Characterization of the herbicide resistance gene bar from *Streptomyces hygroscopicus. EMBOJ* **6**: 2519-2523.

Tonnemaker K A, Auld D L, Thill D C Mallory – Smith C A and Erickson D A (1992). Development of sulfonylurea resistant rape seed wing chemical mutagenesis. *Crop Sci.* **32**: 1387-1394.

Tottman D R, J Holryd, F G H Lupton, R H Oliver, R T Barnes and R M Tysoe (1975), The tolerance of chlorotoluron and isoproturon by varieties of winter wheat. *Proc. Eur. Weed Res. Soc. Symp. Status,* Biol. Control Gramweeds Eur, Pari, 360.

Tranel P and Dekker J (1992). Inheritance of Clomazone resistance in maize, seedlings. Maydica **37**: 137-142.

Truelove B, A M Dinner, D E Davis and J.D. Weete (1979). Metolachlor, membranes and permeability. Abstr. Meeting. *Weed. Sci. Soc. Amer*. 99-100.

Upadyaya V B and C S Bhalla (2002); Efficacy of cultural, mechanical and chemical weed control in chickpea (*Cicer arietnum L.*). *Indian J. Weed Sci.* **34** (1 & 2): 141-142.

Van Heile F J H, Hommes A and Vervelide E J (1970) Cultivar differences in herbicide tolerance and their exploitation. *Proc. Brit. Weed Control. Conf.* **10**: 111-117.

Vasil V, Castillo A M, Fromm M E and Vasil I K (1993) Herbicide – resistant fertile transgenic wheat plants obtained by microprojectile bombardment of regenerable embryonic callus. *Biotechnology* **10**: 667-674.

Vaughn K C and M A Vaughn (1990) Structural and biochemical characterization of dinitroralanien resistance Eleusine. In (*Eds.)* M B Green, H M LeBaron and W K Moberg. Managing resistance to Agrochemicals. *Amer. Chem. Soc.* Washington D.C. **421**: 364-375.

Veerasekaran P R C Kirkwood and W W Fletcher (1977). Studies on mode of asulam in bracken (*Pteridiam aqulinum L Kuhn*) II Biochemical activity in the rhizome buds. *Weed Res.* **17**: 85-92.

Wain R L (1964). Ioxynil – some considerations on its mode of action. *Proc 7th Brit. Weed Conf.* **1**: 306-311.

Walia U S, L S Brar and B K Dhaliwal (1998). Performance of clodinafop and fenoxaprop-p-ethyl for the control of P*halaris minor* in wheat. *Indian J. Weed Sci*. **30** (1 & 2): 48-50.

Wan Y and Lemaux P G (1994). Generation of large number of independently transformed fertile barley plants. *Plant Physiol*. **104**: 37-48.

Warwick S I (1991) Herbicide resistance in Weedy plants Physiology and population biology. *Ann Rev Ecoli Syst* **22**: 95-114.

Watson M C, PG Bartels and K.C. Hamilton (1980). Action of selected herbicides and tween twenty on Oat (*Avena sativa*) membranes. *Weed Sci.* **28**: 122-127.

Wrubel R P and J Gressel (1994). Are herbicide mixtures useful for delaying the rapid evolution of Resistance? A case study weed technol. **8**: 635-648.

Xide Gourmay, J Koller and J L Dufour (1975). Difficulties rencon dans la propsoection degeniteurs pour la reistance a un herbicide; car de latrazine et duble d hiver, proc *Eur weed Res Soc Symp.* Status, Bio. Cartrol Grainweeds Eur, Paru, 388.

Yaduraju N T and M Devender Reddy (2001) All India Coordinated Research Programme on Weed Control, Hyderabad Centre. Fifteenth Annual Progress Report 2000-2001. ANGRAU, Rajendranagar, Hyderabad.

Yaduraju N T, Ramachander Prasad T V (2003): All India Coordinated Research Programme on Weed Control, Bangalore Centre (ICAR) Ann Progress Report 2002, UAS, Hebbal Banglore.

Yoder J I and Goldsbrough A P (1994). Transformation systems from generating marker free transgenic plants. *Biotechnology* **12** : 263-267.

York A C and C J Amtzen (1979). Photosynthetic electron transport inhibition with buthidazole.Abstr. *Weed Sci Soc Amer*, 103.

York A C, C J Arntzen, F W Slife (1978). Effect of buthidazole on the photochemical reactions of isolated pea (*Pisum sativum L*) chloroplast. *Abstr. Weed Sci Soc. Amer* 76.

York A C, C J Arntzen and F W Slife (1981). Photosynthetic electron transport inhibition by buthidazole. *Weed Sci* **29**: 59-64.

Zweig G N Shavit and M Avron (1965). Diquat (1,1-ethylene 2,2-dipyridilium dibromide) in photoreactions of isolated chlorplasts. *Biochem Biophy. Acta* **109**: 332-346.

Colour Plates

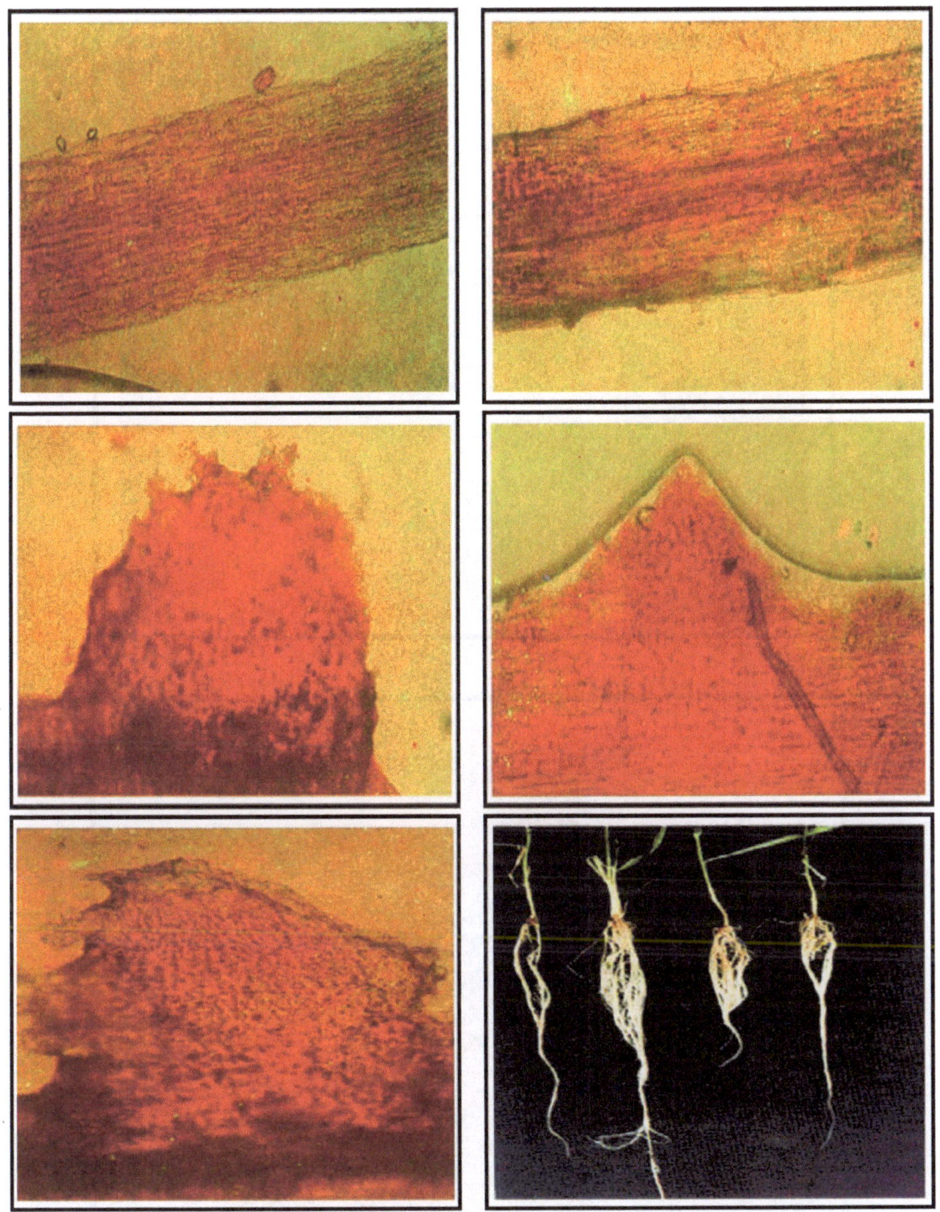

Fig. 6a. Transverse section of roots as affected by 2, 4-D and *Azospirillum brasilense* treatments (200 x magnification) showing untreated control (top left); roots inoculated with *Azospirillum brasilense* (top right); roots treated with 2, 4-D showing nodular outgrowth (middle left); nodule initation in 2, 4-D and *Azospirillum brasilense* treated roots (middle right); well developed nodule with 2, 4-D and *Azospirillum brasilense* inoculation (bottom left); and comparative root growth of different treatments, *viz.,* uninoculated control, *Azospirillum* inoculation, 2, 4-D treatment and *Azospirillum* combined treatment respectively from left to right (bottom right)

Fig. 6b. Nodulation in rice using 2, 4-D and *Azospirillum brasilense* in nitrogen for medium (b) nodules showing the crack entry and presence of bacteria inside the paranodule.

INDEX

α-amylase, 43,44,46
α-glucosidase, 43,45, 46
β-amylase, 43,44,47
β-D-fructofuranoside fructohydrolase, 37
β-ketoacyl ACP reduetase, 25
-synthetase, 25
δ aminolevulinie acid, 131
1-aminocyclopropane-1-carboxylic acid, 9
1,5-biphosphate carboxylase activity, 187
2-(4-carboxyphenyl)-4,4,5,5-tetramethylimidazoline-1-oxyl-3-oxide, 10
2, 4-D, 205,207,222,272
2,3-dichloroisobutyrate, 185
-dichlorapropionate, 185
2,4-DB, 205,222
2,4-diacetlyphloroglucinol produced, 166
2,4-D,NAD, 188
2,4,5-T esters, 235

A

A.caulinodans, 144
Abiotic stress, 95, 169
-resistant fruit cultivars ,190
Abcisic acid, 161
Abutilon theophrasti, 220, 223
ACC (1-aminocyolo propane-1- carboxylates), 161
-activity, 161
-deanminase, 162
-oxidase, 162
-synthase, 162, 187
Acdtyl COA carboxylase, 241
Acetamides, 205, 212, 245, 246, 247
Aceto hydroxy acid synthase, 242
Acetobacter, 149,173
-diazotrophicus, 149
Acetochlor, 205
-(2-Chloro-N-(ethoxy methyl)-N-(2ethyl-6 methyl phenyl), 212
Acetolactate synthase, 242
Acetyl elongase, 241
Acetyl tranasacylase, 25
Acetylene, 188
Acid invertase, 38
Acifluorfen, 218
Acinetobacter, 173
Aconitase, 11
Acrolein, 205, 248
Active iron, 13,14
Acyl units, 24
Acylation, 25
ADP, 38
-glucose pyrophosphorylase, 43,
ADPG pyrophosphorylase, 47
Aegle marmelos, 182
Aeschenomene, 127
Agriculture, 203
Agrobacterium, 161, 162, 191
Agrochemicals, 140
Alachlor, 205, 235
-(2-Chloro-N-(2,6 diethyl phenyl)-N-(methoxy methyl)acetamide), 212
Alanine, 12
Alcaligene, 148
-eutrophus, 272
-faecalis, 148
Alfalfa, 107
-(Medicago Sativa), 127
Aliphatics, 205, 213, 248
Alkaline nutrition, 4
Allantoin, 134
-acid, 134
Allium canadense, 204
Alpha- ketobutyrate, 161
Alternate and erratic bearing, 183
Amaranthus spp., 221
Ambrosia artemisiifolia, 220,223
Ametryn, 230
Amide, 133
Amino acids, 7
Aminobutyric acid, 12
Amitrole, 231
Ammonium assimilation, 55
Amylase, 113

Anthers, 13
Anthocyanin, 9
Antibiosis, 160
Antibiotics, 163
Antifungal activity, 165
Anti-gibberellin activity, 184
Antiport, 54
Aonla, 181
Apical dominance, 186
Apple, 183, 184
Apple fruit, 23
A-pyrrolline-5-carboxylate synthetase(P5CS), 100
Aquaporins, 99
Arabidopsis rice, 98
-mutants, 5
-thaliana, 98, 165, 187
Arginine, 12
Aroline, 213
Aromatic compound biosynthesis, 238
Arsenic, 107
Arsenicals, 205, 213, 249
Arthrobacter, 173
-oxidans 272
Artocarpus spp., 182
Aryloxyphenoxy propionics, 222, 245
Asparagine, 133
Aspartate, 133
Aspartic acid, 12
Asperugo procumbens, 220
Association bacteria, 126
Associative symbiosis, 127
Asulam, 215, 250
Atrazine, 230
Auxin, 9, 161, 171
Avena fatua, 204, 220
Azoarcus, 150
Azolla, 148
-anabaena, 126
-pinnata, 118
Azorhizobium, 127, 146
-bacillus, 113
Azospirillium, 127, 145, 161
-brasilense, 143, 144,162, 167,169
-lipoferum, 168
-strain ORS 571, 146
Azospirillum spp., 171
Azotobacter, 127, 167
-chroococcum, 168

B

Bacillus, 163
-AF1, 167
-amyloliquefaciens, 167
-brevis, 168
-cereus UW 85, 167
-coagulans, 168
-megaterium, 168
-polymyxa, 168
-pumulus, 168
-subtilis, 167
Bacterium, 162
Bacteroids, 133
Bacterial blight, 168
Bael, 181
Bajra, 107
Banana,181,183,192
Barley, 2, 173
Beans, 127, 168
Benefin, 205,217
Bensulfuron, 206, 227
Bensulide, 233,246
Bentazon, 205, 249
Benzamides, 205, 214, 246, 249
Benzoics, 205, 249
Benzothiadiazoles, 214, 249
Benzothia, 205
Ber, 181
Berkholeria, 173
Biennial bearing, 191, 192
Bifenox, 218
Biochemical indicator index(BI), 8
Biofertilization, 171
-control, 171
-remediation, 172
-stimulation, 171
Biofertilizers, 160
Bioinoculants, 173
Biological nitrogen fixation(BNF), 125,159
Biomonitoring, 118
Biosynthesis of chlorophyll, 109
-phosphoglycerides, 22
Biotechnological improvement, 267
-manipulations, 267
Biotic resistant fruit cultivars, 190
-stress, 166
Bipyridiliums, 205,215,249
Bisphosphatidylglycerol (BPG or DPG), 22, 32

Black pepper, 167
Blast, 168
Blue green algae, 127
Bordeaux mixture, 196
Boundary layer resistance, 90
(Brady)rhizobia, 2, 127
Bradyrhizobium, 173
-japonicum, 127, 139
Branched chain amino acid biosynthesis, 238
Branching, 203
Brassica compestris, 267
-kaber, 220
-napus, 98
Bromocil, 232
Bromoxynil, 272
-, ioxynil, 205
Bromus tectorum, 223
Burkholderia, 148
Butachlor, 248
-N-(Butoxymethyl-2-chloro-N-(2,6-diethylphenyl) acetamide, 212
Buthidazole, 205, 219, 251
Butylate, 229
Bypiridiliums (diquat, paraquat), 244

C

Cacodylic acid, 205
Cadmium, 107, 113,117
-acetate, 110
-chloride, 110
Cajanus cajan, 116
Calcareous soils, 3
Calcium carbide, 188
-deficiency, 191
Capacitance, 92
Capsella bursa- pastoris, 220
Carbamates, 205,215, 235,246,250
Carbendazim, 193
Carbohydrate, 186
-metabolism, 11, 46,111
Carbonates, 235
Cardiolipin(bisphosphatidylgycerol),22
Carotenoid, 9,110
-biosynthesis, 240
Carpelloid, 185
Carrot(cell line), 98
Caster bean, 23
-endosperm, 32
Catabolism of phosphlipids, 31
Catalase, 8
Cation-anion balance, 56
-ethanolamine, 28
Cdiclfop, 241
CDP chloine, 28
-1,2-diacylglycrolcholine phophotransferase, 29
-diacylglycerol, 32
-serine-o-phosphatidyl transferase, 30
-myo-inositol phosphatidyltransferase, 30
-ethanolamine: 1,2-diacylglycerol ethanolamine phophotransferase, 29
Cell elongation, 162
-water relation, 86
Cellulose, 84
Cereals, 125
Chara corallina, 119
Chelated Copper, 193
Chelation, 106
Chemical herbicide, 204
-weed control, 207
Chenopodium album, 219,220,223
Chick pea (Cicerarietinum),110,167
Chitinase, 190
Chlamydomonas reinhardtii, 8
Chlophyll a/b ratio, 110
-a, 110
-b,110
Chloramben, 249
Chlorella sorokiniana, 162
-vulgaris, 162
Chlorimuron, 206, 227
Chlormequat Chloride, 184
Chloronerva, 6
Chlorophyll,1,8
-biosynthesis, 240
Chlorophyllase activity, 109
Chloroplasts,1
-envelopes, 23
-lamellae, 23
-spinach, 23
Chlorsulfuron, 206,227,272
Choline, 28
-dehydrogenase(glycine betaine biosynthesis), 98
-oxidase(glycine betaine synthesis), 98
Cinemethylin, 205, 216
Cineoles, 205, 216, 246, 250

Cirsium arvense, 220
Citrulline, 134
Citrus spp., 182,184
-sinensis, 98
Clethodium, 216
Clomazone, 205, 220
Clopyralid, 206, 225
Clovers, 127
Cluster bean, 107
C-N ratio, 187
Coconut, 168, 181
-milk iron, 13
Colletotrichum spp., 190
Colligative properties of water, 83
Combined inoculation, 136
-nitrogen, 131
Compartmentation, 31
Conductanse versus resistance, 92
Contact herbicides, 206
Cotton, 221
Cracking in Citrus, 192
Crop yield, 13,116
Crown rot, 168
CTP: phosphatdate cytidyltransferase, 30
Cucurbita, 70
Cultivation method, 209
-practices, 209
CV, 191
Cyanazine, 231
Cycelohexanediones, 205
Cyclohexanediones, 216,250
Cycloxidim, 216
Cynobacteria, 126, 147
Cynodon dactylon, 221,223
Cytochromes,11
-oxidase,11
Cytokinins,188

D

Dalapon, 213, 248, 272
Damping-off, 168
Dashehari, 195
Datura stramonium, 219
Daucus carota, 220
Debranching enzyme, 45
Density, 83
Descuraina sophia, 220
Desmedipham, 215,250
D-fructose6-phosphate 2-?-C-D-glucosyl tranferase, 38
Diacylglycerol, 32
Diallate, 229
Diazoles, 205
Dicamba, 205,214, 249
Dichlorprop, 222
Diclopfop, 223
Difenzoquat, 206,224
Digitaria spp., 220, 223
Dimethenamid metolachlor, 205
Dinirtoanilines, 205,216,246,250
Dinitrogen fixation, 55
Dinoseb, 252
Diphenylethers, 218,244, 251
Diphenyl ethers, 205
Diquat, 215,244
-paraquat, 205
Disease Resistance, 190
Disodium methanearsonate, 214
Dissociation, 82
Dithiopyr, 206,225
Diuron, 233
D-ononitol, 101
Downey mildew, 168
Driving force, 90
Drought, 137,169
-stress, 191

E

Dwarfing rootstock, 183
Echinocloa Crusgalli, 219,220,223
Efficacy of Herbicides, 211
Eichornia crassipes, 118
Electron acceptors, 238
-transport inhibitors, 238
Elevated atmospheric CO2, 140
Elytrigia repens, 220
Emblica officinalis, 182
Endoplasmic reticulum, 23
-caster been, 23
Enhanced iron availibility, 160
Energy transfer, 238
Enoyl ACP hydrase, 25
-reductose, 25
Enterobacteria, 173
Environmental factors, 210
-strees, 191

Enzymes, 113
EPTC: S- ethyl dipropyl carbamothoate, 229
Erwinia, 161
-ureadovora, 272
Escherichia Coli, 11
-Agrobacterium lumefaciens, 272
Ethanolamine, 28
Ethalfluralin, 205,217
Ethephon, 184,188,197
Ethofumesate, 206,233
Ethylene, 161,162,188
-(ET), 164
-biosynthesis, 238,242
-synthesis, 187
Euphorbia esula spp, 219,220
Experimental compounds, 206

F

Fababean, 167
Fatty acid, 32
-(lipid) Biosynthesis, 240
-Biosyntheis, 238
Fe-deficiency, 1
-III chelate reductase activity, 9
-III hydroxyoxides, 4
-stress-inducible proteins, 5
-stress syndrome, 10
Fenoxaprop, 223,241
Ferric, 7
-chelators, 7
-chelate reductase activity,2,8,9
Ferrous, 7
-(Fe^{2}+), 13
-iron, 14
Flavin adenine dinucleotide (FAD), 58
Flavoproteins, 11
Floral Macformation, 193
Flower meristem, 197
Flowering, 186
Fluazifop- P, 223,241
Fluchloralin, 205,217
Flumetsulam, 232
Flumiclorac, 206
Fluometuron, 233
Fluoride toxicity, 113
Fluoroglycofen, 218
Fluridone, 206,226
Fomesafen, 218
Fosamine, 206,234
Frankia, 126,133,173
Free diazotophes, 130
-living bacteria, 126
-radical, 10
Frost hypoxia, 191
Fructokinases, 42
Fruits, 173
-cracking, 195
-crop, 181,188
-ratio,188
-productivity, 186
Fungal cell-wall, 190
Fusarium oxysporum, 163,166

G

Garlic, 221
GDP, 38
Gene expression, 186
-isolation, 186
-transfer systems, 186
-regulation, 172
Genetic engineering, 95,188
Genetic resource, 181
Germination, 107, 113
Gibberellins, 161
Ginger, 167
Glphosate, 245
Gluconeogenic cells, 41
Glucose 6-phosphate isomerase, 39,43
Glucosephosphate isomerase, 46
Glufosinate, 206,234,242, 245, 272
Glutamate, 133, 169
Glutamic acid, 12
Glutamine, 133
-biosynthesis, 238, 243
-synthetase, 242
Glutathione-S-transferase, 11, 98
Glycerol-3-P, 28
Glyceraphosphate. 30
Glycine betaine, 100,101,169
Glyoxysomes, 23
Glyphosate, 206,234,242,259, 272
Granulation, 195
Grape(Vitis vinifera), 187
Grapevines, 190
Gravitational potential, 86
Green revolution, 141

Groundnut, 167, 168, 221
Growth Substances, 13
Guava, 181
-(Psidium guajava), 188

H

Halosufuron, 206, 227
Haloxyfop, 241
Harpins, 166
Head group, 25
-mould,168
Heat of condensation, 82
-vaporization, 82
Heavy metal pollution, 105,116
Helianthus annuus, 112, 114
Herbicides, 203, 204
-adsorption, 210
-metabolism, 211
-resistance, 271
-resistant crops, 264, 266
-translocation, 210
Hexazinone, 231
Hexokinase, 41,46
Hexose kinase, 39, 43
Hg toxicity, 116
High density technology, 183
High dielectric constant, 83
High surface tension and viscosity, 83
High temperature, 138
Hofler diagram, 87
Homeostasis, 99
Hordeum vulgare, 2, 98
Hormones, 9
Horticulture, 203
Hybridization, 192, 269
Hydration shells, 100
Hydrilla verticullata, 119
Hydrogen bonding, 82
Hydrogen peroxide, 191
Hydrolytic enzyme activities,238,242
Hydroxylamine, 69
Hyptis suaveolens, 112, 114

I

IAA, 161, 185
Illuminated chloroplast, 70
Imazamethabenz, 251
Imazapyr, 219
Imazaquin, 219
Imazethapyr, 205,219
Imidazolidinones, 219,242,251
Imines, 220, 251
Indole-3-acetic acid, 3, 171
Induced resistance, 160
Inducing systmatic resistance, 164
Infection, 128
Inorganic assimilation, 155
-nitrogen, 53
Intensive orchard creation, 182
Internal necrosis, 192, 193
Interveinal chlorosis, 10
Invertase, 37,39
Ioxynil, 220
Ipomoea spp.219
Iron, 13
-chelators, 8
-clorosis, 1,5
-deficiency, 3
-detoxification, 7
-efficient cultivors, 7
-fluxes, 7
-homeostasis mechanism, 7
-insuffuciency, 2
-scavenging, 160
-superoxide dismutase(Fe SOD), 12
Isopropalin, 205
Isoxaben, 214, 205, 249
Isoxazolidinones, 220, 245

J

Jackfruit, 181
Jasmonate(JA), 164
Jowar, 107
Juvenility, 186

K

Klebsiella, 173
-ozone, 272
-mRNA, 132
Kochia scoparia, 219, 220

L

L.pimpinellifolium, 98
Lactofen, 218
Lamellar phospholipo proteins, 8

Laptochloa fusca, 150
Late embryogensis abudant (LEA) proteins, 102
LEA proteins, 98,103
Lead, 107
-acetate, 110
-nitrate, 116
Leaf area, 108
-chlorosis, 2
-curl virus(PLCV), 190
-rot, 168
Leafy gene, 187
Leghaemoglobin, 131
Legumes, 126
Legume-rhizobium symbiosis, 127, 128, 136
Lemna polyrhiza, 118
Lentil, 168
Lignocellulosic materials, 119
Lime, 196
Linuron, 233
Lipopolysaccharide (LPS), 166
Litchi, 181, 192
Lolium perennespp 221,223
Long wave radiation absorbers, 82
Low temperature, 138
L-serine, 28
Lupinus lutens, 112
Lycopersicon esculenutum, 6, 98
Lysine, 12

M

Macformation, 193
Maize, 10, 23,117, 173
-mutants,11
-wheat system, 147
Malformation, 191,193
Malonyl transacylase, 25
Manganese, 106, 115
Mangifera spp, 182
Mangiferin, 193
Mango, 181, 182, 184,193
-(Mangifera indica), 187
-hybridization, 192
Mannitol, 100
-1-phosphate dehydrogenase (mannitol biosynthesis), 98
Matric Potential ($?_{m}$), 84
Matric Potential , 85
MCPA, 222
Medicago sativa, 98
Megapascals (Mpa), 84
Mendelian genes, 63
Mercuric chloride, 110
Mercury, 106,107,115
Metabolism of Sucrose, 37
-strach,42
Metabolic activities, 117
Metflurazon, 206
Methazole, 205
Methi, 107
Methionine, 243
Metolachlor, 213,248
Metribuzin, 231
Metsulfuron, 206,227
MH, 188
Michaelis-Menten kinetics, 38
Micropropagation, 182
Microspore, 269
Mikania micrantha, 204
Millet, 173
Mitochondria, 23
Mitochondrial activities, 238, 239
Molecular Biology, 95, 186
Molinate, 229
Molybdenum, 59
Monosodium methanearsonate, 214
Mosambi, 181
Mugineic acid, 6
-(MA), 8
Mungbean, 106, 107, 167
Mustard, 23,107, 168
Myo-inositol, 28
-O-methyl transferase (D-ononitol biosynthesis), 98

N

NAA, 185,188,193
NADPH, 70
-dehydrogenase, 170
Napropamide, 205,248
Naptalam, 206,224
Near-Isogenic Strain Consortia (NISC), 173
Necrosis, 192
Nediones, 205
Nepropamide, 213
Net photosynthetic efficiency, 189
Neurospora crassa, 63

Nickel, 108, 113
Nicosulfuron, 206,227
Nicotiana plumbaginifolia, 265
Nicotiana tabacum, 98,272
Nicotianamine (NA), 6
Nitriles, 244,252
Nitralin, 205
Nitrate Assimilation, 53, 56, 59
-reductase, 50,58,67
- activity, 53
-reduction, 56
Nitric acid (NO), 10
Nitrite reduction, 69
Nitrofen, 205
Nitrogen, 186
-assimilation, 133
-deficiency, 54
-fixation, 160
-metabolism, 12
Nitrogenase, 131
-activity, 146
-synthesis, 132
NO scavenger, 10
Nodulation in cereal, 140
Nodule formation, 128
-function, 2
-morphology and anatomy, 142
Nonproteinogenic amino acids, 7
Non-selective herbicides, 208
Norflurazon, 206,225,272
N-Phenylphthalamides, 253
N-phenylpthalimides, 223
NR under water stress, 65
Nuclear membrane - 23
Nucleic Acids, 112
-biosynthesis, 238
-Metabolism, 13
Nucleoside bases, 38
Nutrient efficiency, 54

O

Onion, 23,221
Oranges, 181
Organic herbicides, 204
Ornithine decarboxylase, 98
Oryza sativa, 98.113
Oryzalin, 205
Oscimum sanctum, 118
Osmoprotectants, 99
Oxadiazoles, 205,221,252
Oxadiazolidines, 205,221
Oxadiazon, 205
Oxalis spp., 221
Oxidase, 187
Oxyfluorfen, 205,218
Ozone, 191

P

Paclobutrazol, 183, 184, 192
Paelobcetrazol, 186
Paenibacillus Polymyxa, 167
Panicum spp,220,223
-dichotomiflorum, 219
Papaya, 181, 183,190
-(carica papayaL.), 184
Paranodulation, 142
Paraquat, 215,244,249
Parasitism, 160, 164
Parasponia, 127
Pathway for sucrose synthesis, 39
Patterns of Phytosiderophore release, 7
Pea, 107,167
Pebulate, 230
Pendimethalin, 205,217
Pentachlorophenol, 188
PEP carboxylase, 135
Peroxidase, 8,12,113
-(PO),165
Pesticides, 159
Phaseolus aureus, 110,113
-vulgaris, 4,108,109
Phenazine-1-carboxylic acid, 166
Phenmedipham, 215,272
Phenols, 205,221,246,252
Phenotypic expression, 184
Phenoxyalkanoic acids, 205,253
Phenoxys, 245,246
-acetics, 221
-butyrics, 222
Phenyl carbamates, 244
-proparioid pathways, 165
-pyridazinges. 224
-triazinones, 224, 244,254
Phenylalanine ammonia lyase(PAL), 165
Phenylpyridazines, 253
Phloem-loading, 7

Phosphoric acid groups, 21
Phosphatidate phophatase,29
Phosphatidylcholine, 32
Phosphatidylethanolamine, 32
Phosphatidylethanolamine: L- Serine phosphatidyltransferase, 29
Phosphatidylethanolamone,22
Phosphatidylglycerol: CDP diacylglycerol phosphatidyl-transferase, 32
Phosphatidylinositol, 32
-: myo-inositol phosphatidyltransferse, 31
Phosphatidylserine decarboxylase, 30
Phosphatidylserine, 32
Phosphayidylglycerphosphate phosphohydrolase, 30
Phosphobacteria IISR, 167
Phosphoglycerides, 21
Phospholipases, 32
Phospholipid biogenesis, 32
-biosynthesis pathway, 26
-composition, 23
-metabolism, 21
-transfer, 31
Phosphorus solubilization, 160
Phosphorylase, 46
Phosphtidylglycerol: CDP diacylglycerol phosphtidyl- transferase,30
Photoperiod, 186
Photoregulation of nitrate assimilation, 61
Photorhizobium, 127
Photosynthesis, 188,238, 244
-mechanism, 238
Photosynthetic diazotrophs, 130
-efficiency, 188
-rate, 108
Photosystem I, 244
-II, 244
Phthalamates, 224, 246,254
Phtyochrome chromophore synthesis,11
Physiological and biochemical effects of iron, 10
-disorders, 191
Phytoferritin, 7
Phytopathogenic bacteria, 166
Phytosiderophores, 5,6,7
Picloram, 206,221,225
Pigeonpea, 167
Pigment, 109
-biosynthesis, 238,239
Pineapple, 181, 183
-(Ananas comosus),188
Pisum sativum, 6,107
Plant breeding methods, 266
-density, 183
-growth, 160
-promoting rhizobium(PGPR), 136
-meta-bug, 172
-iron nutrition, 1
-morphology, 209
-pathogen interaction, 165
-phenological development, 209
-phospholipids, 23
-resistance, 211
-rhizobacterial interaction, 165
-transformation, 269
-water relationship, 81
Plasmalemma, 4
-bound ferric reductase, 7
Plastid ferritin, 8
Poa annua, 220
Pollen, 203
-grain size, 13
Polyamines, 101
Polyembryonic mango, 187
Polygonum convolvulus, 220
Polygonum spp,232
Polyphenol oxidase (PPO), 165
Pomegranate, 181
Poor orchard efficiency, 185
Portulaca oleracea, 220
Potassium nitrate, 186
Potato, 173,221
-tuber, 23
Predation, 160
Pressure chamber apparatus, 94
-potential, 86
Primisulfurin, 206
Prodiamine, 205,217
Proliferation, 162
Proline, 100,116,169
-synthesis, 98
Prometryn, 231
Promoting rhizobacteria (PGPR), 160
Pronamide, 205,213,248
Propachlor, 205,213,235
Propagation techniques, 182
Propanil, 213,248

Properties of water, 82
Prosulfuron, 206,228
Protease, 113
Protein, 238
-kinase, 33
-metabolism,111
-synthesis, 12
Proteus, 173
Prunus amygdalopersica, 8
Prunus spp., 182
PS II electron transport, 238
Pseudomonas, 161, 162
-aeruginosa, 167
-capacia, 168
-fluorescens, 165, 166, 167
-pseudoalcaligenes, 169
-putid, 272
-putida, 163, 168, 171
-spp, 165, 167,173
Psychrometric tecniques, 93
Pulses, 125
-production, 126
Pyrazoliums, 224, 246,254
Pyrazon, 206,225
Pyridate, 206,253
Pyridaziones, 254
Pyridazionones, 224,244
Pyridine carboxylic acids, 225,243,246,254
Pyridines, 225
Pyridinones, 226, 244,255
Pyrimidinyl thio-benzoates, 226,245,255
Pyrithiobac, 206,226
Pyrophosphorylase to 3-phosphoglycerate, 44
Pyrrolnitrin, 166
Pyrus spp., 182
Pythium, 172

Q

Quinclorac, 206
Quinoline, 255
-carboxylic acids, 226,246
Qunizalofop-P, 206,223

R

R.meliloti, 139
Radish, 107
Recurrent flowering, 192
Redox environments, 6
-pump, 54
Regeneration, 186
Regulation, 186,189
Rejuvenation, 197
Repression, 54
Reproductive physiology, 12
Resistant Crops, 266
Respiration, 11
Rhizobacteria, 160
Rhizobium, 126,138,159,173
-leguminosarum, 127,171
-leguminosarum bv., 127
-tropicii, 127
Rhizoctonia batasticola, 166
-solani, 170
Rhizosphere CO_2, 140
Ribonuclease solubilizable fractions, 8
Ribulose , 187
-biphosphate carboxylose (Rubisco), 189
Rice, 168, 173,221
Richardia scabra, 221
Ricinus, 23
Ring spot virus (PRSV), 190
Root elongation, 162
-exudates, 169
-hairs, 10
-rot, 168
Rootstock, 182
RuBP carboxylase, 135

S

S- Adenosyl- L- methionine: phosphatidylethanolamine N- methyl-transferase, 29
S-adenosylmethionine, (SAM), 243
S-Nitroso-N-acetylpenicillamine, 10
Salicylic acid, 184,197
Salicylic acid (SA), 164
Salinity, 136
Salt Tolerance, 81
-stress determinant, 97
Salvinia,119
-natans, 118
Sapota, 181
Sclerotium rolfsii, 166
Seamum indicum, 112
Seed mutagenesis, 269

Seedling growth, 107
Selective herbicides, 208
Selectivity, 208
Serine, 28
Serratia, 173
Sesbania, 127,169
Setaria spp., 223
Sethoxydim, 216
Sheath blight, 168
-rot, 186
Siderophores, 7, 163
Signal Transduction in Higher Plants, 34
Signalling molecules, 164
Simazine, 231
Sinorhizobium fredi, 127
Siroheme groups, 68
Sisymbrium, 220
Soil Acidity, 139
-Plant Atmosphere Centinuum (SPAC), 87
-Salinity, 169
Solanum nigurm, 232
-spp, 220
-lycopersium, 118
Soluble protein, 112
Solute potential, 85
Sorghum, 173
-bicolor, 219
-halepense, 219
Soybean, 4, 167, 173,221
-(Glycirine nax), 127
Specific heat, 82
Spice crop, 167
Spinach, 23
Spongy Tissue, 191,194
Sprodela Polyrhiza, 118
Stem and root rot, 168
Stomatal Frequency, 108, 109
-Index, 108,109
-Synthase, 47
-Reistance , 90
Strategy I, 5, 7
- II, 5,7
Streptomyces hygroscopicus, 272
Stress tolerance, 98
Succinic dehydrogenase, 11, 12
Succinyl-COA+glycine, 131
Sucrose, 37
-Degradation, 41, 44
-phosphatase, 39
-phosphate phosphatase (SPPase), 38, 46
-synthetase (SPS), 38
-synthase, 38, 39,46, 47
-synthesis, 37
Sufentrazone, 206
Sufficiency value, 2
Sugarcane,221
Sulfentrazone, 206,224
Sulfometuron, 206
Sulfony lureas, 242,245,256
Sulphur dioxide, 191
Sunflower, 167,173
Sunhemp, 107
Superoxide radicals, 191
-dismutose(SOD),191
Sustainable crop productivity, 160
Symbiotic nitrogen fixation, 2
Symplast transport, 7
Symport, 54
Synechocystis PCC 6803, 170
Synorhizobium meliloti, 127
Syzyzium cuminii, 182

T

Taraxacum officinale, 220
TDP, 38
Tebuthiuron, 233
Temperate Fruits, 183
Temperature, 84
Terbacil, 232
Tetrahydropyrimidinones, 229, 244
Tetrahydropyrionidinones, 256
Thermocouple psychrometer sample chamber, 94
Thermotolorant mutant, 147
Thiazopyr, 206,226
Thicarbamates, 245
Thifensulfuron, 228
Thiobencarb, 230
Thiocarbamates, 229,241
Thiourea, 188
Thlaspi arvense, 220
Threonine, 12
TIBA, 185
Tillering, 203
Tissue cluture selection, 268
Tobacco, 98
Tolerance, 211

Tomato, 6
Tralkoxydim, 216
Transcription Factor, 98
Transgenic Plants, 268
Translocated herbicides, 206
Tree, 189
Triasulfuron, 228
Triazines, 206,230,244,256
Triazinones, 231
Triazoles, 184,231,244,257
Triazolopyrimidine sulfon anilide, 232, 245,257
Tribenuron, 228
-experimental compounds, 206
Trichoderma viride, 168
Triclopyr, 206,225
Trifluralin, 205,217
Tungro diseases, 168
Turgor potential ($?_p$), 84
Turmeric, 167

U

UCC - C4243, 206
UDP-, 38
-glucose, 38
-pyrophosphorylase, 39,46
-dehydrogenase, 42
Uncouplers, 238
Uniport, 54
Uptake and distribution of Iron in Plants, 5
Uracils, 232,258
Urdbean, 168
Ureas, 232,258
Ureides, 134, 135
-metabolism, 134
Urticas spp., 221

V

Valine, 12
Vegetative Malformation, 193
Vigna unguiculata, 12, 116, 114

W

Water hyacinth, 118
-Potential, 83,84, 85
-soluble precursors, 28
Weed, 203,204
-management, 203
Wheat, 167, 173
Wild oats, 204
-onion, 204
-rice, 204
Wilt, 168

X

Xanthium strumarium, 220, 227
Xanthomonas, 173
-ozyzae,170

Z

Zea mays, 10
Zinc Phosphomidon, 193
Zinc sulphate, 196
Zn- SOD gene, 191